Confocal Microscopy
Methods and Protocols

METHODS IN MOLECULAR BIOLOGY™

John M. Walker, SERIES EDITOR

METHODS IN MOLECULAR BIOLOGY™

Confocal Microscopy Methods and Protocols

Edited by

Stephen W. Paddock

Department of Molecular Biology, University of Wisconsin, Madison

Humana Press ✳ Totowa, New Jersey

Cover design by Patricia F. Cleary.
Cover legend: Foreground panel: A triple labeled *Drosophila* third instar wing imaginal disc imaged using the laser scanning confocal microscope (see Chapter 1, Fig. 4) and displayed so that color is mapped to the expression of three wing patterning genes using methods outlined in Chapter 21. *(Specimen courtesy of Jim Williams and imaged by Steve Paddock, both of the University of Wisconsin.)* Background image: A confocal Z-series collected from groups of cultured rat BN/MSV cells is displayed so that color is mapped to depth in the specimen using methods outlined in Chapter 21 (Note 3). *(Specimen courtesy of Tim Hammond, Tulane University, and imaged by Steve Paddock.)*

For additional copies, pricing for bulk purchases, and/or information about other Humana titles, contact Humana at the above address or at any of the following numbers: Tel: 973-256-1699; Fax: 973-256-8341; E-mail: humana@humanapr.com, or visit our Website at www.humanapress.com

Printed in the United States of America. 10 9 8 7 6 5 4 3 2

Library of Congress Cataloging-in-Publication Data
Main entry under title:
Methods in molecular biology™
Confocal microscopy methods and protocols / edited by Stephen W. Paddock.
      p.   cm.—(Methods in molecular biology; vol. 122)
   Includes index.
   ISBN 0-89603-526-3 (alk. paper)
   1. Confocal microscopy. I. Paddock, Stephen W. II. Series: Methods in molecular biology (Totowa, NJ); 122.
QH224.P76   1999
570'.28'2 -- dc21   98-38205
                                                                                                              CIP

Preface

The confocal microscope is an established tool in many fields of biomedical research where its major application is for improved light microscopic imaging of cells within fluorescently-labeled tissues. Indeed, before the advent of practical confocal microscopy, most fluorescently-labeled tissues were imaged using a conventional light microscope. Resolution was often compromised by fluorescence from outside the focal plane of interest, especially in tissues made up of multiple cell layers. In order to attain acceptable resolution, microscopists were forced into all manner of tricks to prepare their specimens, including cutting sections of the tissues, growing cells on a coverslip, flattening cells under agar, or squashing embryos between coverslips. Most of these methods had the potential of introducing artifacts and therefore led to questions about the validity of the results, especially images of living cells.

The confocal microscope has now enabled the imaging of discrete regions of tissues virtually free of out-of-focus fluorescence, and with reduced chances of artifacts from the techniques of specimen preparation. Indeed, in some cases virtually no preparation of the tissue is required prior to imaging although it is usual to stain tissues with one or more fluorescent probes for confocal imaging of specific macromolecules within cells. The field of confocal microscopy has grown exponentially over the past ten or so years, and it touches on many areas of contemporary biological research where a light microscope is required for imaging, including applications in cell biology, developmental biology, neurobiology, and pathology. It would therefore be impossible to cover all of the protocols in current use in a single book, and references to other sources including microscopy web pages have been included.

An overall effort has been made in recent years to render confocal imaging systems more user friendly. A relatively short training period is now required before a novice, who has experience with a light microscope and a basic working knowledge of a computer, is able to produce acceptable images. In practice, however, resolution of the images collected with a confocal microscope depends upon various properties of the sample itself, which means that a well-prepared specimen is extremely important for achieving images of the highest quality. The old saying "garbage in, garbage out" is a valuable maxim to keep in mind, not only for the novice, but also for the

v

more-experienced user. Protocols for preparing such specimens are therefore extremely important, and emphasis has been placed on the details of specimen preparation throughout.

The aim of *Confocal Microscopy Methods and Protocols* is to take the researcher from the bench top, through the imaging process, to the journal page. This book is light on the technical details of the microscopes themselves, as these can be found elsewhere and are continually changing as new technology is incorporated into confocal systems. The chapters have been chosen to highlight the biological applications of the confocal microscope and methods for the analysis and the presentation of the images for publication. Protocols for the preparation of tissues from most of the currently popular model organisms, including plants, have been covered, with the addition of chapters on confocal imaging of living cells, three dimensional analysis, and the measurement and presentation of confocal images for publication.

I would like to thank all of the authors, especially those who have persevered through many forms of adversity—both mental and physical (including at least one El-Niño-related incident)—in order to complete their chapters in a timely fashion. I would also like to thank my laboratory colleagues (past and present) for presenting me with such a plethora of imaging questions. A special thank-you goes to Sean Carroll for his encouragement of this project. The series editor, John Walker, has not only provided expert editorial advice, but has also kept me abreast of the cricket scores in the UK! In addition, Tom Lanigan and the staff at Humana Press, especially Patricia Cleary and Fran Lipton, have performed to an extremely high standard of professionalism throughout the project. The color section of the book would not have been possible without the extremely generous sponsorship of Bio-Rad, and I thank Leonard Pulig of Bio-Rad for his help in this matter. Finally, I would like to thank Diana Wheeler for her tolerance and support, especially during the final stages of the project, which happened to coincide with the preparations for our wedding.

Steve Paddock

Contents

Contributors

PAUL R. ADAMS • *Department of Neurobiology and Behaviour, State University of New York, Stony Brook, NY*

AUDREY L. ATKIN • *School of Biological Sciences, University of Nebraska, Lincoln, NE*

ALISON F. BEVEN • *Department of Cell Biology, John Innes Centre, Norwich, UK*

KURT BOUDONCK • *Department of Cell Biology, John Innes Centre, Norwich, UK*

ANDREA H. BRAND • *Wellcome/CRC Institute, University of Cambridge, UK*

NADEAN L. BROWN • *Howard Hughes Medical Institute, University of Michigan, Ann Arbor, MI*

BARRY J. BURBACH • *Beckman Neuroscience Center, Cold Spring Harbor Laboratory, Cold Spring Harbor, NY*

DAVID CARTER • *Genomic Solutions Inc., Ann Arbor, MI*

MARK S. COOPER • *Department of Zoology, University of Washington, Seattle, WA*

GUY COX • *University of Sydney, Sydney, Australia*

SARAH L. CRITTENDEN • *Howard Hughes Medical Institute, University of Wisconsin, Madison, WI*

CHRISTOPHER CULLANDER • *Department of Biopharmaceutical Sciences, School of Pharmacy, University of California, San Francisco, CA*

LEONARD A. D'AMICO • *Department of Zoology, University of Washington, Seattle, WA*

EMMA-LOUISE DORMAND • *Wellcome/CRC Institute, University of Cambridge, UK*

EWA DZIAK • *Department of Anatomy and Cell Biology, University of Toronto, Toronto, Canada*

RACHEL J. ERRINGTON • *Department of Cell Physiology, University of Nijmegen, Nijmegen, The Netherlands*

HELEN FRANCIS-LANG • *Department of Biology, University of California, Santa Cruz, CA*

SCOTT E. FRASER • *Biological Imaging Centre, Beckman Institute, California Institute of Technology, Pasadena, CA*

GEORG HALDER • *Department of Molecular Biology, University of Wisconsin, Madison, WI*

JIM HASELOFF • *Wellcome/CRC Institute, University of Cambridge, UK*

ERIC HAZEN • *University of Minnesota, Minneapolis, MN*

CLARISSA A. HENRY • *Department of Zoology, University of Washington, Seattle, WA*

JON M. HOLY • *Department of Anatomy and Cell Biology, University of Minnesota, Duluth, MN*

JANET HOOGSTRAATE • *Astra Pain Control AB, Sodertalje, Sweden*

SARAH C. HUGHES • *Banting and Best Department of Medical Research, University of Toronto, Toronto, Canada*

DELPHINE IMBERT • *Cellegy Pharmaceutical, Inc., Foster City, CA*

HARVEY J. KARTEN • *Deparment of Neurosciences, University of California at San Diego, La Jolla, CA*

JUDITH KIMBLE • *Howard Hughes Medical Institute, University of Wisconsin, Madison, WI*

HENRY M. KRAUSE • *Banting and Best Department of Medical Research, University of Toronto, Toronto, Canada*

PAUL M. KULESA • *Biological Imaging Centre, Beckman Institute, California Institute of Technology, Pasadena, CA*

JAMES A. LANGELAND • *Biology Department, Kalamazoo College, Kalamazoo, MI*

CLIVE W. LLOYD • *Department of Cell Biology, John Innes Centre, Norwich, UK*

EMMELINE MARTIIN • *Leiden-Amsterdam Center for Drug Research, University of Leiden, The Netherlands*

GABRIEL G. MARTINS • *Department of Anatomy and Cell Biology, State University of New York, Buffalo, NY*

JONATHAN MINDEN • *Department of Biological Sciences, Carnegie Mellon University, Pittsburg, PA*

KAREN OEGEMA • *Department of Cell Biology, Harvard University Medical School, Boston, MA*

DONALD M. O'MALLEY • *Department of Biology, Northeastern University, Boston, MA*

MICHAL OPAS • *Department of Anatomy and Cell Biology, University of Toronto, Toronto, Canada*

STEPHEN W. PADDOCK • *Laboratory of Molecular Biology, University of Wisconsin, Madison, WI*

DENISE L. ROBB • *Institute of Human Genetics, University of Minnesota Medical School, Minneapolis, MN*

ALAN T. STONEBRAKER • *Department of Anatomy and Cell Biology, State University of New York, Buffalo, NY*

WILLIAM SULLIVAN • *Department of Biology, University of California, Santa Cruz, CA*

ROBERT G. SUMMERS • *Department of Anatomy and Cell Biology, State University of New York, Buffalo, NY*

NICK S. WHITE • *Department of Plant Sciences, University of Oxford, Oxford, UK*

CHRIS WYLIE • *Institute of Human Genetics, University of Minnesota Medical School, Minneapolis, MN*

CAROL L. WYMER • *Department of Biological and Environmental Sciences, Morehead State University, Morehead, KY*

Color Plates

Color Plates I–IV appear as an insert following p. 372.

Plate I *Top Sequence*: Multiple-label *Drosophila* embryos. **(A-B)** Structure of double label images, Fig. 21-2 A-B; **(C-D)** a three-color image, Figs. 10-1, 21-1 B, D. *Lower Left Sequence*: Examples of double labeling of a *Drosophila* embryo and wing imaginal disk using FISH, Fig. 5-1 A-B. *Lower Right Sequence*: Four-dimensional visualization of in situ chondrocytes by color-coded relative volume changes, Fig. 18-13 A-F.

Plate II Double and triple antibody labelling of *Drosophila* eye imaginal disks, Fig. 4-1.

Plate III Nuclear and cytosolic calcium dynamics, Fig. 16-8 A.

Plate IV Three-dimensional illustrations prepared for the parallel-eyed viewing method (standard method for publication), Fig. 22-8 A-C.

1

An Introduction to Confocal Imaging

Stephen W. Paddock

1. Introduction

The major application of confocal microscopy in the biomedical sciences is for imaging either fixed or living tissues that have usually been labeled with one or more fluorescent probes. When these samples are imaged using a conventional light microscope, the fluorescence in the specimen away from the region of interest interferes with resolution of structures in focus, especially for those specimens that are thicker than approx. 2 µm (**Fig. 1**). The confocal approach provides a slight increase in both lateral and axial resolution, although it is the ability of the instrument to eliminate the "out-of-focus" flare from thick fluorescently labeled specimens that has caused the explosion in its popularity in recent years. Most modern confocal microscopes are now relatively easy to operate and have become integral parts of many multiuser imaging facilities. Because the resolution achieved by the laser scanning confocal microscope (LSCM) is somewhat better than that achieved in a conventional, wide-field light microscope (theoretical maximum resolution of 0.2 µm), but not as great as that in the transmission electron microscope (0.1 nm), it has bridged the gap between these two commonly used techniques.

The method of image formation in a confocal microscope is fundamentally different from that in a conventional wide-field microscope in which the entire specimen is bathed in light from a mercury or xenon source, and the image can be viewed directly by eye. In contrast, the illumination in a confocal microscope is achieved by scanning one or more focused beams of light, usually from a laser, across the specimen. The images produced by scanning the specimen in this way are called optical sections. This refers to the noninvasive method of image collection by the instrument, which uses light rather than physical means to section the specimen. The confocal approach has facilitated

From: Methods in Molecular Biology, vol. 122: Confocal Microscopy Methods and Protocols
Edited by: S. Paddock Humana Press Inc., Totowa, NJ

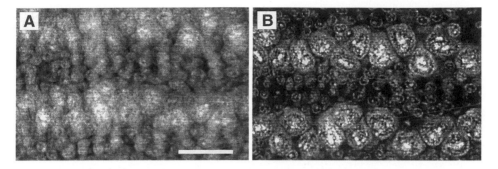

Fig. 1. Conventional epifluorescence image (**A**) compared with a confocal image (**B**) of a similar region of a whole mount of a butterfly pupal wing epithelium stained with propidium iodide. Note the improved resolution of the nuclei in (**B**), due to the rejection of out-of-focus flares by the LSCM.

the imaging of living specimens, enabled the automated collection of three-dimensional (3D) data in the form of Z-series, and improved the images of multi-labeled specimens.

Emphasis has been placed on the LSCM throughout the book because it is currently the instrument of choice for most biomedical research applications, and is therefore most likely to be the instrument first encountered by the novice user. Several alternative designs of confocal instruments occupy specific niches within the biological imaging field *(1)*. Most of the protocols included in this book can be used, albeit with minor modifications, to prepare samples for all of these confocal microscopes, and to related, but not strictly confocal, methodologies that produce perfectly good optical sections including deconvolution techniques *(2)* and multiple-photon imaging *(3)*.

The protocols in this book were chosen with the novice user in mind, and the authors were encouraged to include details in their chapters that they would not usually be able to include in a traditional article. This first chapter serves as a primer on confocal imaging, as an introduction to the subsequent chapters, and provides a list of more detailed information source. The second chapter covers some practical considerations for collecting images with a confocal microscope. Because fluorescence is the most prevalent method of adding contrast to specimens for confocal microscopy, the third chapter contains essential information on fluorescent probes. The next eight chapters cover protocols for preparing tissues from a range of the "model" organisms currently imaged using confocal microscopy. The following six chapters emphasize live cell analysis with the confocal microscope including methods of imaging various ions and green fluorescent protein as well as a novel method of imaging the changes in the 3D structure of living cells. The last section of the book focuses on the analysis and presentation of confocal images. The field of confocal microscopy is now extremely large, and it would be impossible to include every protocol here. This current edition has been designed to give the novice an introduction to confocal imaging, and the authors have included sources of more detailed information for the interested reader.

2. Evolution of the Confocal Approach

The development of confocal microscopes was driven largely by a desire to image biological events as they occur in vivo. The invention of the confocal microscope is usually attributed to Marvin Minsky, who built a working microscope in 1955 with the goal of imaging neural networks in unstained preparations of living brains. Details of the microscope and its development can be found in an informative memoir by Minsky *(4)*. All modern confocal microscopes employ the principle of confocal imaging patented in 1957 *(5)*.

In Minsky's original confocal microscope the point source of light was produced by a pinhole placed in front of a zirconium arc source. The point of light was focused by an objective lens into the specimen, and light that passed through it was focused by a second objective lens at a second pinhole, which had the same focus as the first pinhole, i.e., it was confocal with it. Any light that passed through the second pinhole struck a low-noise photomultiplier, which produced a signal that was related to the brightness of the light. The second pinhole prevented light from above or below the plane of focus from striking the photomultiplier. This is the key to the confocal approach, namely eliminating out-of-focus light or "flare" in the specimen by spatial filtering. Minsky also described a reflected light version of the microscope that used a single objective lens and a dichromatic mirror arrangement. This is the basic configuration of most modern confocal systems used for fluorescence imaging (**Fig. 2**).

To build an image, the focused spot of light must be scanned across the specimen in some way. In Minsky's original microscope the beam was stationary and the specimen itself was moved on a vibrating stage. This optical arrangement has the advantage of always scanning on the optical axis, which can eliminate any lens defects. However, for biological specimens, movement of the specimen can cause wobble and distortion, which results in a loss of resolution in the image. Moreover, it is impossible to perform various manipulations such as microinjection of fluorescently labeled probes when the specimen is moving.

Finally an image of the specimen has to be produced. A real image was not formed in Minsky's original microscope but rather the output from the photodetector was translated into an image of the region of interest. In Minsky's original design the image was built up on the screen of a military surplus long persistence oscilloscope with no facility for hard copy. Minsky wrote at a later date that the image quality in his microscope was not very impressive because

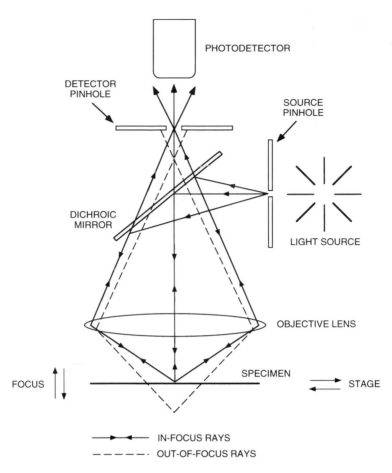

Fig. 2. Light path in a stage scanning LSCM.

of the quality of the oscilloscope display and not because of lack of resolution achieved with the microscope itself *(4)*.

It is clear that the technology was not available to Minsky in 1955 to demonstrate fully the potential of the confocal approach especially for imaging biological structures. According to Minsky, this is perhaps a reason why confocal microscopy was not immediately adopted by the biological community, who were, as they are now, a highly demanding and fickle group concerning the quality of their images. After all, at the time they could quite easily view and photograph their brightly stained and colorful histological tissue sections using light microscopes with excellent optics and high resolution film.

In modern confocal microscopes the image is either built up from the output of a photomultiplier tube or captured using a digital charge-coupled device

(CCD) camera, directly processed in a computer imaging system and then displayed on a high-resolution video monitor, and recorded on modern hard copy devices, with outstanding results.

The optics of the light microscope have not changed drastically in decades because the final resolution achieved by the instrument is governed by the wavelength of light, the objective lens, and properties of the specimen itself. However, the associated technology and the dyes used to add contrast to the specimens have been improved significantly over the past 20 years. The confocal approach is a direct result of a renaissance in light microscopy that has been fueled largely by advancements in modern technology. Several major technological advances that would have benefited Minsky's confocal design have gradually become available to biologists. These include:

1. Stable multiwavelength lasers for brighter point sources of light
2. More efficiently reflecting mirrors
3. Sensitive low-noise photodetectors
4. Fast microcomputers with image processing capabilities
5. Elegant software solutions for analyzing the images
6. High-resolution video displays and digital printers

These technologies were developed independently, and since 1955, they have been incorporated into modern confocal imaging systems. For example, digital image processing was first effectively applied to biological imaging in the early 1980s by Shinya Inoue and Robert Allen at Woods Hole. Their "video-enhanced microscopes" enabled an apparent increase in resolution of structures using digital enhancement of the images which were captured using a low light level silicon intensified target (SIT) video camera mounted on a light microscope and connected to a digital image processor. Cellular structures such as the microtubules, which are just beyond the theoretical resolution of the light microscope, were imaged using differential interference contract (DIC) optics and the images were further enhanced using digital methods. These techniques are reviewed in a landmark book titled *Video Microscopy* by Shinya Inoue, which has been recently updated with Ken Spring, and provides an excellent primer on the principles and practices of modern light microscopy (**6**).

Confocal microscopes are usually classified using the method by which the specimens are scanned. Minsky's original design was a stage scanning system driven by a primitive tuning fork arrangement that was rather slow to build an image. Stage scanning confocal microscopes have evolved into instruments that are used traditionally in materials science applications such as the microchip industry. Systems based upon this principle have recently gained in popularity for biomedical applications for screening DNA on microchips (**7**).

An alternative to moving the specimen is to scan the beam across a stationary specimen, which is more practical for imaging biological specimens. This is the basis of many systems that have evolved into the research microscopes in vogue today. The more technical aspects of confocal microscopy have been covered elsewhere *(1)*, but in brief, there are two fundamentally different methods of beam scanning; multiple-beam scanning or single-beam scanning. The more popular method at present is single-beam scanning, which is typified by the LSCM. Here the scanning is most commonly achieved by computer-controlled galvanometer-driven mirrors (one frame per second), or in some systems, by an acoustooptical device or by oscillating mirrors for faster scanning rates (near-video rates). The alternative is to scan the specimen with multiple beams (almost real time) usually using some form of spinning Nipkow disc. The forerunner of these systems was the tandem scanning microscope (TSM), and subsequent improvements to the design have become more efficient for collecting images from fluorescently labeled specimens.

There are currently two viable alternatives to confocal microscopy that produce optical sections in technically different ways: deconvolution *(2)* and multiple-photon imaging *(3)*, and as with confocal imaging they are based on a conventional light microscope. Deconvolution is a computer-based method that calculates and removes the out-of-focus information from a fluorescence image. The deconvolution algorithms and the computers themselves are now much faster, with the result that this technique is a practical option for imaging. Multiple-photon microscopy uses a scanning system that is identical to that of the LSCM but without the pinhole. This is because the laser excites the fluorochrome only at the point of focus, and a pinhole is therefore not necessary. Using this method, photobleaching is reduced, which makes it more amenable to imaging living tissues.

3. The Laser Scanning Confocal Microscope

The LSCM is built around a conventional light microscope, and uses a laser rather than a lamp for a light source, sensitive photomultiplier tube detectors (PMTs), and a computer to control the scanning mirrors and to facilitate the collection and display of the images. The images are subsequently stored using computer media and analyzed by means of a plethora of computer software either using the computer of the confocal system or a second computer (**Fig. 3**).

In the LSCM, illumination and detection are confined to a single, diffraction-limited, point in the specimen. This point is focused in the specimen by an objective lens, and scanned across it using some form of scanning device. Points of light from the specimen are detected by a photomultiplier behind a pinhole, or in some designs, a slit, and the output from this is built into an image by the computer (**Fig. 2**). Specimens are usually labeled with one or more fluo-

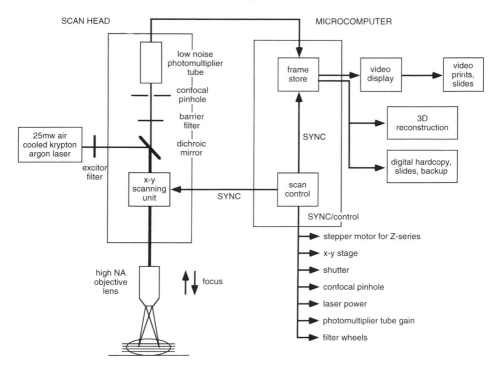

Fig. 3. Information flow in a generic LSCM.

rescent probes, or unstained specimens can be viewed using the light reflected back from the specimen.

One of the more commercially successful LSCMs was designed by White, Amos, Durbin, and Fordham *(8)* to tackle a fundamental problem in developmental biology: imaging specific macromolecules in immunofluorescently labeled embryos. Many of the structures inside these embryos are impossible to image after the two-cell stage using conventional epifluorescence microscopy because as cell numbers increase, the overall volume of the embryo remains approximately the same, which means that increased fluorescence from the more and more closely packed cells out of the focal plane of interest interferes with image resolution.

When he investigated the confocal microscopes available to him at the time, White discovered that no system existed that would satisfy his imaging needs. The technology consisted of the stage scanning instruments, which tended to be slow to produce images (approx. 10 s for one full-frame image), and the multiple-beam microscopes, which were not practical for fluorescence imaging at the time. White and his colleagues designed a LSCM that was suitable for conventional epifluorescence microscopy that has since evolved into an instrument that is used in many different biomedical applications.

In a landmark paper that captured the attention of the cell biology community *(9)*, White et al. compared images collected from the same specimens using conventional wide-field epifluorescence microscopy and their LSCM. Rather than physically cutting sections of multicellular embryos their LSCM produced "optical sections" that were thin enough to resolve structures of interest and were free from much of the out-of-focus fluorescence that previously contaminated their images. This technological advance allowed them to follow changes in the cytoskeleton in cells of early embryos at a higher resolution than was previously possible using conventional epifluorescence microscopy.

The thickness of the optical sections could be varied simply by adjusting the diameter of a pinhole in front of the photodetector. This optical path has proven to be extremely flexible for imaging biological structures as compared with some other designs that employ fixed-diameter pinholes. The image can be zoomed with no loss of resolution simply by decreasing the region of the specimen that is scanned by the mirrors by placing the scanned information into the same size of digital memory or framestore. This imparts a range of magnifications to a single objective lens, and is extremely useful when imaging rare events when changing to another lens may risk losing the region of interest.

This microscope together with several other LSCMs, developed during the same time period, were the forerunners of the sophisticated instruments that are now available to biomedical researchers from several commercial vendors *(10)*. There has been a tremendous explosion in the popularity of confocal microscopy over the past 10 years. Indeed many laboratories are purchasing the systems as shared instruments in preference to electron microscopes. The advantage of confocal microscopy lies within its great number of applications and its relative ease for producing extremely high-quality images from specimens prepared for the light microscope.

The first-generation LSCMs were tremendously wasteful of photons in comparison to the new microscopes. The early systems worked well for fixed specimens but tended to kill living specimens unless extreme care was taken to preserve the viability of specimens on the stage of the microscope. Nevertheless the microscopes produced such good images of fixed specimens that confocal microscopy was fully embraced by the biological imagers. Improvements have been made at all stages of the imaging process in the subsequent generations of instruments including more stable lasers, more efficient mirrors and photodetectors, and improved digital imaging systems (**Fig. 3**). The new instruments are much improved ergonomically so that alignment, choosing filter combinations, and changing laser power, all of which are often controlled by software, is much easier to achieve. Up to three fluorochromes can be imaged simultaneously, and more of them sequentially, and it is easier to manipulate the images using improved, more reliable software and faster computers with more hard disk space and cheaper random access memory (RAM).

4. Confocal Imaging Modes

4.1. Single Optical Sections

The optical section is the basic image unit of the confocal microscope. Data are collected from fixed and stained samples in single, double, triple- or multiple-wavelength modes (**Fig. 4** and Color Plates I and II, following page 372). The images collected from multiple-labeled specimens will be in register with each other as long as an objective lens that is corrected for chromatic aberration is used. Registration can usually be restored using digital methods. Using most LSCMs it takes approximately 1 s to collect a single optical section although several such sections are usually averaged to improve the signal-to-noise ratio. The time of image collection will also depend on the size of the image and the speed of the computer, e.g., a typical 8-bit image of 768 by 512 pixels in size will occupy approx. 0.3 Mb.

4.2. Time-Lapse and Live Cell Imaging

Time-lapse confocal imaging uses the improved resolution of the LSCM for studies of living cells (**Fig. 5**). Time-lapse imaging was the method of choice for early studies of cell locomotion using 16 mm movie film with a clockwork intervalometer coupled to the camera, and more recently using a time-lapse VCR, OMDR, digital imaging system, and now using the LSCM to collect single optical sections at preset time intervals.

Imaging living tissues is perhaps an order of magnitude more difficult than imaging fixed ones using the LSCM (**Table 1**), and this approach is not always a practical option because the specimen may not tolerate the rigors of live imaging. It may not be possible to keep the specimen alive on the microscope stage, or the phenomenon of interest may not be accessible to the objective lens or the specimen may not physically fit on the stage of the microscope. For example, the wing imaginal disks of the fruit fly develop too deeply in the larva, and when dissected out they cannot be grown in culture, which means that the only method of imaging gene expression in such tissues is currently to dissect, fix, and stain imaginal disks from different animals at different stages of development.

For successful live cell imaging extreme care must be taken to maintain the cells on the stage of the microscope throughout the imaging process (*11*), and to use the minimum laser exposure necessary for imaging because photodamage from the laser beam can accumulate over multiple scans. Antioxidants such as ascorbic acid can be added to the medium to reduce oxygen from excited fluorescent molecules, which can cause free radicals to form and kill

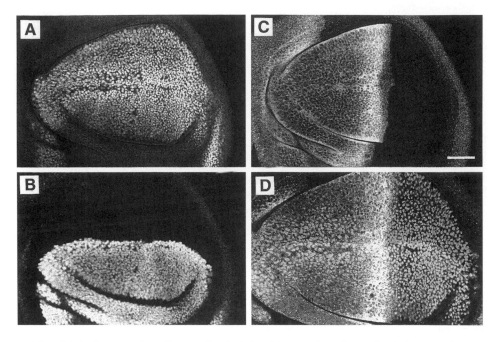

Fig. 4. Single optical sections collected simultaneously using a single krypton/argon laser at three different excitation wavelengths—488 nm, 568 nm and 647 nm—of a fruit fly third instar wing imaginal disk labeled for three genes involved with patterning the wing: **(A)** vestigial (fluorescein 496 nm); **(B)** apterous (lissamine rhodamine 572 nm); and **(C)** CiD (cyanine 5 649 nm); with a grayscale image of the three images merged **(D)**.

the cells. An extensive series of preliminary control experiments is usually necessary to assess the effects of light exposure on the fluorescently labeled cells. It is a good idea to note down all of the details of the imaging parameters—even those that appear to be irrelevant. A postimaging test of viability should be performed. Embryos should continue their normal development after imaging; for example, sea urchin embryos should hatch after being imaged. Any abnormalities that are caused by the imaging process or properties of the dyes used should be determined.

Each cell type has its own specific requirements for life, e.g., most cells will require a stage heating device, and perhaps a perfusion chamber to maintain the carbon dioxide balance in the medium (*see* Chapter 13), whereas other cells such as insect cells usually can be maintained at room temperature in a relatively large volume of medium (*see* Chapter 14). Many experimental problems can be avoided by choosing a cell type that is more amenable to imaging with the LSCM. The photon efficiency of most modern confocal

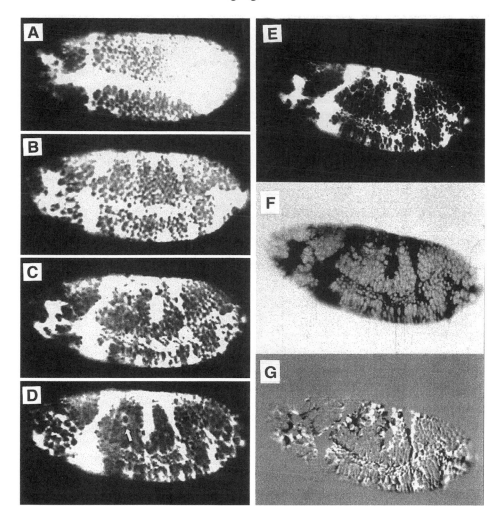

Fig. 5. Time-lapse imaging of a living fruit fly embryo injected with Calcium green (**A–D**). One method of showing change in distribution of the fluorescent probe over time on a journal page is to merge a regular image of one time point (**E**) with a reversed contrast image of a second time point (**F**) to give a composite image (**G**). The same technique can be used by merging different colored images from different time points.

systems has been improved significantly over the early models, and when coupled with brighter objective lenses and less phototoxic dyes, has made live cell confocal analysis a practical option. The bottom line is to use the least amount of laser power possible for imaging and to collect the images quickly. The pinhole may be opened wider than for fixed samples to speed

Table 1
Different Considerations for Imaging Fixed and Living Cells with the LSCM

	Fixed Cells	Living Cells
Limits of illumination	Fading of fluorophore	Phototoxicity and fading of dye
Antifade reagent	Phenylenediamine, etc.	NO!
Mountant	Glycerol ($n = 1.51$)	Water ($n = 1.33$)
Highest NA lens	1.4	1.2
Time per image	Unlimited	Limited by speed of phenomenon; light sensitivity of specimen
Signal averaging	Yes	No
Resolution	Wave optics	Photon statistics

up the imaging process, and deconvolution may be used later to improve the images.

Many physiological events take place faster than the image acquisition speed of most LSCMs, which is typically on the order of a single frame per second. Faster scanning LSCMs that use an acoustooptical device and a slit to scan the specimen rather than the slower galvanometer-driven point scanning systems are more practical for physiological imaging. This design has the advantage of good spatial resolution coupled with good temporal resolution, i.e., full screen resolution of 30 frames per second (near-video rate). Using the point scanning LSCMs, good temporal resolution is achieved by scanning a much reduced area. Here frames at full spatial resolution are collected more infrequently *(12)*. The disk scanning and oscillating mirror systems can also be used for imaging fast physiological events.

4.3. Z-Series and Three-Dimensional Imaging

A Z-series is a sequence of optical sections collected at different levels from a specimen (**Fig. 6**). Z-series are collected by coordinating the movement of the fine focus of the microscope with image collection, usually using a computer-controlled stepping motor to move the stage by preset distances. This is relatively easily accomplished using a macro program that collects an image, moves the focus by a predetermined distance, collects a second image, stores it, moves the focus again, and continues on in this way until several images through the region of interest have been collected. Often two or three images are extracted from such a Z-series and digitally merged to highlight cells of interest. It is also relatively easy to display a Z-series as a montage of images (**Fig. 6**). These programs are standard features of most of the commercially available imaging systems.

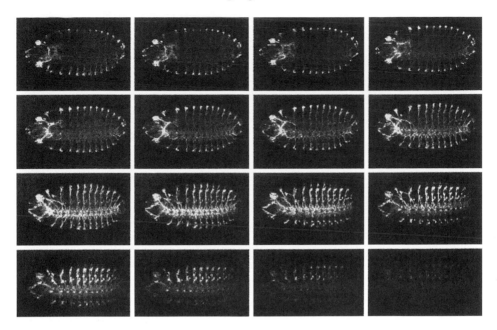

Fig. 6. A Z-series of optical sections displayed as a montage collected from a fruit fly embryo labeled with the antibody designated 22C10, which stains the peripheral nervous system.

Z-series are ideal for further processing into a 3D representation of the specimen using volume visualization techniques *(13)*. This approach is now used to elucidate the relationships between the 3D structure and function of tissues (*see* **Chapter 18**), as it can be conceptually difficult to visualize complex interconnected structures from a series of 200 or more optical sections taken through a structure with the LSCM. Care must be taken to collect the images at the correct Z-step of the motor in order to reflect the actual depth of the specimen in the image. Because the Z-series produced with the LSCM are in perfect register (assuming the specimen itself does not move during the period of image acquisition) and are in a digital form, they can be processed relatively easily into a 3D representation of the specimen (**Fig. 7**).

There is sometimes confusion about what is meant by optical section thickness. This usually refers to the thickness of the section of the sample collected with the microscope and depends on the lens and the pinhole diameter, and not to the step size taken by the stepper motor, which is set up by the operator. In some cases these have the same value, however, and may be a source of the confusion.

The Z-series file is usually exported into a computer 3D reconstruction program. These packages are now available for processing confocal images and run either on workstations at extremely high speeds or using more affordable,

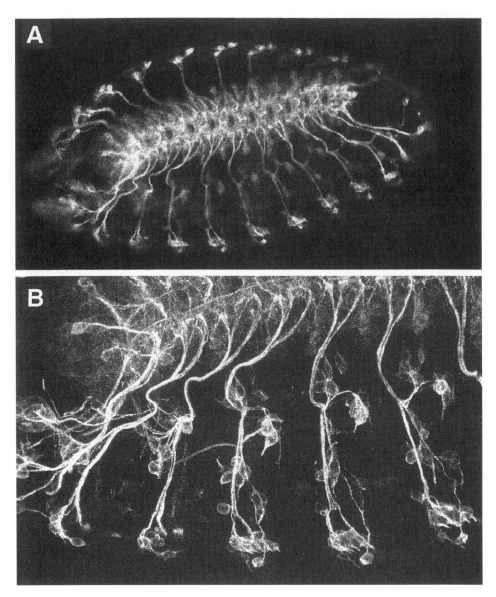

Fig. 7. A single optical section (**A**) compared with a Z-series projection (**B**) of a fruit fly peripheral nervous system, stained with the antibody 22C10.

personal computers. With the introduction of faster computer chips and the availability of cheaper RAM, 3D reconstructions can be produced quite effectively on the workstation of the confocal microscope. The 3D software packages produce a single 3D representation or a movie sequence compiled from different views of the specimen. Specific parameters of the 3D image such as opacity can be interactively changed to reveal structures of interest at different levels within the specimen, and various length, depth, and volume measurements can be made.

The series of optical sections from a time-lapse run can also be processed into a 3D representation of the data set so that time is the Z-axis. This approach is useful as a method for visualizing physiological changes during development. For example, calcium dynamics have been characterized in sea urchin embryos when this method of displaying the data was used *(14)*. A simple method for displaying 3D information is by color coding optical sections at different depths. This can be achieved by assigning a color (usually red, green or blue) to sequential optical sections collected at various depths within the specimen. The colored images from the Z-series are then merged and colorized using an image manipulation program such as Adobe Photoshop® *(15)*.

4.4. Four-Dimensional Imaging

Time-lapse sequences of Z-series can also be collected from living preparations using the LSCM to produce 4D data sets, i.e., three spatial dimensions— X, Y, and Z—with time as the fourth dimension. Such series can be viewed using a 4D viewer program; stereo pairs of each time point can be constructed and viewed as a movie or a 3D reconstruction at each time point is subsequently processed and viewed as a movie or montage *(16,17)*.

4.5. X–Z Imaging

An X–Z section produces a profile of the specimen, e.g., a vertical slice of an epithelial layer (**Fig. 8**). Such X–Z profiles can be produced either by scanning a single line at different Z depths under the control of the stepper motor or by extracting the profile from a Z-series of optical sections using a cut plane option in a 3D reconstruction program.

4.6. Reflected Light Imaging

Unstained preparations can also be viewed with the LSCM using reflected (backscattered) light imaging. This is the mode used in all of the early confocal instruments (**Fig. 9**). In addition, the specimen can be labeled with probes that reflect light such as immunogold or silver grains *(18)*. This method of imaging has the advantage that photobleaching is not a problem, especially for living samples. Some of the probes tend to attenuate the laser beam, and in some LSCMs there can be a reflection from optical elements in the microscope. The problem can be solved using polarizers or by imaging away from the reflection artifact and off the optical axis. The reflection artifact is not present in the slit or multiple-beam scanning systems.

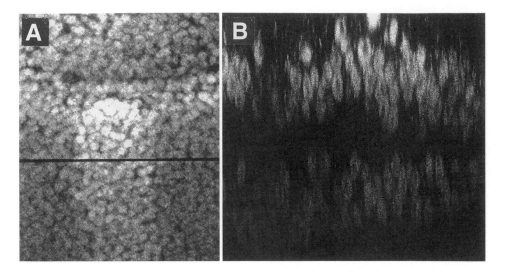

Fig. 8. *X–Z* imaging; the laser was scanned across a single position in the sample (marked by the horizontal black line in (**A**)) at different *Z* depths. An *X–Z* image was built up and displayed in the confocal imaging system (**B**). Note that the butterfly wing epithelium is made up of two epithelial layers, and since the fluorescence intensity drops off deeper into the specimen, only the upper layer is visualized.

4.7. Transmitted Light Imaging

Any form of transmitted-light microscope image, including phase-contrast, DIC, polarized light, or dark field can be collected using a transmitted light detector (**Fig. 9**), which is a device that collects the light passing through the specimen in the LSCM. The signal is transferred to one of the PMTs in the scan head via a fiber optic cable. Because confocal epifluorescence images and transmitted light images are collected simultaneously using the same beam, image registration is preserved, so that the precise localization of labeled cells within the tissues can be mapped when the images are combined using digital methods.

It is often informative to collect a transmitted, nonconfocal image of a specimen and to merge it with one or more confocal fluorescence images of labeled cells. For example, the spatial and temporal components of the migration of a subset of labeled cells within an unlabeled population of cells can be followed over hours or even years *(19)*.

A real color transmitted light detector has recently been introduced that collects the transmitted signal in the red, the green, and the blue channels to build the real color image in a similar way to some color digital cameras. This device is useful to pathologists who are familiar with viewing real colors in transmitted light and overlaying the images with fluorescence.

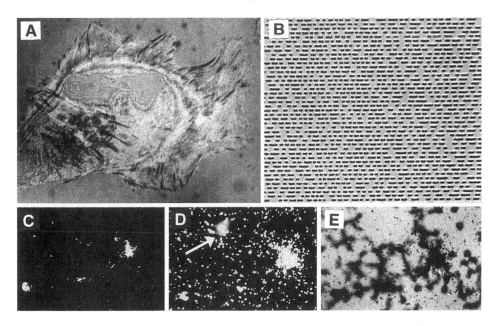

Fig. 9. Examples of reflected light (**A,B,C**) and transmission imaging (**D,E**): Interference reflection microscopy in the LSCM demonstrates cell substratum contacts in black around the cell periphery (**A**); confocal systems are used extensively in the materials sciences—here the surface of an audio CD is shown (**B**); (**C**) through (**E**) show an in situ hybridization of HIV-infected blood cells. The silver grains can be clearly seen in the reflected light confocal image (**C**) and in the transmitted light dark field image (**D**) and bright field image (**E**). Note the false positive from the dust particle [arrow in (**D**)], which is not present in the optical section (**C**).

4.8. Correlative Microscopy

The premise of correlative or integrated microscopy is to collect information from the same region of a specimen using more than one microscopic technique. For example, the LSCM can be used in tandem with the transmission electron microscope (TEM). The distribution of microtubules within fixed tissues has been imaged using the LSCM, and the same region was imaged in the TEM using eosin as a fluorescence marker in the LSCM and as an electron dense marker in the TEM *(20)*. Reflected light imaging and the TEM have also been used in correlative microscopy to image the cell substratum contacts in the LSCM (**Fig. 9A**) and in the TEM *(21)*.

5. Specimen Preparation and Imaging

Most of the protocols for confocal imaging are based upon those developed over many years for preparing samples for the conventional wide field microscope *(22–25)*. A good starting point for the development of a new protocol for the confocal microscope therefore is with a protocol for preparing the samples for conventional light microscopy, and to later modify it for the confocal instrument if necessary. Most of the methods for preparing specimens for the conventional light microscope were developed to reduce the amount of out-of-focus fluorescence. The confocal system undersamples the fluorescence in a thick sample as compared with a conventional epifluorescence light microscope, with the result that samples may require increased staining times or concentrations for confocal analysis, and may appear to be overstained in the light microscope.

The illumination in a typical laser scanning confocal system appears to be extremely bright although many points are scanned per second. For example, a typical scan speed is one point per 1.6 µs so that the average illumination at any one point is relatively moderate, and generally less than in a conventional epifluorescence light microscope. Many protocols include an antibleaching agent that protects the fluorophore from the bleaching effects of the laser beam. It is advisable to use the lowest laser power that is practical for imaging in order to protect the fluorochrome, and antibleaching agents may not be required when using many of the more modern instruments (*see* Chapter 3).

The major application of the confocal microscope is for improved imaging of thicker specimens, although the success of the approach depends on the specific properties of the specimen. Some simple ergonomic principles apply; e.g., the specimen must physically fit on the stage of the microscope and the area of interest should be within the working distance of the lens. For example, a high resolution lens such as a 60× numerical aperture (NA) 1.4 has a working distance of 170 µm whereas a 20× NA 0.75 has a working distance of 660 µm Occasionally, resolution may have to be compromised in order to reach the region of interest, and to prevent squashing the specimen with the lens and risking damage to it.

Steps should be taken to preserve the 3D structure of the specimen for confocal analysis using some form of spacer between the slide and the coverslip, e.g., a piece of coverslip or plastic fishing line. When living specimens are the subject of study it is usually necessary to mount them in a chamber that will provide all of the essentials for life on the stage of the microscope, and will also allow access to the specimen using the objective lens for imaging without deforming the specimen.

Properties of the specimen such as opacity and turbidity can influence the depth into the specimen that the laser beam may penetrate. For example, unfixed and unstained corneal epithelium of the eye is relatively transparent and therefore the laser beam will penetrate further into it (approx. 200 µm) than, for example, into unfixed skin (approx. 10 µm), which is relatively opaque and therefore scatters more light. The tissue acts like a neutral density filter and attenuates the laser beam. Many fixation protocols incorporate some form of clearing agent that will increase the transparency of the specimen.

If problems do occur with depth penetration of the laser light into the specimen then thick sections can be cut using a microtome; usually of fixed specimens but also slices of living brain have been cut using a vibratome, and imaged successfully. The specimen can also be removed from the slide, inverted, and remounted, although this is often messy, and usually not very successful. Dyes that are excited at longer wavelengths, e.g., cyanine 5, can be used to collect images from a somewhat deeper part of the specimen than with dyes excited at shorter wavelengths *(26)*. Here the resolution is slightly reduced in comparison to that attained with images collected at shorter wavelengths. Multiple-photon imaging allows images to be collected from deeper areas within specimens because red light is used for excitation.

5.1. The Objective Lens

The choice of objective lens for confocal investigation of a specimen is extremely important *(27)*, as the NA of the lens, which is a measure of its light-collecting ability, is related to optical section thickness and to the final resolution. Basically, the higher the NA is, the thinner the optical section will be. The optical section thickness for the 60× (NA 1.4) objective lens with the pinhole set at 1 mm (closed) is on the order of 0.4 µm, and for a 16× (NA 0.5) objective, again with the pinhole at 1 mm, the optical section thickness is approx. 1.8 mm. Opening the pinhole (or selecting a pinhole of increased diameter) will increase the optical section thickness further (**Table 2**). These values were measured from the BioRad MRC600 LSCM. The vertical resolution is never as good as lateral resolution. For example, for a 60× NA 1.4 objective lens the horizontal resolution is approx. 0.2 µm and the vertical resolution is approx. 0.5 µm. Chromatic aberration, especially when imaging multilabeled specimens at different wavelengths, and flatness of field are additional factors to consider when choosing an objective lens *(6)*.

The lenses with the highest NAs are generally those with the highest magnifications, and most expensive, so that a compromise is often struck between the area of the specimen to be scanned and the maximum achievable resolution for the area (**Table 3**). For example, when imaging *Drosophila* embryos and imaginal disks a 4× lens is used to locate the specimen on the slide, a 16× (NA 0.5) lens for imaging whole embryos, and a 40× (NA 1.2) or 60× (NA 1.4) lens for resolving individual cell nuclei within embryos and imaginal disks. For large tissues, for example, butterfly imaginal disks, the 4× lens is extremely useful for whole wing disks, and for cellular resolution 40× or 60× is used (**Fig. 10**). Some microscopes have the facility to view large fields at high resolution using an automated *X–Y* stage that can move around the specimen, and collecting images into a montage. Such montages can also be built manually and pasted together digitally.

Table 2
Optical Section Thickness (in microns) for Different Objective Lenses Using the Bio-Rad MRC600 Laser Scanning Confocal Microscope

Objective		Pinhole	
Magnification	NA	Closed (1 mm)	Open (7 mm)
60×	1.40	0.4	1.9
40×	1.30	0.6	3.3
40×	0.55	1.4	4.3
25×	0.80	1.4	7.8
4×	0.20	20.0	100.0

Table 3
Important Properties of Microscope Objective Lenses for Confocal Imaging. An Aid for Choosing the Correct Lens for Imaging.

Property	Objective 1	Objective 2
Design	Plan-apochromat	CF-fluor DL
Magnification	60	20
Numerical aperture	1.4	0.75
Coverslip thickness	170 um	170 um
Working distance	170 um	660 um
Tube length	160 mm	160 mm
Medium	Oil	Dry
Color correction	Best	Good
Flatness of field	Best	Fair
UV transmission	None	Excellent

*a*Objective 1 would be more suited for high-resolution imaging of fixed cells whereas Objective 2 would be better for imaging a living preparation stained with a UV dye.

A useful feature of most LSCMs is the ability to zoom an image with no loss of resolution using the same objective lens. This is achieved simply by decreasing the area of the specimen scanned by the laser by controlling the scanning mirrors and by placing the information from the scan into the same area of framestore or computer memory. Several magnifications can be imparted onto a single lens without moving the specimen (**Fig. 10C,D,E,F**). However, when possible a lens with a higher NA should be used for the best resolution, rather than zooming a lens of lower NA.

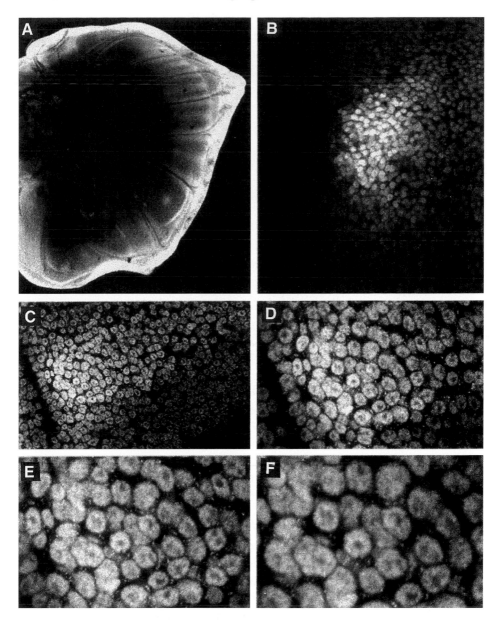

Fig. 10. Different objective lenses and zooming using the same lens. The 4×
lens (**A**) is useful for viewing the entire butterfly fifth instar wing imaginal disk
although the 16× lens (**B**) gives more nuclear detail of the distal-less stain. The
40× lens gives even more exquisite nuclear detail (**C**), and zoomed by progressive
increments (**D,E,F**).

Table 4
Peak Excitation and Emission Wavelengths of Some Commonly-Used Fluorophores and Nuclear Counterstains

Dye	Exc. Max. (nm)	Em. Max. (nm)
FITC	496	518
Bodipy	503	511
CY3	554	568
Tetramethylrhodamine	554	576
Lissamine rhodamine	572	590
CY3.5	581	588
Texas Red	592	610
CY5	652	672
CY5.5	682	703
Nuclear dyes		
Hoechst 33342	346	460
DAPI	359	461
Acridine orange	502	526
Propidium iodide	536	617
TOTO3	642	661

Many instruments have an adjustable pinhole. Opening the pinhole gives a thicker optical section and reduced resolution but it is often necessary to provide more detail within the specimen or to allow more light to strike the photodetector. As the pinhole is closed the section thickness and brightness decrease, and resolution increases up to a certain pinhole diameter, at which resolution does not increase but brightness continues to decrease. This point is different for each objective lens *(28)*.

5.2. Probes for Confocal Imaging

The synthesis of novel fluorescent probes for improved immunofluorescence localization continues to influence the development of confocal instrumentation *(29)* and *see* Chapter 3. Fluorochromes have been introduced over the years with excitation and emission spectra more closely matched to the wavelengths delivered by the lasers supplied with most commercial LSCMs (**Table 4**). Improved probes that can be conjugated to antibodies continue to appear. For example, the cyanine dyes are alternatives to more established dyes; cyanine 3 is a brighter alternative to rhodamine and cyanine 5 is useful in triple-label strategies.

Fluorescence in situ hybridization (FISH) is an important approach for imaging the distribution of fluorescently labeled DNA and RNA sequences in cells *(30)* and *see* Chapter 5. In addition, brighter probes are now available for

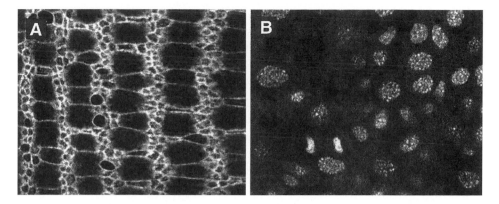

Fig. 11. Examples of dyes used for labeling cellular features. Cell outlines can be labeled with fluorescently labeled phalloidin (**A**) or nuclei using ToPro (**B**). Both samples are whole mounts of butterfly pupal wing imaginal disks.

imaging total DNA in nuclei and isolated chromosomes using the LSCM. For example, the dimeric nucleic acid dyes TOTO-1 and YOYO-1 and dyes such as Hoechst 33342 and 4,'6-diamidino-2-phenylindale (DAPI) have excitation spectra (346 nm and 359 nm) that are too short for most of the lasers and mirrors that are supplied with the commercially available LSCMs, although these dyes can be imaged using a HeNe laser/UV system *(31)* or multiple-photon microscopy. The latter technique does not require specialized UV mirrors and lenses because it uses red light for excitation with a pulsed Ti-Sapphire laser for illumination *(3)*.

Many fluorescent probes are available that stain, using relatively simple protocols, specific cellular organelles and structures. These probes include a plethora of dyes that label nuclei, mitochondria, the Golgi apparatus, and the endoplasmic reticulum, and also dyes such as the fluorescently labeled phalloidins that label polymerized actin in cells *(29)*. Phalloidin is used to image cell outlines in developing tissues, as the peripheral actin meshwork is labeled as bright fluorescent rings (**Fig. 11**). These dyes are extremely useful in multiple labeling strategies to locate antigens of interest with specific compartments in the cell, for example, a combination of phalloidin and a nuclear dye with the antigen of interest in a triple labeling scheme (**Fig. 11**). In addition, antibodies to proteins of known distribution or function in cells, e.g., antitubulin, are useful inclusions in multilabel experiments.

When imaging living cells it is most important to be aware of the effects of adding fluorochromes to the system. Such probes can be toxic to living cells, especially when they are excited with the laser. These effects can be reduced by adding ascorbic acid to the medium. The cellular component labeled can also affect its viability during imaging, e.g., nuclear stains tend to have a more deleterious effect than cytoplasmic stains. One way to overcome this problem is to include a fluorescent dye in the medium around the cells. Probes that distinguish between living and dead cells are also available and can be used to assay cell viability during imaging. Most of these assays are based upon the premise that the membranes of dead cells are permeable to many dyes that cannot cross them in the living state. Such probes include acridine orange; various kits are available from companies such as Molecular Probes *(29)*.

Many dyes, for example, Fluo-3 and rhod-2, have been synthesized that change their fluorescence characteristics in the presence of ions such as calcium. New probes for imaging gene expression have been introduced. For example, the jellyfish green fluorescent protein (GFP) allows gene expression and protein localization to be observed in vivo. GFP has been used to monitor gene expression in many different cell types including living *Drosophila* oocytes, mammalian cells, and plants using the 488 nm line of the LSCM for excitation *(32)*. Spectral mutants of GFP are now available for multi-label experiments and are also useful for avoiding problems with autofluorescence of living tissues *(33* and *see* Chapter 15).

5.3. Autofluorescence

Autofluorescence can be a major source of increased background when imaging some tissues. Tissue autofluorescence occurs naturally in many cell types. In yeast and in plant cells, for example, chlorophyll fluoresces in the red spectrum. In addition, some reagents, especially glutaraldehyde fixative, are sources of autofluorescence, which can be decreased by borohydride treatment. Autofluorescence can be avoided by using a wavelength for excitation that is out of the range of natural autofluorescence. The longer wavelength excitation of cyanine 5 is often chosen to avoid autofluorescence at shorter wavelengths.

The amount of autofluorescence can be assessed by viewing an unstained specimen at different wavelengths and taking note of the PMT settings of gain and black level together with the laser power (**Fig. 12**). Autofluorescence may be bleached out using a quick flash at high laser power or flooding the specimen with light from the mercury lamp. A more sophisticated method of dealing with autofluorescence is using time resolved fluorescence imaging. Autofluorescence can also be removed digitally by image subtraction. Although it is more often a problem, tissue autofluorescence can be utilized for imaging overall cell morphology in multiple-labeling schemes.

5.4. Collecting the Images

The novice user can gain experience in confocal imaging from several sources. The manual provided with the confocal imaging system usually

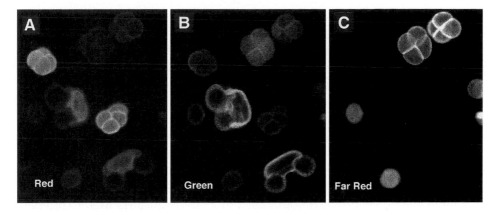

Fig. 12. Example of tissue autofluorescence. Note that different types of pollen fluoresce at different excitation wavelengths; these images were collected simultaneously at the same settings of gain, black level, and pinhole diameter.

includes a series of simple exercises necessary for getting started. The person responsible for operating the instrument may provide a short orientation session, and in most multiuser facilities the manager will usually require a short training session and demonstration of a certain competence level before solo imaging is allowed. The novice should pay particular attention to the house rules of the facility. Other useful sources of information are the training courses conducted by the confocal companies, workshops on light microscopy, and various publications.

It is essential to be familiar with the basic operation of the imaging system before working with experimental slides. It is usually recommended, for the novice at least, to start imaging with a relatively easy specimen rather than with a more difficult experimental one. Some good test samples include paper soaked in one or more fluorescent dyes or a preparation of fluorescent beads, which are both bright and relatively easy specimens to image with the confocal microscope. A particular favorite of mine is a slide of mixed pollen grains that autofluoresce at many different wavelengths (**Fig. 12**). Such slides are available from biological suppliers such as Carolina Biological or can be easily prepared from pollen collected from garden plants. These specimens tend to have some interesting surface features and hold up well in the laser beam. A relatively reliable test specimen for living studies can be prepared from onion epithelium or the water plant *Elodea* sp., using autofluorescence or staining with DiOC6 *(11)*. Many examples of test specimens are covered further in Chapter 2.

The aim should always be to gain the best possible performance from the instrument, and this starts with optimal alignment, especially when imaging with older model confocal instruments. The alignment routine depends on the specific instrument, and is usually best performed by the person responsible for the instrument. Alignment should definitely not be attempted before training and permission from the microscope owner has been granted. This is because the beam can be lost completely, and in the case of some instruments it may require a service visit to rectify the situation.

The basic practices of light microscopic technique should be followed at all times *(6)*. For example, all glass surfaces should be clean because dirt and grease on coverslips and objective lenses are major causes of poor images. Care should be taken to mount the specimen so that it is within the working distance of the objective lens. The refractive index between the lens and the specimen should be matched correctly. For example, use the correct immersion oil for a particular NA and use a coverslip of correct thickness for the objective lens, especially for higher power lenses, which will require a No. 1 or No. 1.5 coverslip, and not a No. 2 coverslip. The coverslip should be sealed to the slide in some way, and mounted flat—use nail polish for fixed specimens, making sure that it's dry before imaging, and some form of nontoxic sealing agent for live specimens; e.g., a Vaseline, beeswax, and lanolin mixture works well. Much time and effort can be saved by taking great care with the simple basics of cleanliness at this stage.

A region of interest is located using either bright field or conventional epifluorescence microscopy, preferably using the microscope of the confocal system. It can be extremely difficult for the novice to find the correct focal plane using the confocal imaging mode alone (*see* Chapter 2). If conventional imaging is not available then structures of interest can be located using a separate fluorescence microscope and their positions marked using a diamond marker mounted on the microscope, a sharpie, or by recording the coordinates from the microscope stage. The ability to preview samples with the actual microscope of the confocal imaging system using the epifluorescence mode is especially useful when attempting to image a rare event such as a gene expressed at a specific stage of development in a sample containing hundreds of embryos of different ages. This can save much time in scanning many specimens using the confocal mode. Many instruments have a low-resolution fast-scanning mode that alleviates some of these problems. It is far easier, however, to scan slides using a conventional microscope when searching for rarely occurring events, and then immediately switching to the confocal mode to collect the images.

The secret to successful confocal imaging is in mastering the interplay between lens NA, pinhole size, and image brightness using the lowest laser power possible to achieve the best image. The new user should vary these parameters using the test specimen and several different objective lenses of different magnifications and NAs to gain a sense of the capabilities of the instrument before progressing to the experimental specimens. Try zooming using the zoom function and compare these images with those obtained using an objective lens of higher NA.

The specific imaging parameters of the microscope should be set up away from the region of interest to avoid photobleaching of valuable regions of the specimen. This usually involves setting the gain and the black levels of the photomultiplier detectors together with the pinhole size to achieve a balance between acceptable resolution and adequate contrast using the lowest laser power possible to avoid excessive photobleaching. Many instruments have color tables that aid in setting the correct dynamic range within the image. Such tables are designed so that the blackest pixels, around zero, are pseudocolored green and the brightest pixels, around 255 in an 8-bit system, are colored red. The gain and black levels (and the pinhole) are adjusted so that there are a few red and green pixels in the image, thus ensuring the full dynamic range from 0 to 255 is utilized. These adjustments can also be made by eye. It is not always practical to collect an image at full dynamic range because full laser power cannot be used or the specimen has uneven fluorescence, so that a bright region may obscure a dimmer region of interest in the frame.

As the specimen is scanned an image averaging routine is usually employed to filter out random noise from the photomultiplier and to enhance the constant features in the image. An image equalization routine can be applied directly after collection of the images so that the image is scaled to the full dynamic range. This routine should not be applied if measurements of fluorescence intensity are to be made unless a control image is included in the same frame as the rest of the experimental images before applying the equalization routine (*see* Chapter 20). If space on the hard disk allows it is often a good strategy to save raw unprocessed images in addition to any processed ones.

The image is usually saved to the hard disk of the computer and later backed up onto a mass storage device. In general it is advisable to collect as many images as possible during an imaging session, and, if necessary, to cull out the unsatisfactory ones in a later review session. It is quite surprising how a seemingly unnecessary image at first sight suddenly becomes valued at a later date after further review—especially with one's peers! It is much harder to prepare another specimen, and often harder still to reproduce the exact parameters of previously prepared specimens.

A strategy for labeling image files should be mapped out before imaging, and during imaging many notes should be taken or placed on the image file along with the image if this facility is available. Users should conduct a test to determine if this information is accessible after imaging and remember that it can be lost when the images are subsequently transferred to image manipulation programs such as NIH Image or Photoshop on other computers. It is hard to replace a well-ordered notebook, or perhaps a laptop computer file, preferably with a table of image file names with facility for comments and details of the objective lens and the zoom factor for calculating scale bars at a later date. Most confocal imaging systems do not automatically keep track of the lens used; this is important for calculating the scale bars for publication. In addition, some computer systems will accept up to nine characters for their file labels; and beware of using periods in the file names that can sometimes be confused by the software. For example, STEVE.NEW.PIC may be read by the imaging system as STEVE.NEW rather than as .PIC image file. Many modern systems incorporate an image database that will keep track of file names and location of the files, and may also include a thumbnail of the images (*see* Chapter 23).

5.5. Troubleshooting

A protocol will sometimes inexplicably cease to work, and there is often an initial reflex to blame the instrument rather than the sample. The authors have been encouraged to include tips on such eventualities in the *Notes* sections of their chapters, and this is covered in more detail in Chapter 2. A good test is to view the sample on a conventional epifluorescence microscope, and if some fluorescence is visible by eye then the signal should be very bright on the confocal system. If this is the case, it might be time to run through some checks of the confocal system using a known test specimen and not the experimental one. A digital file of an image of the test specimen should be accessible to all users together with all of the parameters of its collection including laser power, gain, black level, pinhole diameter, zoom, and objective lens used.

It is advisable to seek help from an expert who may have prior experience of the problem. If all else fails, do not panic, each of the confocal companies should have a good help line whose number is usually posted close to the microscope, and can be accessed through websites listed at the end of this chapter. As a rule, if you are not sure of something ask, or at least step back from the problem before attempting to remedy it.

Problems with the protocols themselves are usually caused by degradation of reagents, and a series of diagnostic tests should be performed. It is usually best to make up many of the reagents fresh yourself or, at least, "borrow" them from a trusted co-worker. Antibodies should be aliquoted in small batches from the frozen stock, and stored in the refrigerator. They should be reused only if absolutely necessary although this is sometimes unavoidable when using expensive or rare reagents, and often does not present a problem.

Bleedthrough can occur from one channel into another in multilabeled specimens. It can be caused by properties of the specimen itself or can result from problems with the instrument. The causes and remedies of bleedthrough have been reviewed in much detail elsewhere *(34)*. A good test of the instrument is to view a test sample with known bleedthrough properties using both the multiple-label settings and the single-label settings. It is advisable to collect an image of the test specimen and record the settings of laser power, gain, black level, and pinhole diameter so that when problems do occur one can return to these settings with a test sample and compare the images collected with those of the stored test images collected when the instrument was operating in an optimal way.

Additional tests include a visual inspection of the laser color and the anode voltage of the laser, e.g., if the beam from a krypton/argon laser appears blue and not white when scanning on a multiple-label setting then this suggests that the red line is weak. If this is the case then the anode voltage will usually be high, and can usually be reduced to an acceptable level by adjusting the mirrors on the laser; this should usually be left to the person responsible for the instrument. If it is not possible to reduce the voltage a new or refurbished laser may be required.

Sometimes the antibody probes may have degraded or need to be cleaned up. Older specimens may have increased background fluorescence and bleedthrough caused by the fluorochrome separating from the secondary antibody and diffusing into the tissue. Always view freshly prepared specimens if at all possible. Changing the concentration and/or the distribution of the fluorochromes often helps. For example, if fluorescein bleeds into the rhodamine channel then switch the fluorochromes so that rhodamine is on the stronger channel because the fluorescein excitation spectrum has a tail that is excited in the rhodamine wavelengths. The concentration of the secondaries can be reduced in subsequent experiments.

5.6. Image Processing and Publication

Confocal images are usually collected as digital computer files, and they can usually be manipulated using the proprietary software provided with the confocal imaging system. One of the most dramatically improved features of the LSCM has been in the display of confocal images. This part of the process is extremely important because although it is good to achieve improved resolution using the LSCM, this improvement is of little value if it cannot be displayed and reproduced in hard copy format.

Even 5 years ago most laboratories used darkrooms and chemicals for their final hard copy. Color images were even harder to reproduce because they were usually printed by an independent printer who had little idea of the correct color balance. For hard copies, images are now exported to a slide maker, a color laser printer, or to a dye sublimation printer for publication quality prints. Photographs are taken directly from the screen of the video monitor. Moreover, movie sequences can be published on the worldwide web.

The quality of published images has also improved dramatically as most journals are able to accept digital images for publication. This means that the resolution achieved within the computer of the confocal imaging system is more faithfully reproduced in the final published article. Some journals also publish their articles on CD ROM, which means that the images should be exactly the same as those collected using the confocal microscope. These technological advances are especially useful for color images where the intended resolution and color balance can be accurately reproduced by the journals, and, theoretically, at a much lower cost to the author.

6. Information Sources

6.1. Websites

6.1.1. Good General Sites

www.videomicroscopy.com   Superb magazine on video and digital imaging; excellent links to many websites that pertain to confocal technology and imaging. Good basic tutorials and sources of instrumentation including hard copy devices

www.ou.edu/research/electron/mirror/web-org.html   A directory of microscopy websites listed by organization

www.patents.ibm.com   The IBM patents webserver is a useful database of patents and contains those patents that pertain to confocal imaging. The entire patent including diagrams can be accessed through this server.

www.bocklabs.wisc.edu/imr/home2.htm   Useful site for basic principles of confocal, two-photon, and four-dimensional imaging. Lists meetings and workshops, and a booking form for reserving time on the instruments in Madison.

6.1.2. Confocal Microsxope Companies

www.microscopy.bio-rad.com   Bio-Rad Microscopes: Information on their laser scanning, real-time, and two-photon systems. Many useful application notes can be downloaded and a database of papers can be accessed here.

www.leica.com   Details on light microscopes including a laser scanning confocal system. Tutorials on confocal imaging

www.nikon.com   Microscopic products and technical information.

www.noran.com   Noran Instruments: Details of a real-time scanning system and image analysis software

www.olympus.co.jp   Microscopes and confocal imaging systems.

www.lasertec.co.jp   Lasertec

www.optiscan.com.au   Optiscan

www.technical.com   Technical Instruments

www.lsr.co.uk   Life Science Resources

www.mdyn.com.   Molecular Dynamics. Good application notes.

www.zeiss.com   Website for Carl Zeiss in the USA with details of light microscopes including real-time and laser scanning systems

6.1.3. Filters

www.chroma.com   Useful handbook of optical filters.
www.omegafilters.com
www.image1.com

6.1.4. Dyes

www.jacksonimmuno.com   Fluorescent probes and antibodies.

www.probes.com   The "Molecular Probes" website is great for details of most of the fluorescent probes used for imaging

6.1.5. Confocal Methodology

www.bioimage.org   Details of 3D microscopy.

www2.uchc.edu/htterasaki   Live cell imaging

6.1.6. Courses and Societies

www.mbl.edu   Marine Biological Laboratory at Woods Hole, which runs two excellent courses on basic light microscopy including sessions on confocal imaging. A good place to see many confocal microscopes at the same site.

www.cshl.org   Cold Spring Harbor laboratory web page; details of various courses and CSH Press publications

msa.microscopy.com   Web site of the Microscopy Society of America contains useful links to other sites and microscopy societies. Details of their annual conference and of other meetings pertaining to all forms of microscopy including confocal microscopy.

www.rms.org.uk   Website of the Royal Microscopy Society of the UK, and links to the *Journal of Microscopy*

6.2. Listservers

One of the most useful confocal resources is the confocal e-mail group based at SUNY, Buffalo and started by Robert Summers. The confocal listserver was set up some years ago as a discussion group for Bio-Rad users, and it has developed into a discussion group on all forms of confocal microscopy and related technologies. An extremely useful aspect of the group is that previous messages are archived for reference purposes.

To join the group send the e-mail message "Subscribe Confocal (your name)" to the confocal microscopy list Confocal@listserv.acsu.buffalo, or for help contact the current listowner, Paddock@facstaff.wisc.edu.

Acknowledgments

I would like to thank Jim Williams (**Fig. 4**), Grace Panganiban (**Fig. 7**), Jane Selegue (**Fig. 8**), Dorothy Lewis (**Fig. 9C,D,E**), and Julie Gates and Ron Galant for allowing me to image their beautiful specimens.

References

1. Pawley, J. B. (1995) *Handbook of Biological Confocal Microscopy*, 2nd edition, Plenum Press, New York.
2. Chen, H., Hughes, D. D., Chan, T. A., Sedat, J. W., and Agard, D. A. (1996) IVE (Image Visualisation Environment): a software platform for all three-dimensional microscopy applications. *J. Struct. Biol.* **116,** 56–60.
3. Potter, S. M (1996) Vital imaging: two photons are better than one. *Curr. Biol.* **6,** 1595–1598.
4. Minsky, M. (1988) Memoir on inventing the the confocal scanning microscope. *Scanning* **10,** 128–138.
5. Minsky, M. (1957) U.S. Patent No. 3013467
6. Inoue, S. and Spring, K. S. (1997) *Video Microscopy: The Fundamentals*, 2nd. edition. Plenum Press, New York.
7. DeRisi, J., Penland, L., Brown, P. O., Bittner, M. L., Meltzer, P. S., Ray, M., Chen, Y., Su, Y. A., and Trent, J. M. (1996) Use of a cDNA microarray to analyze gene expression patterns in human cancer. *Nat. Gene.* **14,** 457– 460.
8. White, J. G., Amos, W. B., Durbin, R., and Fordham, M. (1990) Development of a confocal imaging system for biological epifluorescence application, in *Optical Microscopy for Biology*, Wiley-Liss, New York, pp. 1–18.
9. White, J. G., Amos, W. B., and Fordham, M. (1987) An evaluation of confocal versus conventional imaging of biological structures by fluorescence light microscopy. *J. Cell Biol.* **105,** 41–48.
10. Pawley, J. B. (1995) Light paths of current commercial confocal microscopes for biology, in Pawley, J. B. (ed.) *Handbook of Biological Confocal Microscopy*, 2nd edition. Plenum Press, Plenum Press. pp. 581–598.
11. Terasaki, M. and Dailey, M. E. (1995) Confocal microscopy of living cells, in Pawley, J. B. (ed.) *Handbook of Biological Confocal Microscopy*, 2nd edition, Plenum Press, New York, pp. 327–346.

12. Cheng, H., Lederer, W. J., and Cannell, M. B. (1993) Calcium sparks: elementary events underlying excitation-contraction coupling in heart muscle. *Science* **262**, 740–744.

13. White, N. S. (1995) Visualization systems for multidimensional CLSM, *Handbook of Biological Confocal Microscopy*, 2nd edition (J. B. Pawley, ed.), Plenum Press, New York, pp. 211–254.

14. Stricker, S. A., Centonze, V. E., Paddock, S. W., and Schatten, G. (1992) Confocal microscopy of fertilisation-induced calcium dynamics in sea urchin eggs. *Dev. Biol.* **149**, 370–380.

15. Paddock, S. W., Hazen, E. J., and DeVries, P. J. (1997) Methods and applications of three colour confocal imaging. *BioTechniques* **22**, 120–126

16. Thomas, C. F., DeVries, P., Hardin, J., and White, J. G. (1996) Four dimensional imaging: computer visualization of 3D movements in living specimens. *Science* **273**, 603–607.

17. Mohler, W. A. and White, J. G. (1998) Stereo-4-D reconstruction and animation from living fluorescent specimens. *BioTechniques* **24**, 1006–1012.

18. Paddock, S. W., Mahoney, S., Minshall, M., Smith, L. C., Duvic, M., and Lewis, D. (1991) Improved detection of in situ hybridisation by laser scanning confocal microscopy. *BioTechniques* **11**, 486–494.

19. Serbedzija, G. N., Bronner-Fraser, M., and Fraser, S. (1992) Vital dye analysis of cranial neural crest cell migration in the mouse embryo. *Development* **116**, 297–307.

20. Deerinck, T. J., Martone, M. E., Lev-Ram, V., Green, D. P. L., Tsien, R.Y., Spector, D. L., Huang, S., and Ellisman, M. H. (1994) Fluorescence photooxidation with eosin: a method for high resolution immunolocalisation and *in situ* hybridisation detection for light and electron microscopy. *J. Cell Biol.* **126**, 901–910.

21. Paddock, S. W. and Cooke, P. (1988) Correlated confocal laser scanning microscopy with high-voltage electron microscopy of focal contacts in 3T3 cells stained with Napthol Blue Black. *EMSA Abstr.* **46**, 100–101.

22. Sheppard, C. J. R. and Shotten, D. M. (1997) Confocal laser scanning microscopy. *Royal Microscopical Society Handbook Series No. 38*, Bios, Oxford.

23. Matsumoto, B. (1993) *Cell Biological Appplications of Confocal Microscopy. Methods in Cell Biology*, Vol. **38**, Academic Press, San Diego.

24. Stevens, J. K., Mills, L. R., and Trogadis, J. E. (1994) *Three-Dimensional Confocal Microscopy: Volume Investigation of Biological Systems*, Academic Press, San Diego.

25. Spector, D. L., Goldman, R., and Leinwand, L. (1998) *Cells: A Laboratory Manual*. Vol. **II**: *Light Microscopy and Cell Structure*, Cold Spring Harbor Press, Cold Spring Harbor, NY.

26. Cullander, C. (1994) Imaging in the far-red with electronic light microscopy: requirements and limitations. *J. Microscop.* **176**, 281–286.

27. Keller, H. A. (1995) Objective lenses for confocal microscopy, in *Handbook of Biological Confocal Microscopy*, 2nd edition (J. B. Pawley, ed.), Plenum Press, New York, pp. 111–126.

28 Wilson, T. (1995) The role of the pinhole in confocal imaging system, in *Handbook of Biological Confocal Microscopy*, 2nd edition (J. B. Pawley, ed.), Plenum Press, New York, pp. 167–182.

29. Haugland, R. P. (1996) *Handbook of Fluorescent Probes and Research Chemicals*, 6th edition, Molecular Probes Inc., Eugene, OR.
30. Birchall, P. S., Fishpool, R. M., and Albertson, D. G. (1995) Expression patterns of predicted genes from the *C. elegans* genome sequence visualised by FISH in whole organisms. *Nat. Genet.* **11,** 314–320.
31. Bliton, C., Lechleiter, J., and Clapham, D. E. (1993) Optical modifications enabling simultaneous confocal imaging with dyes excited by ultra-violet and visible-wavelength light. *J. Microscop.* **169,** 15–26.
32. Chalfie, M., Tu, Y., Euskirchen, G., Ward, W. W., and Prasher, D. C. (1994) Green fluorescent protein as a marker for gene expression. *Science* **263,** 802–805.
33. Heim, R. and Tsien, R. Y. (1996) Engineering green fluorescent protein for improved brightness, longer wavelength and fluorescence energy transfer. *Curr. Biol.* **6,** 178–182.
34. Brelje, T. C., Wessendorf, M. W., and Sorenson, R. L. (1993) Multicolor laser scanning confocal immunofluorescence microscopy: practical applications and limitations. *Methods Cell Biol.* **38,** 98–177.

2

Practical Considerations
for Collecting Confocal Images

David Carter

1. Introduction

Conventional microscopy delivers two-dimensional images in real time and real color to the eye of the user. Confocal microscopy adds a third dimension by imaging only one plane within the sample at a time so that variations in depth can be quantified *(1)*. This has both positive and negative aspects. The advantage is that a series of such slices can be reconstructed to give 3D views and enable volume analysis of the sample, and that any one slice is crisper and clearer than a full-field fluorescence image. The disadvantage is that the portion of the sample visible at any one time is so small that finding the most interesting parts of the specimen may no longer be possible (**Fig. 1**).

Computer control makes it easy to explore the temporal dimension, scanning time series instead of Z-series to measure the way a specimen varies over time. A wide choice of laser lines and detection filters delivers up portions of yet another dimension: wavelength. A further benefit of confocal systems is that they produce digital images, which can be manipulated easily and condensed into statistical data.

There are three main objectives to pursue during a series of experiments:

1. Identify new and interesting phenomena.
2. Collect a volume of data that proves that the first impression is real and valuable.
3. Collect a few high-quality images that explain the hypothesis and add sparkle to a publication.

1.1. Identifying New and Interesting Phenomena

It may take the novice user some time to develop skill in collecting and interpreting confocal images, and randomly perusing a sample to obtain a sense

From: *Methods in Molecular Biology, vol. 122: Confocal Microscopy Methods and Protocols*
Edited by: S. Paddock © Humana Press Inc., Totowa, NJ

Specimen Space v Sampled Volume in
Confocal and Bright Field Microscopes

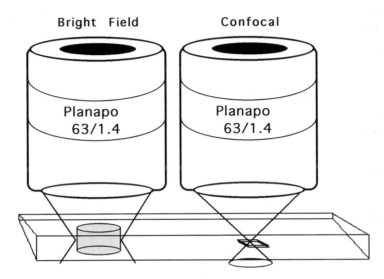

Accessible Sample
For a 22mm cover slip and 50μm useful working distance, specimen space is:
20,000 x 20,000 x 50 = 20,000,000,000 μm^3

Bright Field
Every point within the cylindrical field of view is simultaneously illuminated
and detected at any moment; an instantaneous sample volume of:
3.14 x100 x100 x 50 =1,600,000 μm^3.

Confocal
One point 0.2 x 0.2 x 0.5μm (0.02μm^3) within a square image plane is illuminated
and detected at a time; sample volume for a complete image scan is only:
200 x 200 x 0.5 = 20,000 μm^3.

Fig. 1. Specimen space vs sampled volume in a confocal and a bright field microscope.

for how it all fits together is a necessary precursor to delving in with more
detailed analyses. Most laboratories have stocks of well known animal models
or cell lines, and a supply of reliable staining protocols for fluorescent imaging
(**Subheading 2.0.**). Browsing an interesting sample with a well-designed confocal system is useful, but keep your notebook at hand to jot down chance
discoveries.

1.2. Producing the Supporting Data

When you have seen enough evidence that something interesting is happening, you may need to design rigorous tests to prove or disprove these observations, and to uncover more subtle relationships within the test sample. If a statistical analysis is called for, care should be given to designing the most efficient sampling procedure—one that proves the point without generating unnecessary reams of data.

When the intention is to see the most detail in a sample, pixel spacing should be set at half the separation of the optical resolution. This is the so-called Nyquist criterion, and holds true for all imaging systems at any magnification. Where structures of interest are significantly larger than the maximum resolution of the microscope, a lower magnification can be used so that more structures can be counted in each image. For example, in a study of spatial separation of cell nuclei, a pixel separation of 2 μm may be acceptable; whereas if chromatin spots within each nucleus must be mapped, pixels may have to be 10 times smaller with a consequent reduction in sampled area in each image.

1.3. Producing Convincing Images

The final requirement for a confocal study is the need for really pretty images. Unlike **Subheading 1.2.**, the emphasis here is purely qualitative—pulling out all the stops to get the most attractive, most informative images possible. Exceeding the Nyquist criterion and packing in as many pixels as possible into the region of interest may not gain any information, but the smoother transition across the image from one structure to the next may make it worth the extra exposure time. If available, a good interpolation routine will achieve a similar result, but this may be less satisfying than extracting excellent images directly from the specimen.

Cleaner images come from brighter illumination, slower scan speeds, a smaller confocal aperture, and lower photomultiplier tube (PMT) gains or CCD integration times. Under these circumstances, bleach rates are much higher, so there is a limit to how often a region can be scanned. Practice on unattractive regions of the sample to optimize scan parameters, then go to draft or fast scan mode, frame a good region, then scan it once with the highest resolution parameters.

Most laboratories will have a favorite sequence of postprocessing steps, which may have to be varied considerably between samples, but should be posted on a cue card for the benefit of occasional users. The following is an example:

1. Directional smooth to eliminate random pixel noise
2. Histogram Equalize to improve contrast in the image

3. Threshold to eliminate background in each channel
4. Auto background subtraction to lower the step between threshold and zero
5. Rescale to spread the surviving pixel values across the whole dynamic range
6. Palette change, to exaggerate the most interesting channel

The most attractive confocal images are generated from 3D data sets, where the volume is rendered with simulated shadowing from an apparent light source, and where each channel is given its own opacity value to make it more or less opaque, the so-called SFP reconstruction.

2. Materials
2.1. Stains

Use these stains as test standards for practice and to be sure that everything is working properly before moving up to more challenging material. Make a set of images that describe these samples, and compare new findings with this stock of initial images to hunt for interesting differences.

1. Rhodamine B hexyl ester, 1 to 10 µg/mL live or fixed for membranes
2. Hoechst 33347, 10 µg on fixed material, or 100 µg on live cells for DNA
3. Bodipy phalloidin, 10 µg on fixed material for F-actin
4. Rhodamine 123, 1 µg/mL on live cells for mitochondria
5. Carboxyfluorescein diacetate (CFDA), 7 µg/mL, for live cells

2.2. Lens Choice

The most interesting regions for confocal study are often too small or too subtle to be spotted with low-power, low numerical aperture (NA) objectives, so confocal imaging is nearly entirely done through top-quality oil or water immersion objective lenses. By reducing the sweep angle of the scanning mirrors, pan and zoom controls can be used for homing in on details within a field of view, so the main challenge is finding fields of view that contain promising regions (**Fig. 2**).

The resolution limit of a confocal microscope is set by the NA of the lens and the excitation/collection wavelengths. This means that the ideal lens for routine confocal imaging is the lowest magnification lens with the highest available NA. A 63× oil objective is easier to use than an equivalent 100× lens simply because it offers a larger working area to the user without sacrificing performance. Working distance is also important, but for oil objective lenses performance deteriorates so quickly after 50 µm penetration that provided the sample is mounted flat, a longer working distance gives little benefit.

1. *Water objectives:* There has been a recent trend in confocal microscopy toward the use of water immersion objective lenses, which are significantly more expensive than the oily types and have slightly lower NAs. These disadvantages arc

Available Sample Volume for Typical Objectives

Objective Lens			
100x oil	63x Oil	40x Water	10x Dry
NA 1.3	NA 1.4	NA1.2	NA 0.32

Axial Resolution			
0.45µm	0.35µm	0.56µm	10µm (Reflection)
0.74µm	0.58µm	0.93µm	16µm (Fluorescence)

Viewable Volume			
150x150	200x200	350x350	1400x1400
x90µm	x90µm	x220µm	x1900µm

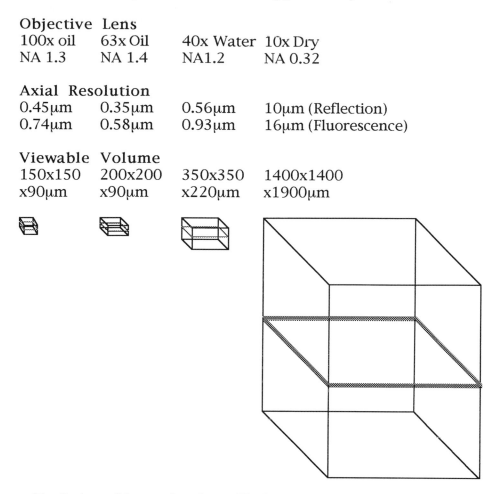

Fig. 2. Accessible sample volume: The key properties of four commonly used objective lenses are shown. Note that there is no benefit in using a higher magnification lens if the NA is not also greater, as the resolution does not increase, but the available sample volume is much reduced.

outweighed by their good compatibility with watery specimens which more than doubles the effective working distance, and the greater convenience of using water as the immersion fluid. They are also UV compatible. A good water objective lens is so versatile that it can replace a whole cluster of dry and oil objective lenses and actually saves money on the cost of a fully equipped system.

2. *Dry objectives:* There are instances when a specimen cannot be exposed to immersion fluid but must be left dry. Dry objective lenses typically have much

lower NAs and hence lower resolution than an equivalent immersion objective. But this is balanced by a longer working distance and cleaner work environment.

3. *Dipping objectives:* Dipping objectives are used where watery specimens need to be viewed directly without a coverslip between lens and specimen. These lenses have a narrow profile for insertion into dishes and are typically encased in either ceramic or Teflon for corrosion resistance in physiological solutions. This is a good type of lens for use with cultured cells on an upright system, as it has a long working distance to allow for addition of reagents, and eliminates the need for perfusion chambers.

2.3. Test Samples

1. Multiline submicron beads: The most universal test samples for confocal microscopes are made from beads that are smaller than the resolution limit of the microscope and have been doped to fluoresce at many wavelengths. Tetraspeck beads (Molecular Probes, Eugene, OR) are excited at all popular laser wavelengths from 354 nm UV to 647nm dark red, and come in a range of sizes from 0.1 to 4 µm.

2. FocalCheck™ beads: Molecular Probes sells a series of 15 µm FocalCheck beads with a visually striking staining pattern that readily shows the accuracy of instrument alignment and are easier to use than submicron multiline beads. Test slides can be made with a 5-µL aliquot of each bead suspension, spread with the pipet tip on a glass slide, and allowed to dry. Apply a drop of mounting medium and cover with a 22-mm No. 1.5 cover glass. Permafluor™ (Lipshaw Inc., Detroit, MI) works well as a mountant, as it sets completely after a few days, but is an aqueous solution that does not dissolve out the fluorophores.

   Both bead types have proved useful for factory alignment and final testing of the Meridian/TR series of instruments, which require no alignment in routine operation. The UV/Blue FocalCheck beads are excellent targets for checking the coalignment of the independent UV and visible lightpaths of the Meridian Ultima, and the Blue/Orange/Red FocalCheck beads are ideal for testing visible-only confocal systems, such as the Meridian InSight.

3. *Inclined mirror:* An easy method for testing axial resolution in reflection is to take a series of images of an inclined surface. If the Z-separation between images is equal to or greater than the resolution, a maximum intensity projection of the images will appear as a series of parallel stripes. If the Z-resolution is less, all the lines will merge together as a single, solid bar. Axial resolution is the step size at which the transition between one bar and the next can just be discriminated as a slight dip in brightness. A front silvered mirror is ideal, but if the lightpath is efficient enough, an ordinary glass slide will give enough reflection. Glue a broken sliver of slide to one edge of another such that it can be held securely on the stage and present an inclined plane to the objective.

4. *Silicon chip:* Good lateral resolution samples are more difficult to devise. The tracks on silicon chips make good test samples for reflection imaging, but can only be used for resolution tests with low NA objective lenses. Chrome on glass

test targets with exact spacing between tracks are commercially available, but are expensive and do not have features with submicron separations.

5. *Diatoms:* Repetitive structures in various species of diatom can be used as resolution tests, and these can be purchased from most biological supply houses, e.g., Carolina Biological Supply, as assortments, or as more expensive individually mounted target sets. Once the system is properly aligned for reflection, take a representative image and keep it for future reference. Diatoms can also be used as fluorescence standards by mounting them in immersion oil doped with fluorophore. This produces negative images, where the specimen appears dark against a bright background.

6. *Pollen:* Pollen grains make notoriously effective demonstration specimens for confocal imaging. The lignified outer skeleton of the grain can be stained with acridine orange to enhance its natural fluorescence and make it stable against bleaching. Lily pollen grains have triangular features along the main ribs that blend together when the system is poorly aligned, and become crisper and more distinct as alignment is optimized.

7. *Fluorescent paper:* Many grades of paper are naturally fluorescent, e.g., linen bond thesis paper, which fluoresces moderately under blue excitation. Special papers, such as bright yellow sticky notes, are considerably brighter. Alternatively soak filter paper in any of the commonly available fluorescent dyes. To reduce the amount of light scatter, the paper can be soaked in immersion oil before it is mounted for observation. This renders it translucent and allows for imaging at great depth.

8. *Fluorescent plastic:* The brightly colored plastic used to make children's toys is uniformly fluorescent, and can be used either as whole sheets or as thin sections for aligning an instrument.

9. *Thin fluorescent dyes:* Extremely thin fluorescent targets can be made by mounting a small quantity of fluorescent ink between slide and coverslip. A suitable mixture of dyes will fluoresce with similar intensity for three or four laser lines simultaneously.

10. *Mixed bead slides:* Populations of fluorescent beads can be mixed and mounted together to make test standards for many experimental procedures. For example, a slide containing a tiny minority of fluorescent beads can be used to test automated search routines, while known proportions of beads with differing size or brightness are useful for confirming population characterizing procedures.

11. *Live cells:* A dish of cells can be stained in just a few minutes with a membrane dye such as Rhodamine B hexyl-ester (hexyl-rhodamine, 10 μg/mL), which labels all membranes throughout the cell, but has very little fluorescence in free solution. Fixed cells can be stained as quickly with a DNA stain such as Hoechst 33342 (10 μg/mL), which will also penetrate live cells at higher concentrations (100 μg/mL, 1 h).

12. *Live plant tissue:* Instant samples can be made from plants or algae, which have chlorophyll that is brightly fluorescent blue to red without any pre-treatment. Slices of spring onion leaf are easy to handle because they are circular and can be

mounted between slide and coverslip without falling over. The outer cuticle of the leaf fluoresces green, and lignified xylem vessels have broad excitation and emission ranges. The epidermis between layers at the root end can be stained with hexyl-rhodamine, and the DNA with ethidium bromide. Provided that the staining solution has low osmolarity, vesicles will be seen parading around the cell due to cytoplasmic streaming.

3. Methods

3.1. Test of Instrument Performance Using Multiline Submicron Beads

1. *Field flatness:* In a perfectly aligned system, the whole field of view will come into focus together as the Z-drive of the confocal system is adjusted. If there is an error in field flatness, it can be characterized by collecting a Z-series of the whole field of view, using a very small step size, then making a depth shaded reconstruciton of the data volume.
2. *Axial resolution:* A monolayer of beads can be used to test axial resolution in fluorescence for any combination of excitation and detection wavelengths. Take an X–Z image through a patch of Tetraspeck™ beads, then take a vertical line query through them. Axial responses are recorded as full width half max (FWHM) values, which is the distance between the two points at which the X–Z image records half its maximum brightness.
3. *Chromatic aberration:* Most objective lenses have excellent color correction at the center of the field, where a line query across a bead image will show each channel rising to a peak at the same location. To test the whole field of view, take a line query across beads at each corner of the field of view and at the middle. Radial errors are more likely than errors tangential to the optical axis, so line queries in a direction that passes through the center of the field of view are more likely to register an error. If an aberration is detected, a second line query perpendicular to the first will fully characterize its magnitude. Alternatively, a "number 4" shaped line query can be used to summarize chromatic separation in a single, folded line query.
4. *Z-drive reproducibility:* Reproducibility of the Z-drive mechanism can be tested by collecting two Z-series through a plane of Tetraspeck beads. The beads should be in sharpest focus and hence brightest at corresponding levels in each series.
5. *Particle analysis:* A mixed population of beads is useful for testing the ability of the whole system to discriminate differences in size brightness. Flow cytometry test kits are available for calibrating a system against a known number of fluorophore molecules per bead. However, they may be too precise and hence too expensive for everyday use.

3.2. Test of Instrument Performance Using FocalCheck Beads

1. *Chromatic aberration:* It is often easier to use FocalCheck beads for characterizing chromatic aberration, as they are more visually striking, and present a larger

target than submicron beads for testing with a line query. The large beads are easier to locate, and it requires less dexterity with the mouse to accurately bisect the bead image with a line query. Colocalization scattergram plots can be used on a single FocalCheck bead to present the accuracy of color registration, and chart its improvement as an alignment procedure is carried out.

2. *Image analysis:* Images of FocalCheck beads should always be circular, and their maximum size should be constant regardless of their position on the field of view. This can be tested by compiling a maximum intensity projection of a field of beads, then querying the size and shape of each bead in the field.

3. *Calibration of Z-drive:* A simple X–Z image of a FocalCheck bead is sufficient to show that the Z-drive (and image rendering software) is indeed performing as expected, as these beads are spherical and should always appear circular in a confocal image. When preparing samples of large beads, be generous with the mounting resin so that it will not compress the beads between slice and coverslip as it shrinks and hardens.

4. *Refractive index correction:* When there is a mismatch between the refractive index of a specimen and that of the immersion fluid, a distortion in the Z-direction is introduced *(2)*. If the two refractive indices are known, and the NA of the objective lens is entered, this error can be compensated for so that the final data set is accurate in all dimensions.

5. *Coalignment of excitation beams:* By taking a multichannel X–Z image of a FocalCheck bead, the coalignment of laser lines can be confirmed. Slight errors in coalignment will appear as a slight vertical shift between channels, producing single-color fringes above and below the circular image of each bead. Standard X–Y images will show any lateral alignment errors between excitation lines, and a Z-series will show up both types of error in a single data set.

3.3. Sampling Strategies: Finding Regions of Interest Under Bright Field

The main benefit of confocal imaging is also its greatest limitation—that only a tiny portion of the specimen is visible at any given time (**Fig. 1**). For example, a standard microscope slide with a 22-mm coverslip viewed through a 100 µm working distance objective lens has a specimen volume of up to $20,000 \times 20,000 \times 100$ or 4×10^{10} µm^3. A single confocal image may be only 200 µm^2 and half a micron thick, or 2×10^4 µm^3 in volume. If each image takes just a second to acquire, it would take 2 million s (46 h) of imaging time to thoroughly scan the whole sample!

Clearly, it is impractical to scan all of a sample at high resolution with a confocal microscope, so strategies must be developed to help navigate through specimen space and find regions of interest quickly.

1. *Ink and scratch:* Simple steps such as mounting the specimen in the center of the coverslip, and marking the slide with a sharpie or scratch mark can save time in

locating your specimen on the confocal instrument. On an inverted microscope, the underside is readily accessible, so regions of interest can be dabbed or circled while the specimen is previewed under low power. These markings can be found easily with the condenser light, and will not be exposed to immersion oil or cleaning fluid that may wash them off.

For thin samples, it may be helpful to mark the specimen side of the slide with indelible ink or a diamond scratch before the sample is mounted, to make it easier to home in on the plane of focus. None of these marks will interfere with confocal imaging, as both the light source and the detection path are on the coverslip side of the specimen.

2. *Full-field flash:* When the sample is moved into the light path, scattered bright field light or fluorescence color changes show up through direct observation, even if the focal plane is not close enough to show any structural details through the ocular lenses. This is easily seen on an inverted microscope which offers a clear view of the underside of a slide-mounted sample, but even on an upright system, enough light is scattered through the slide itself to make its edges brighten when the sample is encountered. This is not the case when using the confocal microscope for locating the specimen because here the image of the specimen is either in focus or out of focus, and it can be extremely hard to find the specimen, especially for the novice user. It is therefore necessary to search for the specimen using the conventional light microscope.

3. *Combing a slide for regions of interest:* Uniform samples, such as cells grown on coverslips, will not flash when a region of interest is swept through full-field illumination, so more careful observation through the ocular lenses may be needed to find interesting locations.

To survey the entire area under a coverslip, use the lowest power objective that will show sufficient detail, and start looking in one corner of the coverslip. While looking through the oculars, scan the stage in one direction until the opposite edge of the coverslip is encountered, then move down one field of view and sweep back over the coverslip. Continue this across then down the pattern until the whole slide is scanned, or until enough locations of interest have been found. Where possible, it is easier to adjust the density of targets such that a random walk starting anywhere on the slide will quickly uncover enough sites for closer examination.

4. *Recording and revisiting regions of interest:* Marks on the specimen itself are easy to find again, and can be related to notes in a lab book. All microscope stages also have a numbered scale to indicate approximate X and Y positions, although reading and reproducing them accurately is often difficult.

If it is necessary to correlate coordinates from one microscope stage to another, a correction table can be made with graph paper glued to a slide. Take stage readings on opposite corners of each stage, then use the coordinates to calculate intermediate values for every location.

Universal grids are available that have a numbered/lettered grid etched on a standard slide. These can be used to correlate positions to a standard scale. Such grids can be constructed using an EM finder grid placed onto a coverslip, and depositing a thin layer of carbon over it onto the coverslip below. Alternatively, when a region of interest has been found and imaged, simply remove the slide without moving the stage, then insert the calibration scale and read off the grid location through the oculars.

5. *Epi-illumination of opaque specimens:* When the specimen is too large or opaque for conventional bright field range finding, locating the right plane of focus within the specimen can be frustrating. If there is any fluorescence, a mercury arc system will suffice, but the beam splitter needed for reflection imaging is not a common peripheral.

   The meter routine used for automatic alignment of the pinhole can also be used as a range finder. Select a very large pinhole and alternately adjust the gain and focus until the brightest plane is found.

   When neither is available, try obliquely illuminating the specimen with a flashlight. The new Meridian/TR system has a narrow gooseneck task light permanently mounted on the benchtop. This can be held very close to the objective lens below the specimen, or shone through an empty arc lamp port and reflected onto the specimen with a plain microscope slide.

6. *Full-field fluorescence from mercury arc or laser:* Most samples can be located and previewed under conventional bright field illumination. Many fluorochromes have enough color under white light to show where a specimen is on the slide, and in water-based samples there is often enough refractive difference between specimen and medium to resolve the sample with bright field or phase contrast. However, it is much easier to home in on promising areas using full field fluorescence from a mercury arc lamp. The color balance will be widely different from that experienced when scanning the sample because the excitation wavelengths are different and the human eye has a natural bias toward detecting greener colors.

   The Meridian ACAS and Ultima systems have beam diffusers that allow the laser frequencies to be used as a full field fluorescence source. This gives a more accurate perception of what the confocal optics will detect, as it uses exactly the same excitation and detection wavelengths.

   Camera-based confocal systems such as the Meridian InSight offer the user direct ocular viewing of the confocal fluorescence image. The human eye has a dynamic range many times greater than that of a CCD or PMT at a given setting, so a slight increase in brightness will be seen as the plane of focus is moved towards the specimen. Alternatively, the aperture can be opened wide until the specimen is found, then stopped back down to return to confocal imaging mode.

3.4. Homing in: Finding Regions of Interest in Confocal Viewing Mode or Matching the Experiment to the Instrument

Every confocal instrument has advantages and disadvantages, and until all confocal facilities are equipped with several different systems, the user will be aided or constrained by whatever instrument is available. To make the best use of available equipment, its benefits and limitations should be understood. Confocal manufacturers have developed a range of user friendly functions to help the user find regions of interest as quickly and easily as possible. When used properly, these capabilities can save a great deal of time and effort, and make the use of the confocal system a more enjoyable than frustrating experience.

1. *Confocal ocular viewing:* InSight and spinning disk systems: Slit scanning laser confocal microscopes, Nipkow and tandem spinning disk systems all have the advantage of delivering a real-time fully confocal image directly to the eye of the user. Focus and stage controls can be used as easily as for a bright field microscope to rapidly survey the specimen and find regions of interest. Moving the prism on the trinocular head of the microscope then sends the selected image to a CCD camera for collection and storage.

   The light throughput of spinning disk systems is so low that they are mostly used for reflection applications, whereas the slit scanner is actually more efficient than a point scanning laser confocal, and can be used to collect video rate images of dim fluorescent samples.

   The human eye boasts a much greater dynamic range and pixel density than any camera or PMT system with the greatest density of 150,000 cells/mm^2 *(3,4)*, and is exquisitely sensitive to detection of movement within the specimen. An image delivered straight to the eye has far more impact than a rendered image on a computer screen, and on good samples can be quite mesmerizing.

   Confocal facilities that have an InSight and a point scanner typically use the InSight on all samples to quickly check the quality of staining, before moving on to the slower, more difficult to use point scanner *(5)*. The newest infinity corrected InSight has a resolution comparable to point scanners in X and Y, and the latest version, the InSight Point, is capable of point scanning as well for optimal Z resolution.

2. *Turbo preview scanning:* When navigating around a specimen, hunting for regions of interest, image quality can be sacrificed for higher scan speeds. A simple way of switching between slow scanning for data collection and fast scanning for navigation is to toggle averaging on and off. For faster scan rates, the beam steering system usually has a speed optimized mode that will allow several images to be scanned in a second, often at the cost of pixel density and field flatness. This feature is worth trying, but generally image quality suffers so much that full-field methods are more productive.

3. *Fast X–Z imaging:* The Leica confocal has an optional fast Z drive whereby the sample is mounted on an arm which is rotated by a galvanometer. This arrangement is so fast that X–Z images can be acquired as quickly as X–Y image. Finding the plane of focus of the specimen is then as simple as scanning a single X–Y image, then moving the stage up or down until the plane of focus enters the specimen.

4. *Kalman averaging:* Averaging of data reduces PMT noise and produces much better images. It can be done on a point-by-point basis, taking many data samples from one pixel before moving to the next; line-by-line, taking a complete line of

data many times, then reading out the average; or frame-by-frame, taking a complete image many times and averaging together whole frames. These approaches are used by the Meridian Ultima, TR, and InSight, respectively.

A variation pioneered by Bio-Rad applies a weighted frame average, where the most recently acquired frame has the most significant effect on the final image. The stage and focus can be moved around while Kalman filtering is in effect, and as soon as the location is left unchanged for a significant number of frames, a high-quality image will appear. This is an interesting solution to the problem of being able to move around, and needing to see a good quality image of promising locations.

5. Locate: Where very large preview scans are possible, as in stage/mirror scanning systems, regions of interest can be homed in on by scanning with a large pixel size, then using a Locate function to specify the center point of the next scan. This can be used to acquire higher resolution $X–Y$ images, or to specify the exact cross-section position for an $X–Z$ image. The Zeiss 410 allows the user to specify a vertical optical section along any line by drawing an oblique Locate line on a preview image.

6. *Pan/Zoom:* Both the zoom factor and the position on a preview image may be specified in one stroke with the Meridian/TR by clicking on the Pan/Zoom tool button, then drawing a rectangle around the previewed image. This may be used repeatedly to home in in several steps until the maximum available zoom value has been reached. A Pan/Zoom page on the scan bar can also be used to alter the position and zoom settings (**Fig. 3**).

7. *Fast Z-series:* A camera-based confocal system such as the Meridian InSight has the advantage that it acquires data from all points in the image simultaneously *(6)*. Acquisition rate is limited only by camera speed, so video rate data acquisition is readily attainable. Alternatively, the camera can be left acquiring data while the focal plane is cycled up and down. In this way, an average 3D projection can be collected through a 50-μm thickness of sample in under a second (**Fig. 4**).

Some confocal systems can acquire maximum intensity in a similar way, by acquiring a sequence of whole images at different depths, and saving the brightest pixels from each location in the image. This takes much longer, but still succeeds in condensing a 3D volume into a single image that can be saved as a summary. It is a common way of saving *four-dimensional* data sets, where a single Z series reconstruction is saved for each time point.

8. *Auto focus, gain, and black level:* In theory, it is possible to have an instrument automatically adjust focal plane, gain and black level to instantly produce a high-quality image of a specimen, in much the same way as a scanning electron microscope optimizes imaging parameters. In practice, there are many difficulties with trying to do this reliably with a confocal system, so these automations are less likely to find universal favor in the confocal field.

An SEM has an extremely thick focal plane, whereas confocal imaging is extremely thin. Automatically adjusting to give the highest contrast works with

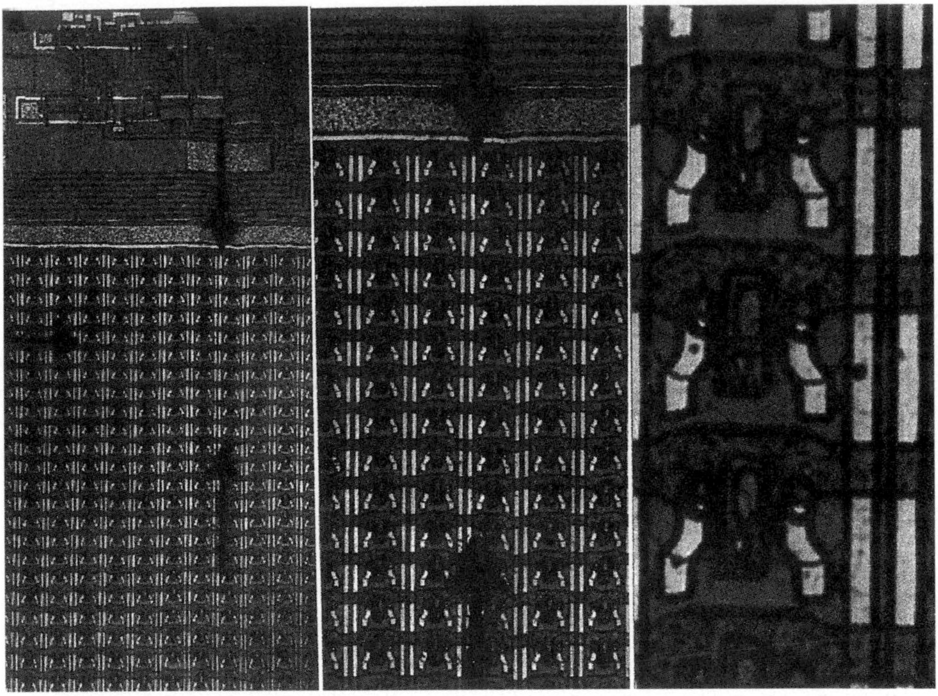

Fig. 3. Automatic zoom. These images were all taken through the same objective lens, but with different areas within the sample being scanned by the laser beam. By taking a wide-field image, then zooming in, locations of interest are identified much more rapidly than could be done at a constant, high magnification. The bright tracks on this silicon chip are 5 µm thick.

EM, but the optical plane with the highest contrast is not necessarily the most interesting to the investigator. Because confocal filtering is so very effective, a system presented with an unfocused sample is lost in blackness, and searching in the wrong direction could crash the objective into the sample.

Automatically setting gain and black level is somewhat easier, and can be done by randomly sampling the image while varying settings until the dimmest pixels fall to zero and the brightest to saturation, and this capability has been available on the Zeiss 410 for many years.

3.5. Systematic Screening Procedures

Examining a sample on a microscope is prone to subjective biases that may not be apparent to the casual user. The more time spent looking at a particular slide, the more likely it is that features of interest or positive results will be discovered *(7)*. Also, when a phenomenon has been noticed and described, it

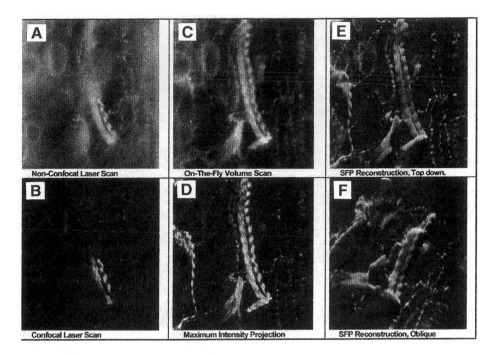

Fig. 4. Rapid 3D reconstructions. **(A)** Image of insect midgut taken on a Meridian InSight Plus using laser scanning at a single level, with the confocal aperture wide open. This indicates the level of optical sectioning achieved by concentrating the light source at a single plane within the specimen, rather than bathing the whole specimen in Kohler or full-field illuminaiton. **(B)** The same location in the specimen scanned with the confocal aperture in place. Note the absence of out-of-focus blur. **(C)** Single-second exposure taken at the same location, with the Z-drive cycling the plane of focus through a 10-μm volume. This image has the sharpness of a confocal image, but samples a larger volume of tissue than the nonconfocal laser scan shown in (A). **(D)** Maximum intensity projection representing 30 video frames of data taken at different levels such that the brightest pixel in each location is retained in the final image. Collecting an image at each level results in a sharper reconstruction than that shown in (C) but takes slightly longer to collect. **(E)** and **(F)** If individual slices are saved, the volume data set can be reconstructed from any vantage point, including top-down and oblique; and can be rendered with a variety of visual enhancements, in this case SFP shadowing.

becomes much more apparent in later studies because the human brain is phenomenally good at pattern recognition. Fortunately, most confocal applications do not require the stringent screening criteria used by clinical pathologists, but there are cases in which strictly objective sampling procedures are needed. Adherent cell cytometry systems have built-in routines that will automatically identify, list, visit, and image all regions of a sample that obey given criteria of fluorescence intensity.

1. *Quick Look:* The Quick Look program on the ACAS and Ultima instruments sweeps a diffuse laser cursor in a raster pattern over the entire specimen and displays a fluorescence map on the screen. Selecting a range of pixel intensities reduces the scanned area to a list of locations, any of which can be visited to test the accuracy of the chosen criteria. Once satisfied with the stringency of the threshold criteria, a complete cell list is compiled, from which the system can be asked to automatically scan and analyze all locations.

    This procedure is used to find rare events, but can also be used to make subcultures of positive cells. Cells are grown on a heat absorbing film in Cookie Cutter dishes. The cell population is queried and each positive cell is visited in turn. Healthy positive cells are then cut around with a high-intensity laser beam that welds a disk of membrane onto the bottom of the dish. Peeling away the backing removes all the negative cells and leaves the positive cells to repopulate the exposed floor of the dish.

2. *Pattern list:* Rather than querying the sample itself to collect a list of locations of interest, a random or uniformly spaced pattern of locations can be generated for objectively and repeatably sampling the specimen. First, the edges of the sample are entered, then a predetermined number of scan locations is generated, which will fit inside this area without any overlap (**Fig. 5**).

3. *Sampling the whole specimen:* If sampling choice locations or spaced locations within a sample are insufficient for a particular study, a grid pattern of locations where all images butt against each other will scan the entire sample at whatever resolution is necessary. This is useful for collecting and displaying the morphology of histological sections where the sample varies enormously from location to location (**Fig. 6**).

    Stage scanning the whole section is a more economical way of summarizing a large specimen, with additional high resolution images being taken at choice locations. The ACAS and Ultima can scan any area up to 10×8 cm with a pixel density of anything up to 1536 square.

3.6. Troubleshooting

The most common difficulty encountered with a confocal microscope is what to do when the image screen remains completely blank during scanning. This may be due to a lack of data at the current position, something blocking the lightpath, incorrect instrument settings, or a problem with the hardware or software. When this happens, check that there is laser light reaching the specimen; change to a false color palette and increase detector gain to see that some PMT signal is being displayed on the screen; and go back to bright field or fullfield fluorescence to see that something is visible under conventional optics that should be detected with the confocal scanner. If these quick checks fail to

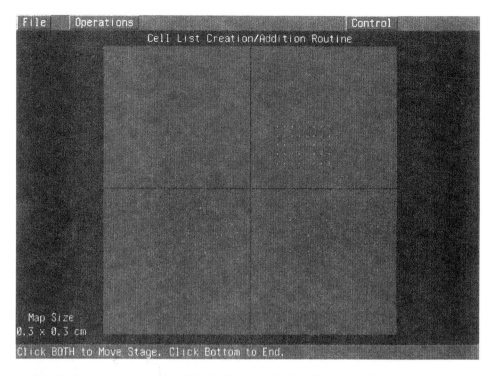

| File | Operations | | Control |

Cell List Creation/Addition Routine

Map Size
0.3 x 0.3 cm

Click BOTH to Move Stage. Click Bottom to End.

Fig. 5. Automatic imaging. **(Top left)** Manual identification of locations of interest using the ocular lenses of the microscope and clicking on each found location for future automatic confocal imaging. **(Bottom left)** Random Cell List of 10 randomly assigned locations within an area of interest designated by the user. **(Bottom right)** Grid Cell List of 12 regularly assigned locations, laid out in a grid pattern within an area of interest defined by the user. **(Top right)** Montage Cell List of 40 fields of view automatically calculated to butt against each other and completely sample the area of interest defined by the user.

reveal the problem, change to using a well-characterized, reliable sample and go through every step in the imaging system until the problem area is identified.

1. *Is laser light reaching the sample?*

Unless the sample is completely opaque, it should be easy to determine what laser lines are reaching the specimen by glancing at the scattered light from the specimen on the stage, or moving the turret to an open port and looking through a frosted glass surface held against the opening. Use fluorescent plastic or paper to detect UV, and take care to avoid direct exposure to the eye.

2. *Is the laser lasing?*

Large lasers typically require a cooldown period during which they are off, but their cooling system and power supply is still active. The sound of the cooling system does not therefore indicate that the laser is active.

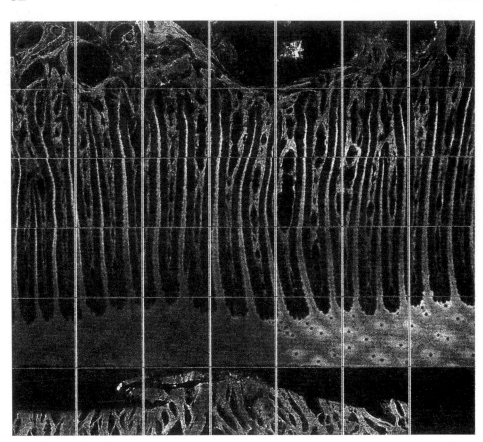

Fig. 6. Montage. Example of Montage Cell List of a histological sample of horse hoof lamellae. The size of each image was selected so that the resolution was just sufficient to determine the length of the lamellar junction between bone and hoof.

3. *Are all shutters and filters correctly set?*

All research grade microscopes have a surplus of knobs and levers, which can block access to the confocal port or the oculars. Make sure the correct positions are clearly marked and correctly set.

4. *Is the confocal aperture set too small?*

Avoid the temptation of routinely using extremely small confocal apertures for fluorescence imaging. Very small pinholes are of value only in reflection, or with extremely bright fluorescent samples, when signal intensity is not a limitation.

5. *Is the sample properly stained and positioned for viewing?*

Full field fluorescence may yield widely different illumination intensities than laser excitation, but if an image is seen under full field, the confocal microscope should also be able to detect it. Try a well-stained slide to confirm that it gives a good image. If so, the staining protocol is suspect and fresh specimens with brighter staining may be necessary.

6. *Is the detector gain or black level set too low?*

Monitor displays are sometimes turned down or turned off to reduce glare in the room and enable a user to get dark adapted for hunting down a dim specimen under full-field fluorescence. If no image appears on the screen, try loading a saved image to ensure that monitor presentation is active and working properly.

7. *Are detection dichroics and filters excluding all signal?*

Make sure that the right cube and filters are being used for the fluorophore of interest. On automated systems, try switching to a less stringent detector filter (say 515LP instead of 530/30) to increase detection efficiency.

8. *Is the computer displaying data from the right channel?*

Set the software to display all channels on the screen, and choose a palette that clearly shows the difference between very low intensity signals and zero output. Increase gain until some signal is seen, then try to increase it by adding condenser light to the signal. If bright field illumination makes no difference on a transparent sample, signal is not getting through to the detectors.

9. *Are all components turned on and properly connected?*

When running a video signal through a video tape recorder, it is usually necessary to keep the VCR turned on, even when data are being acquired directly to computer.

Make sure that all cables are secure and connected to the right ports, that the monitor is turned on and has normal values for brightness and contrast. When a monitor is on and receiving a video signal, its indicator light is green. This either turns yellow when the signal is lost, or goes out completely.

3.7. Ergonomics: High Comfort Equals Low Stress

Time available on a confocal instrument can be both limited and expensive, but working under a perpetual sense of urgency can be distracting and stressful. To be as productive as possible, the user should be comfortable and relaxed. Pieces of technology that cost hundreds of thousands of dollars should be a pleasure to use.

1. Instrument layout: The very occasional user has little control over how the room and the instrument itself are laid out, but if you are using an instrument regularly, make a mental note of any aspect of your time on the instrument that was distracting or annoying. Spend time between confocal sessions thinking up solutions. The chances are that if it bothers you, a suggested change will go down well with other users. Are the room lights too bright or inaccessible? Does the keyboard give you cramps? Is the chair high enough/soft enough? Is there enough workspace? Is the hard drive always full? Is the user manual close at hand? Do the cooling fans broil your feet?

Take a look around a well established EM facility, and note the dedicated room for each instrument, the subdued lighting, the big comfortable armchairs, and the neat banks of drawers for specimens, plates, and useful paraphernalia. Such is the comfort level that is demanded for flagship instrumentation, and high-performance laser scanning systems should be no exception.

2. *Illumination:* Microscopists do it in the dark. But they need light for taking notes and finding their equipment. To avoid having to keep getting up and down to use the main room lights, a desk lamp near the microscope is essential. A lamp with a dimmer switch is even better, as the user can then match ambient light intensity with the brightness through the oculars. The Meridian TR has an integral goose-neck task light, which has both a dimmer switch and a removable red shade. This lets the user work in comfort and makes it easy to maintain dark adapted vision for locating dim specimens through the oculars.

3. *Bench space:* Early confocal microscopes were placed on standard size anti-vibration tables, which help improve the scanned image, but are most uncomfort-able for performing the nonconfocal functions of finding the sample through the microscope. Most modern instruments are now configured on custom-built antivibration tables, which are less restrictive. However, immunity to bench vibrations can be built in to a microscope so that it is either used on a regular desk top or is enclosed in its own integral chassis.

   Very few systems are equipped with adequate work space, so it is often neces-sary to place a workbench on one side of the instrument for holding notes, immersion fluid, tissues, and specimens. Printers and storage devices also need space near the instrument, and can have surprisingly large footprints.

4. *Storage space:* The media used for storage need to be kept in an organized fash-ion. Many multiuser facilities provide backup storage to minimize the risk of useful data being permanently lost. Date stamped files can be accessed by delv-ing into the booking log to see when a particular run of work was completed.

   Users should always keep their own data, usually organized by subject rather than date. Some facilities offer no long-term storage because they are too busy to administer it. There is no point having an archive copy somewhere if the archive is not readily accessed. The onus is then entirely on the user to take good care of their data, and make their own backup copies.

5. *Noise:* Laser power supplies and cooling systems can be noisy, but the sound of rushing air is rarely intrusive to the user. If the room itself suffers from external noise, have a pair of headphones handy and a good collection of CDs. The data CD of most computers is audio compatible, a creature comfort that is often overlooked.

3.8. Software Design

Most menu-driven software is laid out as a simple toolbox, which groups commands under a number of headings, and grays out those that are tempo-rarily unavailable, but otherwise offers very little help to the user in deciding which commands to use in what order. A more productive alternative is to have software modules, each of which walks the user through a known experimental procedure. This logical software design is much more difficult to implement because it requires intimate knowledge of how it might be used so that it can guide the user rapidly to the final result, without being too restrictive.

A promising compromise is to have toolbox type software that contains in it extensive use of wizards, macros, templates and cue cards. With all these aids, the complete palette of tools is always available to the user, but streamlined pathways are made available that guide the user through the mass of options and allow for rapid repetition of experimental procedures.

1. *Macros:* A macro is simply a recorded sequence of commands that is accessed with a programmed keystroke or keyword. They are useful to speed up tediously repetitious procedures, or to ensure that exactly the same instructions were applied to each sample.
2. *Templates:* Most confocal programs save the parameters used in acquiring a data set, either as a separate parameter file, or as a custom header within the image file itself. If an identical scanning procedure is required, the scan parameters can be reset by loading the header or parameter file as a template. This is especially useful for quantitative analyses where all data must be acquired under identical scan conditions.
3. *Wizards:* Wizards are a standard feature of Windows® programs, which guide the user through an involved procedure, asking for appropriate choices to be made at each step of the process. For example, the TR software has a Dye Wizard that uses an internal database of fluorophore spectra to calculate the most efficient optical layout for simultaneous detection of multiple dyes. It lists the dyes in its database and allows users to add their own. Then it asks which dyes are to be used in combination, then it tests whether this combination is feasible and determines the best layout for optimal detection. Finally, it reconfigures the hardware to match these predictions, and changes the labels on channels to reflect what is being imaged in each channel.
4. *Cue cards:* Cue cards offer information and short cuts to performing different commands and can be used like macros for consistently repeating user defined procedures, or like Wizards to sequentially inform the user about appropriate options that carry out the selected commands. For example, to make a good SFP reconstruction, a Z-series must be scanned, the data should be filtered to minimize noise effects, then thresholded to choose a range of interest. A preliminary reconstruction should be made to confirm image quality, then a complete series of reconstruction views assembled to make a movie loop. An SFP cue card could contain action items for each of these steps, with descriptive details of what settings would be appropriate. But unlike a Wizard, any of the steps can be undertaken at any time, and all the other menus on the main screen are still available so the power user can mix and match at will.

3.9. Economics: Time Management, Billing, and Laser Conservation

1. *Laser life:* Lasers all have a rated longevity, expressed in mean hours of lifetime. Ion gas lasers are the shortest lived but most popular lasers for confocal work,

mainly because of their high intensity and good color. Helium/neon lasers are now available with very good colors, but are barely bright enough unless the delivery system is very efficient. A typical 100 mW argon laser is rated about 2000 h at full power, whereas a 5 mW red helium/neon laser has a 25,000 life expectancy. Water-cooled UV argon lasers have an intermediate life span, about 5000 h.

Longevity of ion lasers can be extended considerably by running them at less than full power, and by switching them to standby mode when not in use. If the laser is not going to be used for an hour or more, it can be powered down to further extend operating life, but care must be taken to leave the cooling fan on for at least 10 min, as a hot shutdown can damage the tube. Omnichrome now has laser power supplies with built-in control of cooling fan shutdown, and new confocal systems such as the Meridian/TR have integrated one-button shutdown. Large water-cooled argon lasers should not be turned on and off repeatedly because startup stresses the tube and reduces life more than an additional few hours of constant operation. They should be turned on once and kept running until the day's work utilizing them is finished.

2. *Mercury arc life:* Mercury arc bulbs have much shorter lifetimes, about 200 h, although the cost of a replacement is comparatively trivial. Again, the latest versions have better power regulation which can vary light intensity and extend bulb lifetime.

3. *Booking time on an instrument:* Billing and time management are important aspects of running a multiuser facility. Every user needs adequate access to the instrument, but demand for instrument time will often outstrip availability. In a university setting, 4-h time slots may be appropriate, morning, afternoon, and evening. High-productivity systems such as the InSight may be booked in 1 h or even half-hour slots because they are so much faster and easier to use than a typical point scanner.

A good self-correcting booking scheme is to allow each user to sign up for a maximum number of time slots. This allows some users to book the same time each week, while others who need a whole day to scan samples can also allocate appropriate time. As soon as a time slot is used up, more time can be booked so each user never has more than three slots on the sign up sheet. If the instrument is not in use, any user has access, so early morning and lunch times may be free for users to quickly preview samples without using up a time slot.

Keeping track of printer paper, specific laser usage, and mercury arc burn time is more difficult, as the hourly charges cannot adequately reflect frugal or excessive usage. Inkjet printers are now good enough to generate draft images for lab notes, allowing the facility to keep its dye sublimation printer separate for publication quality printing. Alternatively, an occasional check on the amount of unused paper may be enough to keep all users honest. As a last resort, the ink ribbon can be unwound to reveal negative images of every print taken.

References

1. Matsumoto, B., ed. (1993) Cell Biological Applications of Confocal Microscopy, Academic Press, San Diego.

2. Hell, S., Reiner, G., Cremer, C., and Stelzer, E. H. K. (1993) Aberrations in confocal fluorescence microscopy induced by mismatches in refractive index. *J. of Microscop.* **169,** 391–405.
3. Inoue, S. and Spring, K. R. (1996) Video Microscopy—The Fundamentals. 2nd edition, Plenum Press, New York.
4. *Gray's Anatomy*, 37th edition (1992), p. 1197.
5. Carter, D. G. (1995) InSight User's Guide, Meridian Instruments, Okemos, MI.
6. Carrington, W. A., Fogerty, K. E., Lifschitz, L., and Fay, F. S. (1989) Three dimensional imaging on confocal and wide-field microscopes, in *Handbook of Biological Confocal Microscopy*, 1st edition (James Pawley, ed.), IMR Press, Madison, WI, pp. 137–146.
7. Jiwa, A. H. and Wilson, J. M. (1991) Selection of rare event cells expressing beta-galactosidase. *Methods (A Companion to Methods Enzymol.)* **2,** 272–280.

3

Fluorescent Probes for Confocal Microscopy

Christopher Cullander

1. Introduction

This chapter differs from others in this volume in that it does not describe protocols *per se*. Rather, it consists of checklists, cautions, tips, rules of thumb, and advice related to fluorescent probes in general and probes for confocal microscopy in particular.

An excellent overview of fluorescence and fluorescence techniques can be found in the Introduction to the Handbook of Fluorescent Chemical and Research *(1)*. The company's Web pages include probe spectra, application notes, and Material Safety Data Sheets (MSDSs), as well as an up-to-date version of their catalog *(2)*. There are several lists of probes used in confocal laser scanning microscopy *(3–6)*, and new dyes are constantly being developed (e.g., *7,8,20*).

When specific probes or classes of probes are used as examples in the following sections, the information provided is necessarily incomplete; for detailed handling protocols, please consult the literature. Finally, some of this discussion pertains to single-photon excitation (confocal) microscopy only; in particular, the multiphoton excitation spectra of fluorophores are similar but not identical to their single-photon spectra (*see* **Note 1**).

1.1. General Considerations

1. Obtain a copy of the enclosures that come with the probe from the supplier before you order it (*see* **Note 2**).
2. Consult the published literature carefully *before* purchasing the probe (*see* **Note 3**).
3. To locate probe literature, utilize the specific indices (e.g., the CAS NO in Medline, *see* **Note 4**) available in electronic searches of bibliographic databases (*see* **Note 5**).

From: *Methods in Molecular Biology, vol. 122: Confocal Microscopy Methods and Protocols*
Edited by: S. Paddock © Humana Press Inc., Totowa, NJ

4. Overall, the idea is to make sure you can in fact use the probe for the purpose you intend before you buy it.

1.2. Definitions

1. Fluorophores, fluorochromes, fluorescent dyes, and fluors all refer to molecules that fluoresce.
2. A fluorescent probe often implies a fluorophore attached to another molecule, although it can also mean the fluorophore itself.
3. The estimated fluorescence intensity of a fluorophore is the product of its quantum yield (Q_f, no units, typically 0.05–1.0) and its molar extinction coefficient (ε, $M^{-1} \cdot cm^{-1}$, typically $10^4 - 2 \times 10^5$; *see* **Note 6**) of the probe (assuming subsaturation excitation). However, both quantities are measured under specific conditions that do not usually pertain in a given experiment (*see* **Note 7**).

2. Materials
2.1. Selecting a Probe

In general, a useful fluorescent probe is one that associates specifically with the structure(s) of interest, provides a strong signal, photobleaches slowly, and is nontoxic. Important additional considerations in the selection of a probe for confocal microscopy are the laser lines available, the excitation and emission spectra of the probe, and the confocal filter sets that can be used to collect the fluorescence.

1. Choose a fluorophore that can be excited by the available laser lines; the absorption maximum of the probe (*see* **Note 8**) should be at or near one of the laser wavelengths available (*see* **Note 9**).
2. It is often useful to check the probe's emission spectrum in a solvent system that approximates the one in which it will be used (*see* **Note 10**).
3. Choose a probe with high fluorescence intensity (*see* **Note 11**); use of a low photometric gain will minimize noise (*see* **Note 12**).
4. When using more than one probe in the same preparation, try to choose fluorochromes with well-separated excitation and emission spectra and that have minimal emission overlap (*see* **Note 13**).
5. Verify that the fluorescence filters available can be used to collect the emission spectra of the probes with reasonable efficiency (*see* **Note 14**).
6. If the tissue will be fixed after exposure to the probe, choose a probe that can be fixed as well. Otherwise, it may be washed out during the fixation process (*see* **Note 15**).

2.2. Storing the Probe Upon Arrival

Fluorophores can be air, water, temperature, and/or light sensitive. Some are shipped as mixtures that will degrade over time (e.g., by hydrolysis).

1. Maintain a database of your probes (*see* **Note 16**).
2. Determine the necessary storage conditions before ordering the probe; finding room in a freezer to accommodate a desiccator is not always easy (*see* **Note 17**).
3. Do not automatically store all fluoroprobes in a freezer! Some probes (and certain probe components) should never be frozen (*see* **Note 18**).
4. If a dry fluor is hydroscopic, then minimize its exposure to the water vapor in air by dividing it into aliquots and storing these appropriately (e.g., under dry nitrogen).
5. It is often wise to check the purity of a probe upon arrival, especially when ordering from a new supplier (*see* **Note 19**).

2.3. Getting the Probe into Solution

As a rule, probes in aqueous solution are unstable (e.g., subject to slow hydrolysis), as well as being more light and heat sensitive than the dry form. It is usually best to store hydrophilic probes dry, and to put them into solution just before using them. Always protect a probe in solution from light!

1. Make a habit of using double gloves when handling probes (*see* **Note 20**).
2. Always use freshly prepared buffer; old or used buffer often gives poor staining.
3. For certain compounds (e.g., the salt form of some peptides and proteins, or of nucleotide derivatives), an exact molecular weight is not known, and the concentration of the probe solution should be determined post-preparation (e.g., by measuring the absorbance and using the extinction coefficient to calculate concentration) (*see* **Note 21**).
4. To get a probe into aqueous solution, try changing the pH slightly, adding a small amount of a mild detergent, and/or using strong mechanical agitation (*see* **Note 22**).
5. Remove dissolved oxygen from a solution that has been shaken or agitated by exposing the solution to vacuum for 1–2 min (*see* **Note 23**).
6. Check the probe solution for the presence of insoluble particles (*see* **Note 24**).
7. Calibration solutions for ion indicators must be free of heavy metal ions (*see* **Note 25**).
8. The buffer used for loading acetoxymethyl (AM) probes should be serum free, and should also not contain any primary or secondary amines (*see* **Note 26**).

2.4. Storing Probe Solutions

1. Lipophilic probes can often be safely stored frozen for a long time as a concentrated stock solution (*see* **Note 27**).
2. Allow a stock solution to thaw completely, rather than using the first portion to liquefy and then refreezing the remainder (*see* **Note 28**).
3. Proteinaceous probes (e.g., antibody conjugates) can aggregate during storage, and these aggregates can result in nonspecific staining. Similarly, some probes (e.g., propidium iodide) can precipitate when frozen (*see* **Note 29**).
4. The AM esters of many probes (e.g., fluo-3 and rhod-2) are very susceptible to hydrolysis (*see* **Note 30**) and should be assayed for decomposition before use if they have been stored improperly, or for longer than 6 months (*see* **Note 31**).

5. Check the purity of the probe before use, e.g., by TLC with an appropriate solvent system, particularly if the application requires that no free dye be present (*see* **Note 32**).
6. If you store probe solutions in scintillation vials, be aware that some dyes can react with the metal foil inside the cap (plastic-lined caps are available). Also note that glass binds some dyes more than plastic does, and thus glass storage containers are not always best for probe solutions (*see* **Note 33**).

3. Methods
3.1. Do a Background Check

Autofluorescence can be a significant source of artifact in confocal microscopy, and reduces the signal-to-noise ratio in an image. Biological autofluorescence is usually elicited by excitation wavelengths less than 500 nm; however some specimens (e.g., those that contain significant amounts of chlorophyll or porphyrins) will be autofluorescent in the far red. The nature and extent of autofluorescence can also depend on how the sample was prepared *(9)*.

1. Avoid aldehyde fixation, particularly if glutaraldehyde is used, as this will induce autofluorescence. The same is true for picric acid and periodate (*see* **Note 34**).
2. It is often possible to reduce or eliminate autofluorescence by careful choice of filter cutoffs, or by use of high-wavelength probes (*see* **Note 35**).
3. Reduce or remove the signal from free or nonspecifically bound probe ("reagent background") as much as possible before imaging (*see* **Note 36**).
4. When using ratiometric probes, background fluorescence corrections must be made in order to calculate the ratio at any point.
5. Always check a test sample for autofluorescence after all preparative steps, with the setup that will be used for imaging (*see* **Note 37**).

3.2. Labeling

This section provides a few illustrations of precautions to be taken with fluors of interest in confocal microscopy. It is not intended to be inclusive.

1. The best temperature, exposure time, and probe concentration for probe loading must be determined by trial and error; however, in general, it is best to use as little of the dye as possible, and to load at room temperature (rather than at the optimal physiological temperature for the sample) (*see* **Note 38**).
2. Ion probes (e.g., pH, calcium) must be calibrated in a medium that closely mimics the experimental medium. An in situ calibration (permeablizing the cells) is usually the best way of accomplishing this (*see* **Note 39**).
3. When using AM esters of ion indicators (such as BCECF, fluo-3, and rhod-2), be sure to check for probe compartmentalization (*see* **Note 40**).
4. Increasing the concentration of an ion-indicator dye can buffer the ion of interest, and interfere with cell function (*see* **Note 41**).

5. Be very careful to ensure the complete mixture/dissolution of the probe in the working solution before it is used (*see* **Note 42**).
6. Conventional cell and tissue stains can interfere with the staining of the fluoroprobe and/or quench fluorescence.
7. After dye loading, be sure to wash the sample with buffer to remove any dye that is nonspecifically associated with cell membranes.
8. Finally, keep in mind that the probe may participate in the specimen's biochemistry in unexpected ways (*see* **Note 43**).

3.3. Detecting the Fluorescence Signal

1. Always use the minimum laser intensity possible to obtain a useable image.
2. Always collect an image using the full dynamic range available, but take care not to exceed it (saturation) or to fall below the detection threshold (*see* **Note 44**).
3. Increasing the amount of label present (i.e., adding more of the probe, but below the quenching concentration) can increase the detected intensity, but be aware that the effect is not linear.
4. Do not overexcite the probe with too high an illumination level (*see* **Note 45**).
5. When using multiple probes, check for bleedthrough from the lower channel into the upper. If bleedthrough is unavoidable, determine where it occurs (that is, which features labeled by the lower probe are detected in the higher channel) (*see* **Note 46**).
6. Be aware that when you are using dual (or multiple) probes, and the emission spectrum of the lower wavelength probe overlaps the excitation spectrum of the higher wavelength one, then you can have FRET (fluorescence resonant energy transfer) *(10)* (*see* **Note 47**).
7. Be alert to the possibility of label leaving the probe, or of the existence of artifactual fluorescent fragments (*see* **Note 48**).
8. Do not confuse detection with resolution (*see* **Note 49**).

3.4. Photobleaching

The total fluorescence signal (the brightness of the image) can become larger if the intensity of excitation and/or the collection time per pixel are increased, but in practice, photochemistry (including photobleaching) sets a limit to the amount of light that the specimen can be exposed to. Photobleaching is the irreversible destruction of a fluor by light over time, and is associated with the presence of molecular oxygen.

1. Check the rate of photobleaching on a noncritical portion of the sample before beginning imaging (*see* **Note 50**).
2. Use an antioxidant (antifade) agent in fixed preparations (*see* **Note 51**).
3. To minimize photobleaching in living preparations, reduce the partial pressure or concentration of oxygen (if the preparation can tolerate it) (*see* **Note 52**).
4. Software magnification ("zoom") in some confocals will increase the rate of photobleaching (*see* **Note 53**).

5. To obtain a Z-series in a thick specimen in which probe distribution is believed to be more or less uniform, but intensity diminishes rapidly with depth, consider collecting the images "backwards"; that is, taking the first section at the maximum depth, and then moving up toward the surface (*see* **Note 54**).
6. Phototoxicity considerations can be the determinant of which probe is best to use (*see* **Note 55**).

3.5. Environmental Effects

Fluorescence is strongly affected by the probe's environment, including solvent polarity (*see* **Note 56**), pH, and the presence of substances that quench fluorescence (including the probe itself). Quenching is a molecular interaction that reduces quantum yield.

1. The emission spectra of many probes depend on solvent polarity; in some cases, the probe is virtually nonfluorescent in water (*see* **Note 57**).
2. The excitation spectrum (as well as the emission spectrum) of certain probes (e.g., Nile red) depends on the polarity of the solvent.
3. For maximum fluorescence, fixed specimens stained with a pH-sensitive probe (e.g., fluorescein and some fluorescein derivatives) should be mounted in aqueous media with an appropriate pH (8 or greater for fluorescein) (*see* **Note 58**).
4. Do not assume that the pH of either a buffered probe solution or culture medium is stable over time — test it (*see* **Note 59**).
5. The pH of the probe solution and medium can affect staining efficiency and the stability of the probe as well as its fluorescence emission.
6. Proteins with aromatic amino acid residues can quench some fluors, notably fluorescein and 7-nitrobenz-2-oxa-1, 3-diazole (NBD) (*see* **Note 60**).
7. Self-quenching (or concentration quenching) occurs when the concentration of fluor is too high, and can also take place when multiple fluorophores label the same moiety (*see* **Note 61**).

3.6. Instrumental Concerns

Signal-to-noise ratio, resolution, and the overall image quality can also be affected by instrumental factors, some of which are discussed below.

1. Use a mounting medium with an index of refraction close to that of your sample; otherwise image quality will deteriorate rapidly as a result of the index mismatch (*see* **Note 62**).
2. Use an objective lens with a lower magnification and a high numerical aperture (NA), and employ software magnification ("zoom" on Bio-Rad confocals) rather than an objective lens with higher magnification and similar NA (*see* **Note 63**).
3. Verify that the microscope optics and detector to be used are appropriate for the probe (*see* **Note 64**).
4. The confocal filters should match the probe(s) used (*see* **Note 65**).
5. Check the optical alignment using your sample, and with all imaging elements (objective lenses, filters, etc.) to be used in place (*see* **Note 66**).

6. If you have an unexpectedly dim (or no) image, follow the lightpath from the sample in a stepwise fashion back as far as you can go, and look for mechanical sources of trouble (*see* **Note 67**).
7. If you check your sample using conventional excitation (e.g., looking for features of interest), don't be surprised if a high red label looks very dim (*see* **Note 68**).
8. Be careful to avoid the inadvertent inclusion of fluorescent material during the preparation or observation (e.g., as part of a sample holder) of the specimen (*see* **Note 69**).

3.7. An Example: FITC as a Confocal Probe

Fluorescein isothiocyanate (FITC) is arguably the most widely used fluorescent probe for the preparation of conjugates. Evaluating it in terms of the criteria listed previously:

It has the advantages of a high quantum yield, good absorption, and thus good fluorescence; an emission spectrum that does not shift significantly upon conjugation; excellent water solubility; low nonspecific binding; its maximum excitation (490 nm) is near the 488 nm line of the argon laser; its chemistry is well described, and many protocols for its use appear in the literature; and it is well detected using standard confocal filter sets.

It does, however, have significant disadvantages: its fluorescence is sensitive to pH and solvent polarity; its emission spectrum overlaps the biological autofluorescence range; the dye photobleaches readily, and is subject to quenching upon conjugation; and it has very long emission tail, which often bleeds through to the higher wavelength channel in multiprobe experiments.

FITC is, of course, a useful probe in confocal microscopy, but it is important for the investigator to be aware of its limitations.

4. Notes

1. Multi-photon excitation laser-scanning fluorescence microscopy is a promising new technique in which the specimen is illuminated by short, intense pulses of infrared (IR) light. At the point of focus, the photon density is sufficiently high so that two or more of these photons can be adsorbed essentially simultaneously by a fluorophore. When this occurs, it is as if a single photon having approximately the summed energy of the IR photons had been absorbed, and the fluorophore then returns to its ground state by the emission of a photon. In contrast to confocal (single-photon) imaging, excitation is restricted to the focal volume, and thus the volume in which photobleaching and other photochemistry take place (including the release of caged compounds) is greatly reduced. In addition, because all emitted light must originate from the focal volume, no aperture is required to eliminate out-of-focus light and the amount of light that can in principle be collected is greatly increased. For a short description of the technique, see the IMR web page *(11)*.

   The multiphoton excitation spectra of fluorophores are similar but not identical to their single-photon spectra *(12)*. Note also that UV fluorophores (e.g., indo-1) can be excited by this method without using either UV illumination or optics corrected for UV. The method does have certain limitations: For a given fluorochrome, the spatial resolution using multiphoton imaging is slightly lower than that obtained using confocal imaging. Furthermore, if there exists a UV chromophore in the sample that absorbs at the excitation wavelengths, then there is a possibility of thermal damage to the specimen. However, the major obstacle to the wider use of the technique at present is the expense of the instrumentation, particularly the mode-locked laser.

2. The probe enclosure often provides more detail about use and handling than the description in the catalog does.

3. Don't depend exclusively on the references cited by the supplier. These are a good start, but are not necessarily the most relevant references for your application, and are often not up to date.

4. The CAS NO refers to the Chemical Abstracts number, a "unique" five- to nine-digit number that identifies a chemical substance. The number is usually assigned to the native form of a compound, but occasionally a salt form will receive a separate CAS NO. CAS numbers are also given to classes of substances, e.g. dextrans. In the case of enzymes, a number beginning with EC is assigned by the Enzyme Commission of the International Union of Pure and Applied Chemistry. In older entries, the RN (registry number) field is the CAS number.

5. An index is an identifier that is part of the database record entry for the citation (e.g., AU for author). Useful indexes in Medline (in addition to the CAS NO) include CH (chemical name) and XCH (exact chemical name). EC numbers can be searched as if they were CAS numbers, e.g., "FIND CAS NO EC2.7.1.2" retrieves citations about glucokinase. Displaying a Medline citation in full Medline format will show the contents of all the indices, including CAS NO and EC. Other databases, such as Biosis, can be searched using the chemical name as a keyword, but do not have specific chemical name indices.

6. The phycobiliproteins, with multiple fluorophores on each protein unit, typically have extinction coefficients on the order of 10^6 $M^{-1} \cdot cm^{-1}$.

7. Q_f, which may be taken as the ratio of photons emitted to photons absorbed by a fluorophore, integrates total photoemission over the fluorescence spectrum of the probe, whereas ε (the absorptivity of a $1M$ solution of the fluorophore through a 1-cm lightpath) refers to absorption at a single wavelength, usually the absorption maximum, in which case it may be identified as ε_{max}.

8. In nearly all instances, the fluorescence excitation spectrum for a probe is identical to its absorbance spectrum.

9. The excitation wavelength does not need to be exactly at the absorption maximum of the dye, as long as the extinction coefficient for the fluor at that wavelength is sufficiently large.

10. Published spectra and excitation/emission maxima are often measured using unphysiological solvents, and the emission spectrum of many probes will change as a function of the probe's environment. The excitation spectrum is less likely to do so, but there are exceptions; e.g., the excitation AND emission

spectra of Nile red are shifted to shorter wavelengths with decreasing solvent polarity.

11. The importance of using fluorescence intensity as opposed to quantum efficiency alone as a measure can be seen by noting that fluorescein ($\varepsilon \approx 70,000$, $Q_f \approx 0.9$) and Cy5 ($\varepsilon \approx 200,000$, $Q_f \approx 0.3$) have very different quantum efficiencies, but almost the same fluorescence intensity.

12. Although a probe's actual Q_f (and to some extent, its ε as well) are dependent on experimental conditions, probes tabulated as having high Q_f and high ε under analytical conditions are generally preferable to those with low numbers.

13. "Bleedthrough" — detection of some fluorescence from the shorter-wavelength probe in the longer wavelength channel — can to some extent be compensated, e.g., by weighted subtraction of one image from the other, but this is rarely completely successful *(3)*.

14. General purpose ("one size fits all") filtersets may not be a good match to the spectra of your probes. Note also that if optical elements (including filters) must be moved or exchanged to image different fluorophores, this may create alignment problems — that is, the images may not be in perfect register.

15. For example, standard dextrans are not readily fixable, and may be washed out during the fixation process, but lysine-fixable versions of some dextran flow tracers can be obtained.

16. Include where the probe is stored in your laboratory, when it was received, the chemical name under which the MSDS is filed, the supplier, and the catalog number, as well as the molecular weight, storage conditions, etc.

17. To store probes, first wrap silver foil around the container, but not the around the cap. Make a separate "hat" for the cap that extends at least 1/2-inch past the top of the foil enclosing the container. This lets you open the container without the need to unwrap the foil each time. Repeated unwrappings and rewrappings will eventually result in the foil seal becoming damaged and no longer light-tight. Wrap a ring of white freezer tape (packaging tape intended for use in a freezer) around the foil enclosing the container, and use a permanent marker to record the name of the probe, the date of receipt, etc.

18. Examples include fluorescent low-density lipoprotein (LDL) complexes and streptavidin–alkaline phosphatase conjugate.

19. Probe purity varies widely between manufacturers, and sometimes between lots from the same supplier. TLC with an appropriate solvent system is a convenient way to access the purity of many probes. Note, however, that some molecules that are typically listed with a single molecular weight are in fact polydisperse, and normally have a range of molecular weights rather than a single molecular weight. Labeled dextrans are a good example, with '3000 MW' dextran having polymers with molecular weightss predominantly (but not exclusively) in the range 1500–3000.

20. Many are known (e.g., 4', 6-diamidino-2-phenylindole [DAPI], propidium iodide) or potential (e.g., those that bind with nucleic acids) mutagens, and should be handled with care. In addition, solvents such as dimethyl sulfoxide (DMSO) enhance the penetration of molecules into skin and other tissues.

21. That is, by using Beer's Law, $c = A/\varepsilon b$, where $A \equiv$ absorbance (unitless), $\varepsilon \equiv$ absorptivity (extinction coefficient, in $M^{-1} \bullet cm^{-1}$), $b \equiv$ path length (1 cm), and $c \equiv$ concentration of solution (M).

22. In more detail, try changing the pH slightly (e.g., using a mild alkaline buffer, such as 0.1 M bicarbonate); adding a small amount of detergent (e.g., Pluronic™ F-127, a non-ionic detergent; or Triton X-100); however, note that this may reduce staining efficiency, and that long-term storage in a detergent solution is not advisable; or mechanically agitating the solution strongly, e.g., by flushing the pipet several times, vortex-mixing, sonicating, or mildly heating (40–50°C) the solution.

23. Be sure to keep the solution protected from light, and make sure that the solution is exposed to the vacuum (e.g., the cap of the vial is loose). This is not advisable with a volatile solvent.

24. If any are present, remove them by centrifugation (e.g., in a microfuge at 12,000g for 5 min), by filtration through Whatman No. 2 filter paper, or by forcing the solution through 0.22-µm membrane filters.

25. Heavy metal ions (such as manganese) will affect both the indicator's affinity for the ion and its fluorescence. Metal chelators (e.g., tetraks-(2-pyridylmethyl) ethelenediamine [TPEN] for calcium) can sometimes be used to remove unwanted metal ions. Note also that significant amounts of metal impurities have recently been found in buffer components; make certain that the source certifies the salts as being high purity or effectively metal-free.

26. Serum often contains esterases, and aliphatic amines can cleave the AM esters and thus prevent loading.

27. The stock solution is typically 1–5 mM in an organic solvent; a working solution can then be made up just prior to use by diluting the stock (typically to 1–5 µM) in an appropriate buffer or solution. This saves time and avoids repeated exposure to moisture, air, etc. and the need to repeatedly freeze and thaw the parent probe solution to obtain a working solution. Smaller aliquots will also thaw faster.

28. The liquid that thaws first will often have a higher concentration of dye than the portion that is still frozen. When thawing is complete, the solution should be remixed and inspected for precipitate or evidence of aggregation (such as cloudiness). Always allow a cold solution to warm to room temperature before opening the container.

29. Remove aggregates by centrifugation, using only the supernatant for labeling, and check for the presence of precipitate after the solution reaches room temperature, either sonicating or vortex-mixing the solution to redissolve the dye.

30. Some probes, including the AM esters of fluo-3 and rhod-2, are sold in very small aliquots to avoid this problem.

31. To determine whether an AM ester is intact, dilute an aliquot of the DMSO stock solution to 2–10 µM in a cuvette containing 2.5 µM calcium test solution in buffer, measure the fluorescence, and then increase the calcium concentration to 5 µM and measure the fluorescence again. Because only the hydrolyzed form of the fluor will bind calcium, any increase in fluorescence indicates that partial

hydrolysis of the ester has taken place. The acid (hydrolyzed) form of the probe will not diffuse across the cell membrane.

32. Probes can also break down during storage; for example, labeled dextrans release small amounts of free dye over time, and must be repurified before use if the presence of free dye is not acceptable. Isothiocyanates such as FITC can also deteriorate during storage.

33. One example is SYBR Green, for which Molecular Probes recommends the use of Rubbermaid Servin' Saver® sandwich boxes.

34. If aldehyde fixation is necessary, the induced autofluorescence can be reduced by washing the specimen with 0.1% sodium borohydride in pH 8.0 phosphate-buffered saline (PBS) for 30 min before staining *(9)*.

35. Narrowing the bandpass also reduces the fluorescence intensity detected; thus the use of high-red fluors is preferable. The use of long wavelengths does, however, raise questions about resolution and optical response *(13)*.

36. For example, reagent background due to the leakage of fluo-3 (and other fluorescein derivatives) can be quenched by adding anti-fluorescein antibody to the external medium. The cells or tissue should also be washed with an appropriate buffer (e.g., the solution that the probe was made in) before imaging to remove any probe that is loosely or nonspecifically associated with the sample.

37. As noted in the text, fixation can induce autofluorescence, as can heating, addition of certain buffers, prolonged storage, and many other factors..

38. Some examples: dextran conjugates of ion indicators are less subject to compartmentalization and leakage, but must be loaded into the cell by invasive procedures, and often have a lower affinity for the ion being measured. The appropriate concentration of antibody conjugates needs to be determined by trial and error, but is usually between 5 and 20 μg/mL in most cases *(14)*.

39. For SNARF-1, a ratiometric pH probe, an in situ calibration is performed by using an ionophore or other permeabilizing agent to equilibrate the intracellular pH with that of the known extracellular medium. Agents used to permeabilize cells include A-23187, ionomycin, nigericin, digitonin, and saponin.

40. Ion indicators are usually loaded into cells in the form of AM esters, which are subsequently cleaved by cytoplasmic esterases to form the fluorescent indicator, in this case BCECF*. If the AM form translocates to an intracellular compartment, it may still convert to the fluorescent form, but the sequestered indicator will not be responsive to changes in cytoplasmic ion levels.

41. A "dose–response" curve for probe loading in the particular system can be constructed to determine the maximum allowable loading concentration. Typical intracellular concentrations of probe range between 30 and 100 μM.

42. Remove dissolved oxygen from a nonvolatile solution by subjecting it to a vacuum for 1–2 min, and check for the presence of insoluble particles.

43. For example, rhodamine 123 is a P-gp substrate. The point here is that the probe may not act solely as a probe in the system of interest.

---

*2',7'-*bis*-(2-carboxyethyl)-5-(and 6)-carboxyfluorescein

44. Although contrast adjustment is possible after the image has been obtained (and can make some features easier to see), the amount of information in the image remains the same. Similarly, all signal strengths out of range will be recorded as having the top level (e.g., 255), no matter what their actual intensity is.

45. The laser intensities in spot-scanning confocal microscopy can easily cause excitation saturation of a probe, especially those with long fluorescent decay times, such as the pyrenes. The result is that the detected intensity is no longer primarily a function of fluor concentration *(15)*, and the signal-to-background ratio is reduced. A higher illumination level also means a higher rate of photobleaching.

46. Do so by taking (on the same area, if possible) a first image using the lower wavelength, a second image using the upper wavelength, and a third using both lines simultaneously. Merge images 1 and 2 and compare the result with image 3 (it may be necessary to manipulate intensities slightly in the merged image).

47. If the probes are sufficiently close together (10–100 Å, depending on the probe spectral characteristics), then the excited state of the lower wavelength probe can couple to the ground state of a proximate higher wavelength probe and excite it. In other words, the lower wavelength probe is excited, but it is the upper wavelength probe that emits.

48. High-intensity illumination (as well as cellular esterases) can cause some probes to break down, and the breakdown products themselves can be fluorescent. For example, the mitochondrial stain hexyl ester rhodamine B, which is excited by green wavelengths and has maximal emission at 578 nm, can be broken down by excitation light, and the breakdown products are fluorescent in the fluorescein range.

49. Resolution is limited by the imaging wavelength used — that is, a 200 nm liposome cannot be resolved using visible light. It can, however, be detected (if the intensity of fluorescence emission is sufficient). In analogy, the naked eye can detect stars at night, but cannot resolve a stellar disk — they appear as point sources. Note that for the liposome, this also means that you cannot distinguish between an intact liposome and a labeled fragment.

50. Adjustment of laser intensity and gain is far easier before the critical images are taken.

51. Although it is by no means the only good photoprotectant *(16,17)*, 0.1% *p*-phenylenediamine (Sigma Chemical Co.) in 10% PBS and 90% glycerol (or in a buffered solution) is an excellent antifading agent for confocal microscopy *(18)*. It is, however, carcinogenic and should be handled with extreme care. For living specimens, antioxidants such as ascorbic acid (0.1–1.0 mg/mL, or greater) have been used in the medium as reducing agents. Keep in mind that the primary label and the counterstain (if any) can affect the performance of an antifade reagent.

52. For example, by bubbling nitrogen or argon over it. The use of an enzyme that depletes oxygen in the medium (Oxyrase, Oxyrase Inc., P.O. Box 1345, Mansfield, OH 44901) has also been reported.

53. If the scan of the smaller ("zoomed") image takes the same amount of time as scanning a full-frame, then the total energy delivered per unit area in the small

image is larger than that for a full frame, and there is consequently more photobleaching. One method of avoiding this effect is to first collect a full-frame image and store it, and to then to magnify the digitized image.

54. In a point-scanning confocal, photobleaching occurs throughout the specimen, not just in the optical section being imaged. If the series begins at the top of the specimen, then the deepest section may be photobleached before it can be imaged. By starting deep, and moving progressively back toward the surface, it is the more superficial sections, with their higher intensity, that undergo the most photobleaching.

55. For example, the long-wavelength calcium indicator Calcium Green-1 has been shown to be less phototoxic than fluo-3.

56. "Solvent" here is interpreted to include the interior of any membrane or membrane-bound structure, as well as regions inside molecules.

57. Nile red, ethidium homodimer, the carbocyanines, NBD, propidium iodide, ethidium bromide, and diphenylhexatriene (DPH) are all strongly polarity sensitive. They have a low quantum yield in aqueous solution, and become strongly fluorescent when they are shielded from water by entering a hydrophobic environment or binding to a biomolecule.

58. Fluorescein has a pK_a of about 6.4, and its fluorescence emission drops rapidly as the medium becomes more acidic. Nearly all of its derivatives display some pH sensitivity as well.

59. For example, the pH of Tris buffers increases when they are refrigerated; if an 8.0 buffer is prepared at room temperature, the pH will increase to 8.5 at 4°C. Culture medium pH is affected by both temperature and CO_2 level.

60. This effect can be partially overcome in some cases; e.g., incubating fluorescein–avidin conjugates with free biotin at the end of the staining results in significantly brighter signals from the conjugates *(19)*.

61. For fluorescein, the maximum brightness is obtained with two to four fluorophores per antibody.

62. Use a water lens to image aqueous (living) specimens, and a mounting medium with an index near that of the immersion oil used for fixed specimens. Matching optical indices is critical for high-resolution applications.

63. The intensity of the light collected increases as the square of the NA, but decreases as the square of the magnification. Specifically, a 60×/1.4 lens is preferable to a 100×/1.4 lens. Higher magnification lenses also tend to be "darker," due primarily to the large number of optical elements within them that the light must pass through.

64. UV probes require UV objectives and UV-reflective mirrors *(4)*. Many older systems use photomultiplier tubes (PMTs) with very poor high red sensitivity, and optics that block or do not adequately correct high red wavelengths *(13)*.

65. Ideally, the bandwidth of the emission filter should span 80–90% of the emission spectrum of the probe.

66. For dual or triple imaging, microspheres that are fluorescent at the wavelengths of interest are a convenient means of checking the registration of the channels

relative to each other. In brief, the microsphere images from the different channels are merged to see if they exactly superimpose. This procedure is described in more detail in the Handbook of Fluorescent Probes and Research Chemicals *(2)*.

67. In our setups, these have included: internal diaphragm in objective stopped all the way down, partially or complete closed shutters and/or apertures, and presence of light-blocking filters in the lightpath.

68. The upper limit of human vision is about 750–770 nm, and retinal sensitivity above 650 nm is poor. A substantial part of the fluorescence emitted by some far-red fluors (e.g., the cyanines, such as Cy-5) is thus effectively invisible to the unaided eye. In other words, it is not possible to accurately judge the intensity or, in some cases, the presence of a high-red label by looking through the eyepieces of a conventional fluorescence microscope. However, most confocals can "see" farther into the red range than can the human eye.

69. Saran Wrap™, Canada balsam, various brands of clear fingernail polish, some epoxy resins, and certain machine oils (used to prevent the corrosion of sharp tools, such as razor blades) are fluorescent.

Acknowledgments

Support for C. C. from the AACP New Investigator program, the UCSF Academic Senate, and NIDR-DE11275 is gratefully acknowledged.

References

1. Johnson, I. (1996) Introduction to fluorescence techniques, in *Handbook of Fluorescent Probes and Research Chemicals*, 6th edition (R. Haugland, ed.), Molecular Probes, Inc., Eugene, OR, pp. 1–6.

2. URL for Molecular Probes: http://www.probes.com/lit/

3. Brelje, T. C., Wessendorf, M. W. and Sorenson, R. L. (1993) Multicolor laser scanning confocal immunofluorescence microscopy: practical application and limitations, in *Cell Biological Applications of Confocal Microscopy* (B. Matsumoto, ed.), *Methods in Cell Biology*, Vol. **38**, Academic Press, San Diego, pp. 98–182..

4. Wells, S. and Johnson, I. (1994) Fluorescent labels for confocal microscopy, in *Three-Dimensional Confocal Microscopy: Volume Investigation of Biological Specimens* (J. K. Stevens, L. R. Mills, and J. E. Trogadis, eds.), Academic Press, San Diego, pp. 101–131.

5. Tsien, R. and Waggoner, A. (1995) Fluorophores for Confocal Microscopy, in *The Handbook of Biological Confocal Microscopy* (J. Pawley, ed.), Plenum Press, New York, pp. 267–280.

6. A list of probes that can be excited by laser wavelengths: http://flosun.salk.edu/fcm/fluo.html

7. Patterson, G. H., Knobel, S. M., Sharif, W. D., Kain, S. R., and Piston, D. W. (1997) Use of the green fluorescent protein and its mutants in quantitative fluorescence microscopy. *Biophys. J.* **73**, 2782–2790.

8. Millard, P. J., Roth, B. L., Thi, H. P., Yue, S. T. and Haugland, R. P. (1997) Development of the FUN-1 family of fluorescent probes for vacuole labeling and viability testing of yeasts. *Appl. Environ. Microbiol.* **63,** 2897–2905.

9. Bacallo, R., Kiai, K. and Jesaitis, L. (1995) Guiding principles of specimen preservation for confocal fluorescence microscopy, in *The Handbook of Biological Confocal Microscopy* (J. Pawley, ed.), Plenum Press, New York, pp. 311–326.

10. Chapple, M. R., Johnson, G. D. and Davidson, R. S. (1988) Fluorescence quenching of fluorescein by R-phycoerythrin. A pitfall in dual fluorescence analysis. *J. Immunol. Methods* **111,** 209–218.

11. A short description of multi-photon microscopy: http://www.bocklabs.wisc.edu/imr/facility/2p.htm

12. Xu, C., Zipfel, W., Shear, J. B., Williams, R. M., and Webb, W. W. (1996) Multiphoton fluorescence excitation: new spectral windows for biological nonlinear microscopy. *PNAS* **93,** 10,763–10,768.

13. Cullander, C. (1994) Imaging in the far-red with electronic light microscopy: requirements and limitations. *J. Microscop.* **176,** 281–286.

14. (1995) Short protocols in molecular biology: a compendium of methods, in *Current Protocols in Molecular Biology*, 3rd edition (F. M. Ausubel, ed.), John Wiley & Sons, New York.

15. Pawley, J. (1995) Fundamental limits in confocal microscopy, in *The Handbook of Biological Confocal Microscopy* (J. Pawley, J., ed.), Plenum Press, New York, pp. 19–38.

16. Biloh, H. and Sedat., J. W. (1982) Fluorescence microscopy: reduced photobleaching of rhodamine and fluorescein protein conjugates by n-propyl galate. *Science* **217,** 1252–1255.

17. Johnson, G. D., Davidson, R. S., McNamee, K. C., Russell, G., Goodwin, D., and Holborow, E. J. (1982) Fading of immunofluorescence during microscopy: a study of the phenomenon and its remedy. *J. Immunol. Methods* **55,** 231–242.

18. Schuman, H., Murray, J. M., and DiLullo, C. (1989) Confocal microscopy: an overview. *BioTechniques* **7,** 154–163.

19. Emans, N., Biwersi, J., and Verkman, A. S. (1995) Imaging of endosome fusion in BHK fibroblasts based on a novel fluorimetric avidin-biotin binding assay. *Biophys. J.* **69,** 716–728.

20. Griffin, B. A., Adams, S. R., and Tsieng, R. Y. (1998) Specific covalent labeling of recombinant protein molecules inside live cells. *Science* **281,** 269–272.

4

Imaging Gene Expression Using Antibody Probes

Nadean L. Brown

1. Introduction

The merger of multiple immunoflurescent labeling and the laser scanning confocal microscope (LSCM) has greatly enhanced experimentation in many areas of biomedical research. Recently, genetic tools have become available for marking individual cells, or cells within a tissue of genetically mosaic animals *(1,2)*. These tools have been primarily developed in *Drosophila* and utilize protein epitopes for which commercial antibodies are available, such as the *myc* epitope *(2*, ATCC and Oncogene Research) or an epitope in the CD-2 protein *(3*, Serotec). Similarly, cell lineage experiments have been conducted in *Drosophila (4,5)*, as well as in other organisms such as *Caenorhabditis elegans (6)* or zebrafish *(7,8)*, in which cells express either β-galactosidase or green fluorescent protein (GFP), are physically labeled with dextran coupled to a fluorophore *(8)* or are labeled with fluorescent lipophilic dyes, such as diI *(9)*. Multiple labeling experiments are then performed comparing the expression of these cell markers with endogenous protein(s) of interest.

In this chapter I present methods for sample preparation, single- and multiple-antibody labeling of such samples, and subsequent analysis using the LSCM. I have focused primarily on using whole-mounted *Drosophila* imaginal discs as an example throughout. *Drosophila* imaginal discs are the larval precursors to adult tissues such as the eyes, antennae, wings, and legs. During larval life the imaginal discs are located internally, and therefore need to be dissected out for antibody staining. However, these techniques can be widely applied to a variety of other specimens, namely whole-mount embryos (invertebrate or vertebrate) or sectioned material. The recently published chapter by Patel *(10)* on imaging neuronal cell types in *Drosophila* embryos and larvae contains excellent details on preparing embryos for antibody staining. In addi-

From: *Methods in Molecular Biology, vol. 122: Confocal Microscopy Methods and Protocols*
Edited by: S. Paddock © Humana Press Inc., Totowa, NJ

tion, it is quite likely that most immunohistochemical or immunofluorescent labeling techniques can be adapted for use with the LSCM.

Primary antibodies can be generated against a protein of interest in a variety of ways, and in several different animal species *(11–13)*. Polyclonal antisera are most often purified in some manner (i.e., affinity purification), but may also work adequately in immunofluorescent labeling as crude whole sera *(12)*. Monoclonal antibodies are produced as either tissue culture supernatants or ascites fluid *(11,12)*. Even though antibody reagents may be quantitated to a given protein concentration (or may be distributed from commercial sources or other research laboratories with a suggested working dilution), the optimal working concentration of each primary antibody often needs to be determined experimentally for a particular application, such as the LSCM. Some antibody reagents can be used at lower concentrations (or higher dilutions) than recommended, simply because of the increased sensitivity of the confocal microscope. For multiple labeling experiments (**Subheading 4.**) determining the optimal antibody concentration for each antibody is especially important. As with all protein solutions, antibody reagents are susceptible to bacterial growth and degradation. If a reagent is not supplied with either sodium azide or thimerisol, add sodium azide to a final concentration of 0.02% as soon as possible after receiving an antibody in solution (or upon rehydrating a lyophilized antibody). If it is unclear whether these bacterial inhibitors are present, add azide as a precaution, as excess azide is preferable to none. Harlow and Lane *(12)* and Scopes *(14)* provide additional detailed methods for storing protein solutions, such as antibodies, to retain maximal protein activity.

Perhaps the single greatest advantage of employing the LSCM to antibody labeling experiments is the ability to analyze multiply labeled samples. This chapter details methods for multiple antibody labeling when primary antibodies are produced in different animal species (e.g., rabbit and mouse), as well as methods for multiple antibody labeling using primary antibodies raised in the same species (mouse). In addition, the use of fluorescent dyes that bind to particular subcellular structures are also described and incorporated into multiple labeling protocols. Specific examples will be provided for these types of labeling experiments that will also highlight the comparison of two coexpressed proteins in the *Drosophila* eye imaginal disc, and the analysis of protein expression within genetically marked (with the myc epitope) eye disc cells.

Samples for multiple antibody labeling are isolated and prepared as described in **Subheading 3.1.** Before carrying out multiple labeling experiments, working experimental conditions for each antibody reagent should be determined individually (*see* **Subheading 3.2.**). However, it may still be necessary to match antibody signals to each other in trial multiple labeling experiments. This is especially true if one protein is expressed more strongly than another or if the titer of a primary antibody is much better than another. Be sure that individual antibody experiments are identical in such parameters as fixation, blocking and buffer conditions, increasing the odds that the multiple labeling will work well and making it easier to troubleshoot any problems that do arise. Potential problems such as incompatibility of epitope fixation or reagent crossreactivity are addressed in detail in **Subheading 4**.

Owing to the large number of antibodies made in either rabbits or mice, it is sometimes advantageous to do multiple labeling using two or more antibodies from the same host species (i.e., two monoclonals). This can sometimes be accomplished by doing antibody labeling steps sequentially. Although this technique is not successful for all combinations of antibodies, it is definitely worth trying. In additional, performing multiple antibody labeling sequentially, rather than simultaneously (**Subheading 3.4.**), can circumvent some reagent crossreactivity problems. An example of sequential labeling is provided in this section using two monoclonal antibodies, anti-myc (ATCC) and anti-22C10 (a *Drosophila* neuronal antigen, *15*); both monoclonals are tissue culture supernatants. Sequential monoclonal antibody labeling has also been successfully utilized in standard immunohistochemical antibody labeling *(16)*.

2. Materials

1. 10× Phosphate-buffered saline (PBS) (for 1 L): 18.6 mM NaH_2PO_4 (2.56 g), 84.1 mM Na_2HPO_4 (11.94 g), 1.75 M NaCl (102.2 g). This stock solution is stable at room temperature. For 1× PBS, dilute 1:10 with deionized (d) H_20 and pH to 7.4 with NaOH or HCl. IX PBS is also stable at room temperature for long periods.
2. PEMF fixative: 0.1M PIPES (pH 6.9), 1 mM EGTA, 2 mM $MgSO_4$, 1.0% Nonidet p-40 (NP-40). These ingredients can be made as a stock solution and stored at room temperature. Add formaldehyde just prior to each experiment (final concentration 2–4% depending upon the antigen). Formaldehyde (37% w/v) from Sigma Chemical Co. (Catalog No. F1635) is commonly used but contains methanol. For methanol-sensitive epitopes, use formaldehyde from Polysciences, Inc. (Catalog No. 04018) that is methanol-free.
3. PBS–PFA fixative: 4.0% paraformaldehyde (PFA), 1× PBS. Weigh 4 g of PFA (Polysciences, Inc. Catalog No 00380) and add to 90 mL of dH_20 that has been heated to 50–60°C and is stirring as PFA is added (this is best done in a fume hood). Add 10 μL of a 10N NaOH solution as PFA does not go into solution unless it is basic. It may be necessary to wait 10–20 min for the solution to clear completely. When all of the PFA appears to be in solution, add 10 mL of 10× PBS stock solution. If a bit of insoluble material remains, filter the solution by pouring it through Whatman filter paper. Check the pH of the solution with litmus paper, rather than a pH meter, to avoid PFA contamination of the electrode. If necessary, adjust the pH to 7.4 with HCl. Store this solution at 4°C, in a glass container, protected from the light. This solution is good for up to a month.

4. PLP fixative: 2% PFA, $0.01M$ NaIO$_4$, $0.075M$ lysine, $0.037M$ NaPO$_4$ (pH 7.2). Make an 8% PFA solution as described for PBS-PFA fixative but omit the addition of PBS (make it up in just water). While the 8% PFA is cooling, dissolve 0.36 g of lysine in 10 mL of H$_2$0, 7.5 mL of $0.1M$ NaPO$_4$ (pH 7.2), 2.5 mL of $0.1M$ Na$_2$HPO$_4$ on ice. Immediately before use, mix 15 mL of buffered lysine solution, 5 mL PFA, and add 50 mg of NaIO$_4$. This solution is made fresh each time, and chilled prior to sample addition.

5. Block buffer*: 50 mM Tris (pH 6.8), 150 mM NaCl, 0.5% NP-40, 5 mg/mL bovine serum albumin (BSA). Store at 4°C. This solution is good as long as no bacterial growth occurs. The shelf life can be extended to months by the addition of sodium azide to a final concentration of 0.02%.

6. Wash buffer*: 50 mM Tris (pH 6.8), 150 mM NaCl, 0.5% NP-40, 1 mg/mL of BSA. Store at 4°C. This solution is usable as long as no bacterial growth occurs. To prevent bacterial growth, thereby extending the shelf life add sodium azide to a final concentration of 0.02%.

7. Mounting buffer*: 50 mM Tris (pH 8.8); 150 mM NaCl; glycerol to desired percentage (10–90%). Store at room temperature. Glycerol can support bacterial growth, so take good care of glycerol stocks by autoclaving them or by dedicating a bottle to tissue mounting.

8. 5× *p*-Phenylenediamine (PDA) stock: 30 mg of PDA dissolved in 4 mL of dH$_2$0; add 6 mL of 100% glycerol and mix well. Store aliquoted at –20°C and keep in the dark as much as possible. For 1× solution, make fresh for each use by diluting the 5× stock with 50 mM Tris (pH 8.8), 150 mM NaCl five-fold. PDA solutions (either 5× or 1×) are not stable over time (weeks to months), especially in glycerol solutions less than 80%. Therefore it is advisable to make fresh 5× PDA stocks often (weekly for 10% glycerol mounting). PDA can be obtained from Sigma (Catalog No. P1519).

9. *n*-Propyl gallate mounting solution: 0.5% *n*-propyl gallate, 70–80% glycerol in 1× PBS. Dissolve in PBS, then add glycerol. Mix well. Store aliquoted at –20°C. More stable than PDA solution (which discolors after several weeks to a month at 4°C or room temperature). Order from Sigma (Catalog No. P3130).

10. DABCO mounting solution: 2.5% DABCO, 70–80% glycerol in 1× PBS. Dissolve in PBS, then add glycerol. Mix well. Aliquot and store and –20°C. Also more stable than PDA. DABCO can be obtained from Sigma (Catalog No. D2522).

11. Suppliers
    a. American Type Culture Collection (ATCC), Maryland, USA
    b. Molecular Probes, Oregon, USA
    c. Amicon, Massachusetts, USA
    d. Oncogene Research Products, Massachusetts, USA

---

*These solutions could also be adapted to a PBS-based buffer system. Block buffer: 1× PBS, 0.5% NP-40, 5 mg/mL of BSA. Wash buffer: 1× PBS, 0.5% NP-40, 1 mg/mL of BSA. Mounting Buffer: 1× PBS/10% glycerol.

    e. Corning Costar Corporation, Massachusetts, USA
    f. Polyscience Inc., Pennsylvania, USA
    g. Fine Science Tools Inc., California, USA
    h. Serotec Ltd., Oxfordshire, England
    i. Jackson Immunoresearch Laboratories, Pennsylvania, USA
    j. Sigma Chemical Company, Missouri, USA
    k. Ted Pella Inc., California, USA
    l. Vector Laboratories Inc., California, USA

3. Methods

3.1. Tissue Preparation

1. Third instar larvae or pupae are briefly washed in 1× PBS to remove any food matter that may be stuck to their outer cuticles.
2. Several larvae at a time are transferred to a depression well slide containing fresh PBS buffer. Individual larvae are bissected along the anterior/posterior axis using fine forceps (Dumont No. 5, Fine Science Tools, Catalog No. 11250-20) and the tail portion is discarded. The head portion is inverted inside out. Eye/antennal discs and four of the six leg discs are attached to the brain and central nervous system (CNS) by ganglia. The wing discs, haltere discs and the remaining two leg discs are attached both to the body wall and to the CNS by ganglia. Unwanted fat, digestive tract tissue, and salivary glands are carefully removed with the forceps. What remains are the imaginal discs, the CNS and the head cuticle. This complex is transferred to 1 mL of ice-cold fixative already present in a well of a 48-well tissue culture plate (Corning Costar Corp. Catalog No. 3548). Up to 20 disc complexes can be placed in one well, using the cuticle as a "handle" to grasp each complex with forceps.
3. Disc complexes are fixed for varying amounts of time depending upon the age of the animal dissected (20–30 min for larvae; up to 2 h for pupae) on ice (*see* **Notes 1–3**).
4. After fixation, tissues are incubated in a blocking buffer to prevent nonspecific binding of the primary antibody. Imaginal discs are blocked for at least 45 min either on ice or at 4°C. This is accomplished by transferring the disc complexes to another well of the 48-well plate containing 1 mL of blocking buffer (*see* **Notes 4–8**).

3.2. Single Immunofluorescent Labeling

1. After *Drosophila* imaginal disc complexes have been fixed and blocked, they are transferred to a well containing appropriately diluted primary antibody in wash buffer. Because primary antibodies are often of limited quantity, this step is done at a lower volume (400 µL) and overnight at 4°C. For antibodies that are used at very low concentrations, it may be advantageous to first make an intermediate dilution (in wash buffer). This intermediate dilution should be stable at 4°C for at least a few weeks, provided sodium azide (0.02%) has been added to it (*see* **Notes 9** and **10**).
2. Imaginal discs are washed 4 × 20 min by transferring them with forceps through four wells, each containing 1 mL of Wash Buffer. Washes are done at 4°C. Prior

to beginning these washes, recover the diluted primary antibody for reuse. Antibodies can often be used three or more times (one particular polyclonal antibody was used five times before a noticeable decrease in signal was observed). Store diluted primary antibodies for reuse in the same manner as intermediate dilutions (*see* **Notes 11** and **12**).

3. Transfer imaginal discs to a well containing the appropriate diluted secondary antibody in 400 μL of wash buffer. Incubate discs at 4°C for several hours (or overnight for large or thick samples). If the secondary antibody is conjugated to a fluorochrome, keep samples in the dark as much as possible by covering the plate or tube with foil (*see* **Notes 13–15**).

4. Discs are washed 4 × 15 min at 4°C.

5. If a directly conjugated secondary antibody is used, proceed to **Subheading 3.5.**, for equilibration of stained material in a mounting medium. If the secondary antibody is biotinylated or if a triple antibody sandwich is being used, transfer discs to a well containing the appropriately diluted (in wash buffer) tertiary reagent (*see* **Note 16**).

6. Incubate for 1–2 h at 4°C. Reduce the volume of the tertiary antibody solution as described for primary antibody incubation. Keep the samples in the dark as much as possible; this is most easily accomplished by wrapping the 48-well plate with foil.

7. Wash discs 4 × 15 min at 4°C also keeping the samples in the dark as much as possible. Proceed to **Subheading 3.5.** for equilibration of imaginal discs in mounting medium.

3.3. Double and Triple Antibody Labeling in Parallel

1. Two primary antibodies raised in different host animal species, such as rabbit polyclonal anti-*Drosophila* Atonal (*17* and **Fig. 1A**) and mouse monoclonal anti-Daughterless (*18* and **Fig. 1B**) are added simultaneously to their predetermined final dilutions in 400 μL of wash buffer, e.g., mouse anti-Daughterless to 1:50 dilution and rabbit anti-atonal to 1:5000. Imaginal discs are incubated in this solution at 4°C overnight (*see* **Notes 17–19**).

2. As with single immunofluorescent labeling, mixtures of primary antibodies can also be saved for reuse by adding sodium azide (0.02%) and storing at 4°C.

3. Imaginal discs are washed 4 × 20 min by transferring them through four wells each containing 1 mL of wash buffer at 4°C.

4. A mixture of secondary antibodies is made in 400 μL of wash buffer. In the double label experiment with anti-atonal and anti-daughterless, the secondary reagents are a 1:200 dilution of goat anti-rabbit biotin (Vector Labs) and a 1:200 dilution of rat anti-mouse IgG (Jackson Immunoresearch). Note that both immunofluorescent signals are being built up employing the two different methods (*see* **Notes 20–23**).

5. Incubate imaginal discs at 4°C for 1–2 h.

6. Discs are washed 4 × 15 min at 4°C.

7. Likewise a mixture of tertiary antibodies is made in 400 μL of wash buffer. For the given example, 1:200 streptavidin–rhodamine (Vector Labs) and 1:100 goat anti-rat-fluorescein isothiocyanate (FITC) (Jackson Immunoresearch) are used.

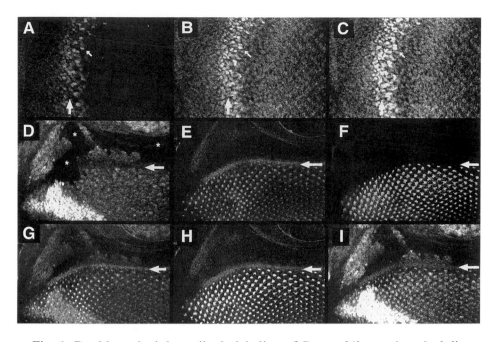

Fig. 1. Double and triple antibody labeling of *Drosophila* eye imaginal discs. **(A–C)** Confocal images of a wild-type eye disc stained with rabbit anti-Atonal (red in A,C) and mouse anti-Daughterless (green in B,C). Single images are shown in A,B and the confocal merge in C. The **small arrow** in all three panels points to a nucleus stained by both antibodies. **(D–I)** Confocal images of a third instar eye disc containing both mutant and mosaic *hairy* tissue. This disc was triple labeled with anti-myc (D,G,I), rhodamine–phalloidin (E,G,H,I), and 22C10 (F,H,I). Single label images are presented in D–F, two different merges of double labeling are shown in G,H, and the triple merger is in panel I. In D, three populations of cells are marked by the myc epitope. The brightest green cells, to the lower left, are homozygous wild-type and all lower level green cells are heterozygous for the *hairy* mutation. Those cells not expressing myc (black areas marked with asterisks) are homozygous mutant for *hairy*. Myc expression is green in D,I and red in G. Phalloidin staining is red in E,H,I and green in G. 22C10 expression is pink in F, green in H, and blue in I. In all panels, anterior is up and the **arrow** marks the position of the MF (*see* Color Plate II).

8. Incubate imaginal discs in this solution for 1 h at 4°C, keeping the 48-well plate wrapped in foil.
9. Wash imaginal discs 4 × 15 min at 4°C, keeping the plate wrapped in foil as much as possible. Proceed to **Subheading 3.5.** for preparation of labeled discs for LSCM analysis.
10. For triple-labeled samples simply add the third set of reagents to each antibody incubation step, as long as the primary antibodies have been made in three differ-

ent host species (e.g., rabbit, rat, mouse). If triple sandwiches are used, remember that the biotin–streptavidin system can be used only once and check for crossreaction between all secondary and tertiary reagents by performing all the pairwise double labelings in practice experiments. Multiple labeling experiments using more than one primary antibody from the same host species are addressed in **Subheading 3.4.**

3.4. Double and Triple Antibody Labeling in Series

1. Samples are isolated and prepared as described in **Subheading 3.1**. The first monoclonal (in this example, anti-myc) is diluted 1:5 in 400 µL of wash buffer and imaginal disc complexes are incubated in this solution overnight at 4°C.
2. Imaginal discs are washed 4×20 min by transferring them through four wells, each containing 1 mL of wash buffer. Washes are done at 4°C. The primary antibody can be saved for reuse.
3. Secondary antibody preparation and incubation are as described in **Subheading 3.2., step 3**. For mouse anti-myc, donkey anti-mouse biotin (Jackson Immunoresearch) at a dilution of 1:50 was used.
4. Disc complexes are washed 4×15 min at 4°C.
5. Tertiary antibody dilution and sample incubation were performed as described in **Subheading 3.2., step 5**. Streptavidin–fluoroscein (Jackson Immunoresearch) at a dilution of 1:200 was used. Rhodamine–phalloidin (Molecular Probes) was included with this last antibody incubation step at a concentration of 250 nM for a triple label. Incubation was extended for 3 h at 4°C, as this was previously determined to provide optimal rhodamine–phalloidin signal (*see* **Notes 24** and **25**).
6. Wash discs 4×20 min at 4°C, keeping the samples protected from the light by wrapping the 48-well plate in foil in this and all subsequent steps.
7. Imaginal disc complexes are taken from the last wash step into the second primary antibody dilution (in this case, 22C10 diluted 1:200 in a volume of 400 µL). Incubation is also done overnight at 4°C.
8. Discs are washed 4×20 min at 4°C.
9. Secondary antibody dilution and sample incubation are performed as described in **Subheading 3.2., step 5**. For detecting 22C10 labeling, goat anti-mouse IgG-Cy5 (Jackson Immunoresearch) was used at a dilution of 1:50.
10. Wash discs 4×20 min at 4°C.
11. Proceed to **Subheading 3.5.** for mounting imaginal discs and **Subheading 3.6.** for merging confocal images. The imaginal discs in the provided example are triply labeled for myc epitope expression using fluorescein (**Fig. 1D**), 22C10 expression using Cy5 (**Fig. 1F**), and phalloidin conjugated to rhodamine (**Fig. 1E**), which labels the actin cytoskeleton, thereby highlighting the morphology of eye imaginal disc cells (**Fig. 1I**).

3.5. Preparing Labeled Samples for LSCM Analysis

1. After the last washing step in each of the labeling protocols, transfer imaginal disc complexes to 1 mL of Tris/glycerol solution (*see* **Subheading 2.1.**). For

very flat mounted preparations of imaginal discs (especially the eye disc) use 10% glycerol. For other specimens, e.g., embryos, use 50–90% glycerol (*see* **Note 26**).

2. Allow samples to equilibrate in glycerol solutions for at least 1 h at room temperature (for immediate mounting) or at 4°C overnight. Keep samples protected from the light by wrapping them in foil.

3. Glycerol-equilibrated samples can be stored at 4°C for several weeks, or at –20°C for months protected from the light. For –20°C storage, the glycerol content must be greater than 10% to prevent samples from freezing. Do not conduct longer term sample storage in the presence of anti-photobleaching agents (*see below*) as they can be unstable.

4. Free radical scavengers reduce or eliminate rapid fluorescent signal fading or photobleaching, which results from either epifluorescent or laser light illumination. Therefore, just prior to LSCM viewing and analysis (within a day or so), incubate imaginal disc complexes in an anti-photobleaching agent for 30–60 min at either room temperature or 4°C, wrapping them in foil.

5. Imaginal disc complexes are dissected apart and arranged for final mounting on a microscope slide. To accomplish this, one well of imaginal discs at a time is transferred to a clean microscope slide in a puddle of mounting buffer. If the puddle of mounting buffer begins to dry up during the mounting procedure, keep adding small amounts to discs to ensure that they do not dry out.

6. Using small tungsten wire dissection needles [Ted Pella Inc. Catalog No. 27-11; see Patel *(10)* for a detailed protocol of needle preparation] dissect imaginal discs apart from the remaining cuticle, brain and other unwanted larval body parts. If necessary, dissect imaginal discs apart from each other.

7. For flat mount preparations (preferable with eye imaginal discs) remove with forceps all unwanted body cuticle (as it is much thicker than the discs) and place all other debris off to one side (this will act to "trap" unwanted air bubbles; *see below*). Arrange discs in their final positions on the slide.

8. Carefully lower a 22 × 22 No. 1 coverslip (thickness should be matched to the objective lens used) onto the imaginal discs. To avoid having the discs move around during this process, use only a small amount of mounting buffer.

9. Once the coverslip is in place, add mounting buffer dropwise from one side of the coverslip, if there is too little buffer. Excess buffer can be carefully wicked away, using a tissue paper labwipe gently touching one side of the coverslip. Coverslips should be lowered at an angle to avoid trapping air bubbles around the discs. Often bubbles that do appear will congregate around debris left to one side of the discs. Bubbles trapped adjacent to specimens can sometimes be freed by gently tapping on the coverslip.

10. Coverslipped slides should be sealed to avoid evaporation of mounting media. Fingernail polish works well for this, but it should be colorless (clear) as many pigmented nail polishes (auto)fluoresce.

11. Slides are ready for microscope/LSCM analysis when the nail polish is dry or can be stored at 4°C protected from the light.

3.6. LSCM Collection, Merging and Analysis of Antibody Labeled Samples

1. Fluorescent images shown in **Fig. 1** were collected using a Bio-Rad MRC600 Laser Scanning Confocal Microscope (Bio-Rad Microscience, Hercules, CA) equipped with a krypton/argon laser (American Laser Co., Ltd., Salt Lake City, UT), which has three major lines at wavelengths 488 nm (fluorescein), 568 nm (rhodamine), and 647 nm (Cy5).

2. Eight to ten single optical planes were Kalman averaged, normalized, and collected at a resolution of 8 bits.

3. The images were transferred to a Macintosh Quadra computer with a 24-bit color card, assigned specific colors, then merged into a single image using the Adobe® Photoshop 3.0 (Adobe Software) program. Details of image collection and storage will depend upon the microscope, LSCM, and computer(s) used.

4. **Figure 1** contains two examples of *Drosophila* eye imaginal disc multiple antibody labeling. In the first example, double labeling with two nuclear antibody proteins, rabbit anti-Atonal (**Fig. 1A**) and mouse anti-Daughterless (**Fig. 1B**) was used to demonstrate a stripe of cells (the merging of red and green creates yellow) in **Fig. 1C** that coexpress both proteins. This type of data presentation is not different from standard immunofluorescence with one important exception. The cells coexpressing both proteins are located within a physical depression of the eye imaginal disc known as the morphogenetic furrow (MF). In addition, because some cells within the MF undergo differentiation into neuronal photoreceptor cells, their nuclei are located in different focal planes from each other and/or from nuclei of undifferentiated cells. By analyzing this double label preparation with the LSCM, it was possible to optically section through the MF region of the eye disc at different depths and determine unambiguously that all MF cells expressing Atonal (**Fig. 1A**) also express high levels of Daughterless (**Fig. 1B**).

5. The second multiply labeled eye disc (**Fig. 1D–1I**) is a mixture of homozygous wild-type cells (brightest green cells in **Fig. 1D**), cells heterozygous for a mutation in the *Drosophila hairy* gene (intermediate green cells in **Fig. 1D**), and homozygous mutant (*hairy–*) cells (those cells not fluorescing, asterisks in **Fig. 1D**). The green fluorescence indicates the presence of a myc epitope, whose expression in this larva marks cells containing one or two copies of the normal *hairy* gene product. This mosaic eye disc was also assayed for expression of two commonly used eye development markers, monoclonal antibody 22C10, which labels differentiating photoreceptor cells (pink cells in **Fig. 1F**), and rhodamine–phalloidin (red labeling in **Fig. 1E**), which binds to actin in all eye disc cells and highlights the position of the MF (arrow in **Fig. 1C–1I**). Two combinations of these labeled images are presented in **Fig. 1G,H** and the triple merger of all three images is shown in **Fig. 1I**. Note the color substitution in these panels.

6. Different image combinations can be built using the Photoshop program to facilitate analysis of the overlap of different antibody signals (**Fig. 1G,H**) or to consolidate this information into a single panel, conserving space for journal presentation (**Fig. 1I**). The final color used for each image is a matter of choice in

the Photoshop program. The myc epitope is green in **Fig. 1D**, accurately reflecting its labeling with fluorescein, but red in **Fig. 1G**. Likewise, 22C10, which was labeled with Cy5, is pink in **Fig. 1F** but green in **Fig. 1H** and blue in **Fig. 1I**.

4. Notes

4.1. Tissue Preparation

1. The choice of fixative also contributes substantially to proper preservation of the antigenic site(s) of many proteins. Several different fixatives can be used for preparing LSCM samples, including PEMF, often used for nuclear proteins; PLP, often used for cell surface or extracellular proteins; and PBS-PFA, which I have used with some success for nuclear, cytoplasmic, or extracellular proteins (*see* **Subheading 2.2.** for detailed buffer recipes). These choices of fixatives are provided only as rough guidelines. Some extracellular proteins are fixed quite well in PEMF and some nuclear antibody experiments work adequately using PLP. Exact fixation conditions often need to be determined by trial and error for each antigen of interest. A good rule of thumb is to choose one fixative that works well for preserving the majority of epitopes of interest and try this protocol first whenever testing new antibodies (e.g., fix *Drosophila* imaginal discs in PEMF for 20 min on ice), then vary this experimental parameter only as needed. This is especially important for multiple labeling experiments.

2. An important consideration in determining the time of fixation is the epitope itself. Tissue or cellular morphology may not be preserved adequately by short fixation times; alternatively epitopes can be destroyed if the fixation is too long. Therefore, the time of fixation may need to be determined empirically. This and subsequent steps could also be performed on embryos or tissues in 1.5-mL microcentrifuge tubes or, by reducing the volumes, could be used to stain sectioned materials on slides.

3. Underfixation can cause high background and overfixation can result in low or no signal. The type of fixative, the concentration of fixative, and the length of time samples are fixed can all be varied to correct this problem.

4. The blocking step can be extended to several hours (I have blocked tissues for up to 8 h with no adverse effect), but this is done only as a matter of convenience. No improvement in the quality of antibody staining has been noted when extended blocking steps were performed. Either BSA or whole animal sera can be used to block nonspecific antibody binding, and some antibody protocols use a combination of both. From 1 to 5 mg/mL of good quality BSA (fraction V, Sigma Chemical Co.) works as well as animal sera (1–10% is often used), but either is adequate

5. If animal sera is used as a blocking agent, it is commonly from the same species that was used to make the secondary antibody. For example, use normal goat serum if the secondary antibody is goat anti-mouse IgG (in this instance the primary antibody would be a mouse monoclonal antibody made against a *Drosophila* protein). Although it is probably not critical that the blocking sera be species matched to the secondary antibody host animal, this can be a source of nonspecific background if the blocking sera contains an IgG molecule that the secondary

antibody can recognize. Heat sera at 56°C for 30 min to inactivate complement. Filter warm serum through a 0.22-μm filter and store aliquoted at –20°C in a nondefrosting freezer.

6. In these *Drosophila* experiments, washing steps between fixation and blocking have not been necessary, most likely because imaginal discs are only several cell diameters thick, facilitating rapid equilibration when discs are transferred between solutions. However, thicker and/or larger tissues may require several (two to three) brief (5–15 min) washing steps in between fixation and blocking. Likewise, imaginal discs do not need agitation in any of the steps described, but good quality antibody labeling of larger tissues and/or embryos may require gentle agitation on a shaking or rocking table or end-over-end rotation within a closed tube.

7. *Drosophila* imaginal discs are prepared (dissected, fixed, and blocked) fresh for each labeling experiment, but other samples are very amenable to being prepared ahead of time and stored. For example, both *Drosophila (10)* and zebrafish embryos *(19)* can be fixed and stored in a nondefrosting –20°C freezer in ethanol or methanol for several weeks to months (I have successfully used both of these type of embryos in antibody staining experiments after storage of longer than 1 year at –20°C). For those epitopes that are sensitive to alcohol incubations and insensitive to long fixation times, it may also be possible to store fixed material in fixative at 4°C. All stored samples are rehydrated, washed, and blocked just prior to primary antibody application. In addition, sectioned material is often prepared ahead of time and stored at either room temperature (e.g., paraffin embedded) or at –20°C (e.g., cryosectioned material) prior to antibody labeling. As for fixation conditions, the ability to use tissues and samples prepared in advance will need to be determined experimentally for each antibody of interest.

8. Larger tissues or embryos may require the addition of 1% dimethyl sulfoxide (DMSO) (Sigma Catalog No. D5879) to fixation, blocking, and all antibody incubation steps. DMSO facilitates large molecules passing through the cell membrane.

4.2. Single Immunofluorescent Labeling

9. The concentration of primary antibody used is also an important factor for the success of antibody labeling experiments. To determine empirically the optimal concentration to use, pick a broad dilution range for initial experiments. Polyclonal antibodies (and ascites) are often used in the dilution range of 1:100 to 1:50,000. Tissue culture supernatants from hybridoma cell lines typically are used in the range of undiluted to 1:500. If a hybridoma supernatant is still weak when it is used undiluted, concentrate the supernatant by using Centricon protein microconcentrators (Amicon, Catalog No. 42409). At least a 50-fold concentration can be achieved in this manner.

10. The length of time that samples are incubated with primary antibodies can also be widely varied. If the signal is low, try lengthening the incubation time to up to 24 h. If the background is a problem, antibody binding can be shortened to as little as 1 hour at either 4°C or room temperature.

11. Nonspecific antibody signal may also be reduced by preabsorption of primary antibodies. This may particularly help polyclonal antibodies. Dilute an antibody four- or five-fold in wash buffer, add fixed experimental samples, and incubate for 1 hour up to overnight with gentle mixing. For *Drosophila* imaginal discs, this could be either an extra well of fixed disc complexes or fixed embryos. After preabsorption, remove the antibody solution and dilute to the desired final concentration or store at 4°C with 0.02% azide added. Discard the material used for preabsorption.

12. Background signal may also be lowered by the addition of a second blocking step after excess primary antibody has been washed away and just prior to the addition of a secondary antibody reagent.

13. Secondary antibodies are made by immunizing host animals against either whole immunoglobulin proteins or portions of these proteins [e.g., F(ab')$_2$ fragments]. If it is known whether a primary antibody is IgG or IgM, be sure to select a secondary antibody that is appropriate. If this information is unknown, many vendors offer secondary reagents that recognize both. For most purposes, reagents made against whole immunoglobulin molecules work well, but those antibodies made against the variable F(ab')$_2$ fragment sometimes lower nonspecific background.

14. The optimal concentration of a secondary reagent may need to be determined empirically, using the manufacturer's recommended dilution as a starting point. Many antibody reagents arrive lyophilized and can be stored indefinitely in this form at 4°C. Reconstitute them according to the accompanying instructions, centrifuge the solution (to remove insoluble particles), and transfer the supernatant to a fresh tube. Antibody reagents should be stored according to the manufacturer's recommendations, which often suggest aliquoting a reagent and storing all but one working aliquot at either –80°C or –20°C in a nondefrosting freezer. The working aliquot is kept at 4°C and is stable for weeks to months. In addition, some researchers store their antibody reagents solely at 4°C with no apparent loss of activity. The most important aspect of reagent storage is to avoid repeated freeze-thawing, which can rapidly lower the titer of antibodies.

15. Many antibody labeling experiments work well using a secondary antibody directly conjugated to a fluorochrome. The fluorochrome used is usually a matter of choice, or based upon the filters available on fluorescent microscopes and/or LSCMs. Commonly used fluorochromes are FITC, lissamine–rhodamine, Texas red, and more recently Cy5 and Cy3.

16. A weak fluorescent signal can often be improved by building up the signal through the subsequent addition of another antibody reagent that specifically recognizes the secondary antibody. In this scenario, the secondary antibody would be unconjugated and the tertiary antibody would contain the fluorophore of interest. An example of such signal building is: primary antibody = mouse anti-*Drosophila* antigen; secondary antibody = rabbit anti-mouse IgG (Jackson Immunoresearch); tertiary antibody = donkey anti-rabbit-FITC (Jackson Immunoresearch). Another antibody labeling system that can also amplify a signal employs a secondary antibody that is biotinylated and a fluorophore conju-

gated to either avidin or streptavidin. Biotin and streptavidin/avidin have a very specific affinity for one another, in which multiple biotin molecules bind to each avidin or streptavidin. The high specificity of this binding can lower or prevent background fluorescence due to secondary reagents binding nonspecifically. These amplification reagents may facilitate multiple labeling experiments when employed to match the signal strength of different antibody reagents.

4.3. Double and Triple Antibody Labeling in Parallell

17. Crossreactivity occurs when part or all of another antibody signal is added to a known labeling pattern. This is caused primarily by secondary or tertiary antibody reagents not being specific enough, usually by recognizing more than one species of immunoglobulin molecule. The more evolutionarily related the host species are to each other (primary and secondary antibodies), the more likely that crossreaction problems will arise with the addition of the next set of reagents (secondary or tertiary antibodies). To help minimize such problems try using secondary or tertiary reagents that have been crossreacted with IgG fractions from multiple animal species. For example, in a double label experiment with primary antibodies made in both rat and mouse, use reagents such as goat anti-rat IgG and donkey anti-mouse biotin from Jackson Immunoresearch that have been specifically immunodepleted of mouse or rat IgG reactivity, respectively. The drawback to using these types of reagents is that they often work at a lower titer than equivalent reagents that have not been cross-species purified.

18. More rarely, antibody crossreaction occurs when primary antibodies against different members of the same gene family are used. In this circumstance, one primary antibody recognizes more than one protein of interest. In the example given in **Subheading 4.1.2.**, both Atonal and Daughterless are transcription factors containing highly related basic helix–loop–helix domains *(20–22)*. But because both of these antibody reagents were developed against a less conserved region (that does not contain the basic helix–loop–helix domain) of these proteins *(17,23)*, cross-reactivity is not an issue here. To determine if this is a problem, it is necessary to know the immunogen used to make primary antibody reagents in question.

19. Optimizing all primary antibody and secondary and/or tertiary reagent concentrations helps to minimize crossreactivity as well.

20. Another problem in multiple labeling experiments that looks very much like a crossreactivity problem occurs when one very strong fluorescent signal bleeds through to another wavelength channel on the LSCM. In this case, an image also appears additive for labeling patterns and signals. This is particularly a problem when two fluorochromes emit close together. For instance, strong rhodamine–phalloidin signal can sometimes be detected on the fluorescein channel even though the emission spectra are 80 nm apart. To distinguish between bleedthrough and reagent crossreaction, switch the first fluorochrome protein pattern farther away from the emission wavelength of the other label(s), e.g., substituting Cy5 (emits at 647 nm) for rhodamine.

21. Perhaps the hardest problem to solve occurs when two or more proteins of interest are found to require different fixation conditions. If one signal is optimal but the other is marginal in a particular set of conditions, try boosting the marginal signal by building a triple (or even quadruple) sandwich. It may also be possible to titer back the better signal (by dropping primary or secondary antibody concentrations) to even out signaling intensity. This may help equalize images at the LSCM. If one set of fixation conditions appears harsher than another, do the antibody labeling sequentially (**Subheading 4.2.**) with the addition of a harsher fixation step and a second blocking step prior to addition of the second antibody. Lastly, if the choice of fixative itself it not at issue, try intermediate concentrations of fix or detergent to optimize both signals.

22. Multiple labeling of samples can also take advantage of fluorescing subcellular dyes (e.g., rhodamine–phalloidin (Molecular Probes), which binds to F-actin filaments within the cytoplasm of cells). Detailed descriptions of such dyes, including their optimal binding conditions and excitation and emitting wavelengths, can be found in the Molecular Probes, Inc. *Handbook of Fluorescent Probes and Research Chemicals* (Eugene, OR). The use of these compounds is most simply incorporated by adding them to a predetermined final concentration to the last antibody labeling step (either secondary or tertiary antibody). The excess is then washed away during the last set of sample washes. Strongly staining dyes like propidium iodide, which binds to nucleic acids, should be added instead to the first wash of the last set of washing steps. A specific example is given in **Subheading 3.4.**

23. Some subcellular dyes emit in the low ultraviolet range and are not detected by many LSCMs. Also other dyes, such as propidium iodide bind to both RNA and DNA, requiring additional sample preparation steps such as RNase treatment and the presence of detergent to achieve specific nuclear DNA labeling (*24*). Therefore, working out compatible multiple labeling conditions for these reagents and antibody reagents may need to be done by trial and error.

4.4. Double and Triple Antibody Labeling in Series

24. A blocking step (**Step 3.4**) could be used prior to the addition of the second primary antibody, but did not prove necessary in the given example.

25. For multiple primary antibodies from the same host species (or to minimize cross-reaction problems), try reversing the order of primary antibody addition. For example, in multiple labels with the anti-myc monoclonal and the 22C10 monoclonal, if the myc antibody was used first, no cross-reaction of signals occurred, but if the order was reversed, the 22C10 expression pattern became additive with the myc expression pattern.

4.5. Preparing Labeled Samples for LSCM Analysis

26. Compounds that prevent rapid fading of fluorophores are *p*-phenylenediamine (PDA), *n*-propyl gallate, and DABCO (1,4-diazabicyclo[2.2.2] octane). Several commercially prepared anti-photobleaching agents are also available (e.g., Slowfade, Molecular Probes, Catalog No. S828). Among standard fluorophores,

fluorescein (FITC) is the most susceptible to photobleaching. However, due to the intensity of the laser of the LSCM, it is a good idea to use an antifading agent with all fluorophores.

Acknowledgments

I thank Steve Paddock for teaching me confocal microscopy and Sean Carroll for his support. I also thank Tom Glaser and Jim Lauderdale for comments on this chapter.

References

1. Golic, K. G. and Lindquist, S. (1989) The FLP recombinase of yeast catalyzes site-specific recombination in the *Drosophila* genome. *Cell* **59,** 499–509.
2. Xu, T. and Rubin, G. M. (1992) Analysis of genetic mosaics in developing and adult *Drosophila* tissues. *Development* **117,** 1223–1237.
3. Zecca, M., Basler, K., and Struhl, G. (1995) Sequential organizing activities of engrailed, hedgehog, and decapentaplegic in the *Drosophila* wing. *Development* **121,** 2265–2278.
4. Buenzow, D. E. and Holmgren, R. (1995) Expression of the *Drosophila* gooseberry locus defines a subset of neuroblast lineages in the central nervous system. *Dev. Biol.* **170,** 338–349.
5. Brewster, R. and Bodmer, R. (1995) Origin and specification of type II sensory neurons in *Drosophila*. *Development* **121,** 2923–2936.
6. Chalfie, M., Tu, Y., Euskirchen, G., Wards, W. W., and Prasher, D. C. (1994) Green fluorescent protein as a marker for gene expression. *Science* **263,** 802–805.
7. Peters, K. G., Rao, P. S., Bell, B. S., and Kindman, L.A. (1995) Green fluorescent fusion proteins: powerful tools for monitoring protein expression in live zebrafish embryos. *Dev. Biol.* **171,** 252–257.
8. Kimmel, C. B. and Warga, R. (1986) Tissue-specific cell lineages originate in the gastrula of the zebrafish. *Science* **231,** 365–368.
9. Chitnis, A. J. and Kuwada, J. Y. (1990) Axonogenesis in the brain of zebrafish embryos. *J. Neuroscience* **10,** 1892–1905.
10. Patel, N. (1994) Imaging neuronal subsets and other cell types in whole mount *Drosophila* embryos and larvae using antibody probes, in *Methods in Cell Biology*, Vol. 44 *Drosophila melanogaster: Practical Uses in Cell Biology* (L.S.B. Goldstein and E. Fryberg, eds.), Academic Press, San Diego.
11. Goding, J. W. (1986) *Monoclonal Antibodies: Principles and Practice*, 2nd edition. Academic Press, London.
12. Harlow, E. and Lane, D., eds. (1988) *Antibodies: A Laboratory Manual.* Cold Spring Harbor Laboratory, Cold Spring Harbor, NY.
13. Williams, J. A., Langeland, J. A., Thalley, B., Skeath, J. B., and Carroll, S. B. (1993) Preparation and purification of antibodies to foreign proteins produced in E. coli plasmid expression system vectors, in *DNA Cloning: Expression Systems*, IRL Press, Oxford.

14. Scopes, R. K. (1987) *Protein Purification: Principles and Practice*, 3rd edition, Springer-Verlag, New York.
15. Zipursky, S. L., Venkatesh, T. R., Teplow, D. B., and Benzer, S. (1984) Neuronal development in the *Drosophila* retina: monoclonal antibodies as molecular probes. *Cell* **36,** 15–26.
16. Skeath, J. B., Panganiban, G., Selegue, J., and Carroll, S. B. (1992) Gene regulation in two dimensions: the proneural *achaete* and *scute* genes are controlled by combinations of axis-patterning genes through a common intergenic control region. *Genes Dev.* **6,** 2606–2619.
17. Jarman, A. P., Grell, E. H., Ackerman, L., Jan, L. Y., and Jan, Y. H. (1994) *atonal* is the proneural gene for *Drosophila* photoreceptors. *Nature* **369,** 398–400.
18. Brown, N. L., Paddock, S. W., Sattler, C. A., Cronmiller, C., Thomas, B. J., and Carroll, S. B. (1996) *daughterless* is required for *Drosophila* photoreceptor cell determination, eye morphogenesis, and G_1 progression. *Dev. Biol.* **179,** 65–78.
19. Westerfield, M., ed. (1995) *The Zebrafish Book: A Guide for the Laboratory Use of Zebrafish* (Danio rerio), 3rd edition, University of Oregon Press, Eugene, OR.
20. Cronmiller, C., Shedl, P., and Cline, T. W. (1988) Molecular characterization of *daughterless*, a *Drosophila* gene with multiple roles in development. *Genes Dev.* **2,** 1666–1676.
21. Caudy, M., Vaessin, H., Brand, M., Tuma, R., Jan, L. Y., and Jan, Y. N. (1988) *daughterless*, a *Drosophila* gene essential for both neurogenesis and sex determination, has sequence similarities to *myc* and the *achaete–scute* complex. *Cell* **55,** 1061–1067.
22. Jarman, A. P., Grau, Y., Jan, L. Y., and Jan, Y. N. (1993) *atonal* is a proneural gene that directs chordotonal organ formation in the *Drosophila* peripheral nervous system. *Cell* **73,** 1307–1321.
23. Cronmiller, C. and Cummings, C. A. (1993) The *daughterless* gene product in *Drosophila* is a nuclear protein that is broadly expressed throughout the organism during development. *Mech. Dev.* **42,** 159–169.
24. Witfield, W. G. F., Gonzalez, C., Maldonado-Codina, G., and Glover, D.M. (1990) The A- and B-type cyclins of *Drosophila* are accumulated and destroyed in temporarily distinct events that define separable phases of the G2-M transition. *EMBO J.* **9,** 2563–2572.

5

Single and Double FISH Protocols for *Drosophila*

Sarah C. Hughes and Henry M. Krause

1. Introduction

In situ hybridization within whole-mount *Drosophila* tissues was made routine with the introduction of digoxigenin labeled probes and alkaline phosphatase based detection methods *(1)*. However, this method of detection until recently has been limited by the required use of alkaline phosphatase conjugated secondary antibodies and chromogenic substrates. The use of alkaline phosphatase substrates and their diffusible products limits the resolution of staining, particularly in thick tissues and deep within the embryo. Without additional probes, double-labeling and interpretation of results is also very difficult.

In this chapter we describe techniques for fluorescent in situ hybridization (FISH) in *Drosophila* tissues. Unlike the products of the alkaline phosphatase reaction, a fluorescent signal is nondiffusible and thus allows higher resolution microscopy *(2)*. Resolution and sensitivity of probe detection are enhanced even further when coupled with laser scanning confocal microscopy (LSCM), or deconvolution microscopy. Other advantages include the ability to reconstruct three-dimensional images and to peer deep within the specimen. If more than one probe is available, the ability to assess overlaps in patterns of expression is also enhanced *(3)*. In addition to the increased resolution, overlaps show up as easily discerned novel colors (e.g., red + green = yellow). With LSCM, one can also tell if transcripts do or do not co-localize at the subcellular level.

This chapter describes how to generate multiple probes for double hybridization. Also described are variations of the protocol that allow simultaneous detection of transcripts and proteins and for use of these detection techniques in whole embryos as well as dissected tissues. Underlying the described detection methods is the use of nonradioactive RNA probes and fluorescently conju-

From: *Methods in Molecular Biology, vol. 122: Confocal Microscopy Methods and Protocols*
Edited by: S. Paddock © Humana Press Inc., Totowa, NJ

gated secondary antibodies. Probes labeled with digoxigenin or fluorescein work equally well. Primary antibodies against these molecules are available from Boehringer Mannheim, as are reagents and kits for making the labeled RNA probes. Non-cross-reacting secondary antibodies, conjugated to cyanine fluorochromes, are then used to bind the primary antibodies.

The strength of the signals generated by this procedure, relative to standard chromogenic detection techniques, depends upon the equipment used for detection. We find that the intensities of fluorescent and chromogenic signals are similar when developed using the same RNA probe, and as detected on our Zeiss Axioplan 2E microscope. However, when detected by LSCM, the FISH approach is more sensitive and gives far greater resolution.

2. Materials
2.1. RNA Probe Preparation
1. 1.5-mL Microcentrifuge tubes, autoclaved
2. RNase-free Diethyl Pyrocarbonate (DEPC) treated or double-distilled water
3. 5X T7/T3 Transcription optimized buffer (Promega, Madison WI, USA; Catalog No. P1181)
4. T7 or T3 RNA Polymerase (Promega, Madison WI, USA: Catalog Nos. P2075, P2083; 1000 U)
5. Fluorescein RNA Labeling Mix (Boehringer Mannheim; Catalog No. 1685 619)
6. Digoxigenin RNA Labeling Mix (Boehringer Mannheim; Catalog No. 1277 073)
7. RNAguard (Pharmacia; Catalog No 27-0815-01)
8. 0.5M EDTA
9. 4M Lithium chloride
10. Absolute ethanol
11. Cold 70% ethanol wash

2.2. Initial Embryo Fixation
1. Chlorine bleach; diluted 1:1 with water
2. 40% Formaldehyde solution (prepared fresh from paraformaldehyde as described below)
3. 10× Phosphate-buffered saline (PBS) solution
4. Heptane
5. Methanol
6. 20-mL disposable glass scintillation vials (Fisher)
7. 1.5-mL microcentrifuge tubes, autoclaved

2.3. Post-Fixation and Hybridization of Whole-Mount Embryos
1. PBT solution: 1× PBS plus 0.1% Tween-20
2. 40% formaldehyde solution prepared that day
3. 20 mg/mL Proteinase K (Sigma). Dissolve in sterile H_2O, divide into 50-µL aliquots, and store at −20°C

4. 2 mg/mL glycine in PBS
5. RNA hybridization solution: 50% formamide, 5× SSC, 100 µg/mL heparin, 100 µg/mL sonicated salmon sperm DNA, and 0.1% Tween 20. Filter through a 20-µM filter and store at –20°C in aliquots (stable for at least 6–12 months)
6. Hot block or water bath at 80°C
7. Water bath at 56°C

2.4. Post-Hybridization Washes and Development of the FISH Signal

1. RNA Hybridization buffer
2. PBT Solution: 1× PBS, 0.1% Tween-20
3. PBTB Solution: 1× PBS, 0.1% Tween-20, and 0.5% milk powder
4. Mouse monoclonal anti-fluorescein antibody [IgG; 1:2000 dilution of a 0.1 mg/mL stock solution (*see* **Notes 2** and **3**); Boehringer Mannheim, Laval, QC, Canada; Catalog No. 1426 320]
5. Sheep anti-digoxigenin antibody (IgG; 1:1000 dilution of a 0.2 mg/mL stock solution; Boehringer Mannheim, Laval, QC, Canada; Catalog No. 1333 089)
6. Goat anti-mouse antibody conjugated to CY2 [F(ab')$_2$ fragment of IgG (H+L); 1:2000 dilution of a 1 mg/mL stock solution; Jackson ImmunoResearch Laboratories Inc., West Grove, PA, USA; Catalog No. 115-226-062]
7. Donkey anti-sheep antibody conjugated to CY3 [F(ab')$_2$ fragment of IgG (H+L); 1:2000 dilution of a 1 mg/mL stock solution; Jackson ImmunoResearch Laboratories Inc., West Grove, PA, USA; Catalog No. 713-166-147]
8. Embryo mountant: 70% glycerol, 2% DABCO (1,4-diazabicyclo [2.2.2.] Octane; Sigma, St. Louis, MO, USA; Catalog No. D-2522)
9. Microscope slides
10. Coverslips (22 × 50 mm)
11. Fluorescence (Zeiss Axioplan 2) and/or LSC microscope (Zeiss)

3. Methods
3.1. RNA Probe Preparation

1. To prepare runoff transcripts, the plasmid template is first linearized to completion (*see* **Note 1**) with the appropriate restriction enzyme, and then the enzyme removed by careful phenol and then chloroform extractions. After removal of all chloroform (heating to 65°C for 15 min helps), precipitate the DNA by adding NaAcetate (pH 5.2) to 0.3M, 3 vol of ethanol, and cooling to –70°C for 20 min. Centrifuge 10 min in a cold microcentrifuge and wash with cold 70% ethanol. We generally prepare 5–10 µg of linearized template, resuspended in 20 µL of RNase-free water.
2. RNA probes are prepared as described by Boehringer Mannheim on their digoxigenin and fluorescein RNA labeling spec sheets. On ice add 1 µg of linearized template DNA (3–5 kb), 2 µL of fluorescein or digoxigenin RNA labeling mix, 4 µL 5× transcription buffer (supplied with the RNA polymerase: 0.4 M Tris-HCl (pH 8.0); 60 mM MgCl$_2$, 100 mM dithiothreitol, 20 mM spermidine),

1 µL RNase inhibitor (1 U/mL) and sterile, RNase-free water to make the final reaction volume equal to 18 µL. Add 2 µL of the appropriate RNA polymerase (T7 or T3), mix well, and incubate at 37°C for 2 h.

3. Following the transcription reaction (*see* **Note 2**), the labeled probe is precipitated by addition of 1 µL 0.5 *M* EDTA, 2.5 µL 4 *M* LiCl, and 75 µL of absolute ethanol. Chill to –70°C, centrifuge, and wash the pellet as described above. After drying, resuspend the pellet in 100 µL of RNase-free water (*see* **Note 3**). Check the probe by loading and running 4–5 µL on a conventional agarose gel (~1%). The runoff transcript should easily be detected by ethidium bromide staining *(4)*. Probe should be stored at –20°C. Several freeze–thaw cycles on ice do not impair probe activity.

3.2. Initial Embryo Fixation

1. Prepare 40% formaldehyde stock solution just prior to embryo dechorionation (*see* **Note 4**). Dissolve 0.92 g of paraformaldehyde in 2.5 mL of water containing 35 µL of 1*N* KOH. Heat the mixture at 80°C until dissolved.
2. Collect and rinse the embryos in water.
3. Dechorionate the collected embryos in a 1:1 mixture of chlorine bleach and water for approximately 90 s. When dechorionated, the embryos will either float to the surface of the bleach solution or stick to the sides of the collection basket. Embryos should be rinsed immediately, as over-dechorionation is apparently detrimental. Rinse the collection basket with plenty of water. Fast flowing tap water can help dechorionate partially dechorionated embryos. An optional rinse with 0.7% NaCl, 0.03% Triton X-100 is helpful for removing residual bleach and for washing embryos down from the sides of the basket.
4. Transfer the embryos to a 20-mL glass scintillation vial (*see* **Note 5**) containing a two-phase mixture of 8 mL of heptane, 2.5 mL of 1× PBS and 250 µL of 40% formaldehyde. Shake for 20 min.
5. Using a 1-mL pipetteman, draw up embryos (which are at the interphase), taking care not to suck up any of the lower aqueous phase (*see* **Note 6**). Transfer to a 1.5-mL microfuge tube containing 0.5 mL of heptane and 0.5 mL of methanol for devitilinization. Shake vigorously until the majority of the embryos sink to the bottom (about 30 s). Carefully remove about 75% of the heptane and methanol and replace with 1 mL of methanol. Shake once more. All or most embryos should have now sunk to the bottom of the tube. Remove all liquid along with any embryos at the interphase, and then rinse two or three times with methanol. Embryos can be stored in methanol at –20°C for several months.

3.3. Post-Fixation and Hybridization of Whole-Mount Embryos

The following steps are optimized for ~50 µL settled embryos in a 1.5-mL microfuge tube.

1. Rinse the embryos once in methanol.
2. Rinse the embryos twice in PBT (1× PBS, 0.1% Tween 20).

3. Post-fix the embryos for 20 min in 0.5 mL of PBT containing 50 μL of freshly prepared 40% formaldehyde. Place tubes on a rocking platform to ensure even fixation.
4. Rinse embryos 3× in PBT. Washes should be approximately 2 min in duration.
5. Add approximately 0.5 mL of PBT containing 50 μg/mL of nondigested proteinase K. Incubate for ~1 to 1.5 min (*see* **Note 7**). Mix by drawing up some of the solution with a pipetteman and gently jetting the embryos back into suspension. Repeat once, allow embryos to settle, and then remove the solution at least 30 s before the end of the incubation period.
6. Stop the proteinase K digestion by immediately adding 1 mL of PBT containing 2 mg/mL of glycine. After about 2 min, remove and rinse for another 2 min in the same solution.
7. Rinse embryos twice in PBT to remove the glycine.
8. Post-fix the embryos once again (as in **step 3**) for 20 min in PBT containing 4% formaldehyde.
9. Wash the embryos extensively in PBT to remove all traces of fixative.
10. Rinse the embryos in 1 mL of 50% PBT, 50% RNA hybridization solution. Replace the mixture with 100% hybridization solution and prehybridize the embryos at 56°C for a minimum of 2 h. If required, embryos can be stored overnight at −20°C in the HYB solution prior to the 2-h heating step.
11. After prehybridization, place the embryos in a sterilized 0.5-mL microfuge tube, remove prehybridization solution and add probe. Optimal probe concentration needs to be determined empirically but generally 1 μL of probe in 100 μL of RNA hybridization solution works well. Diluted probe is heated to 80°C for 3 min, cooled briefly on ice and then added to the embryos.
12. Hybridizations are carried out at 56°C for 12–16 h. Mix embryos two to three times during the course of the incubation, either by quickly inverting the tube, or by using a pipetteman to gently jet the embryos into suspension.

3.4. Post-Hybridization Washes and Development of the FISH Signal

1. Remove any hybridization solution and embryos from the upper walls and cap of the microfuge tube by spinning the tube for ≈10 s at 1500 rpm in a microcentrifuge.
2. Remove the probe solution and rinse the embryos once with 400 μL of prewarmed hybridization buffer. Repeat the wash with another 400 μL of prewarmed hybridization buffer, this time incubating at 56°C for 20–30 min. Invert the tube several times during the course of the wash.
3. Wash embryos for another 20–30 min with a 1:1 mix of hybridization buffer and PBT and then with four 5 min washes of PBT. All washes should be done with preheated solutions at 56°C.
4. Cool to room temperature and incubate for 10 min in 400 μL of PBTB (1× PBS, 0.1% Tween 20, 0.5% milk powder). The use of milk powder in this and subsequent steps helps to reduce background.

5. Hybridized RNA probes are detected by first incubating the embryos with the appropriate primary antibodies diluted in PBTB. For double-labeling, both anti-digoxigenin and anti-fluorescein antibodies are added. Dilutions (*see* **Note 8**) that were optimal in our hands are given in the Materials section, but batches may vary, as may optimal activity given the many variations that exist in a particular laboratory's reagents, equipment, and methodology. Incubate with primary antibodies for 2 h (optionally overnight at 4°C), with constant mixing on a rocking platform or rotating mixer.

6. Wash for 1–2 h (optionally overnight) with four or five changes of PBTB.

7. Add the appropriate (*see* **Note 9**) secondary antibody(s) diluted in PBTB, and incubate with constant mixing for 2 h. Carry out this step and all subsequent steps in dim light with tubes covered or wrapped in foil.

8. Wash for 2 h with four or five changes of PBTB and then finally with PBT.

9. Resuspend embryos in DABCO-containing mountant. Allow the embryos to settle to the bottom of the tube (1–3 h or overnight at 4°C) before resuspending and mounting.

10. Transfer the embryos to a clean slide in ~80 μL of mountant and cover with a 25×50 mm coverslip. Seal the edges with nail polish. Slides can be stored for weeks at 4°C in the dark. Background levels will often decrease over the first few days.

11. Embryos can be viewed by either conventional fluorescence microscopy, LSCM, or deconvolution microscopy. Basic LSCM techniques are discussed elsewhere in this book. An example of double FISH labeling of a *Drosophila* embryo is shown in **Fig. 1A**.

3.5. RNA–Protein Double-Labeling

1. Collect and fix embryos as described previously for FISH, with the exception of the proteinase K step. The proteinase K concentration may have to be lowered to preserve integrity of protein epitopes (*see* **Note 7**).

2. After performing the hybridization and washes, as described, add the primary antibody for the protein of interest, along with the anti-fluorescein or anti-digoxigenin antibody. To obtain non-cross-reacting signals, the protein-specific antibody must have been raised in a host other than the host(s) used for the probe-specific antibodies (i.e., not mouse or sheep).

3. After primary antibody incubation and washes, the primary antibodies are detected using appropriate secondary antibodies (*see* **Note 9**). Wash and mount as with single or double FISH staining. With careful choice of antibodies, triple-staining a combination of transcript and protein targets is possible. However, secondary antibodies conjugated to CY5 are necessary *(3)*, as are the microscope excitation and detection components required for visualization.

3.6. Performing FISH on Dissected Tissues

1. Dissect tissues such as imaginal disks or salivary glands in PBS. Dissected tissues can be stored briefly (up to 30 min) on ice in a microfuge tube containing PBS while collecting enough tissues for analysis.

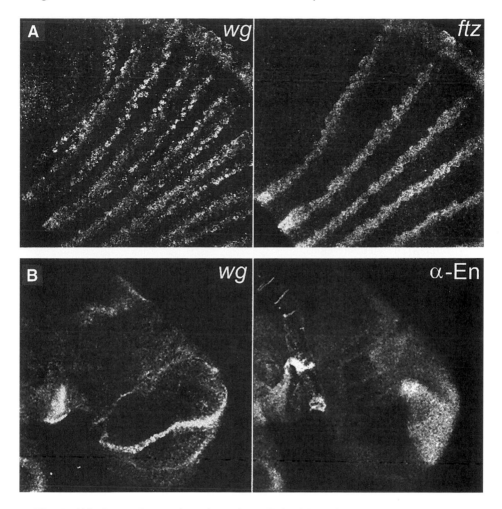

Fig. 1. (**A**) shows the trunk region of a cellular blastoderm embryo with *wg* transcripts detected on the left (CY3 channel) and *ftz* transcripts on the right (CY2 channel). (**B**) A wing imaginal disk with wingless transcripts detected on the left (CY2) and engrailed protein detected on the right (CY2).

An RGB image of the superimposed channels is shown in Color Plate I Examples of double-labeling using FISH. (**A**) A whole-mount embryo at 4 h after egg laying that is double-labeled for *fushi-tarazu* mRNA in green and *wingless* mRNA in red. (**B**) A third instar imaginal wing disk double-labeled for *wingless* mRNA in red and *engrailed* protein (antibody 4D9; *see* **ref. 6**) in green. In both panels, anterior is to the left and ventral is down. (*See* Color Plate I, lower left sequence, appearing after p. 372.)

2. Remove the PBS and add 50 µL of 10× PBS, 325 µL of water, 500 µL of heptane and 125 µL of 40% formaldehyde (freshly prepared as described previously). Shake gently for 45 s.

3. Remove the heptane and most of the fixative, and replace with PBT containing 4% formaldehyde. Continue fixation for another 20 min with gentle mixing.
4. Wash 4× with PBT and proceed to the proteinase K and subsequent steps, as described previously for embryos. Use the appropriate reagents for single or double FISH, or FISH/protein double-staining. **Figure 1B** shows an example of FISH/antibody double labeling in dissected imaginal disks.

4. Notes

1. Template DNA should be chosen and linearized such that runoff transcripts correspond to unique portions of the gene's coding region. So far, we've found that runoff transcripts ranging from about 0.4 to 1 kb work well as probes. Cutting to completion generally takes 2–4 h, and should be confirmed by agarose gel electrophoresis.
2. Removal of template with DNaseI subsequent to the transcription reaction was found to be unnecessary. Precipitation of the probe with LiCl removes most unincorporated nucleotides.
3. Previous protocols (e.g., Protocol 6) used carbonate degradation to reduce the size of probe RNA. In our hands this was found to be unnecessary and in fact was usually detrimental.
4. Freshly prepared formaldehyde appears to be required with the high temperatures used for RNA hybridization. Commercially prepared formaldehyde solutions, even ultrapure, generally yield ruptured embryos.
5. Vessel sizes used here are optimized for small collections (<250 μL of settled embryos). For greater collection sizes, larger vessels should be used, keeping approximately to the same relative ratios. 50-mL Falcon tubes work well for fixing and devitillinizing settled embryo volumes from ~0.25 to 2 mL. Care should be used as some tubes or plastics appear to interfere with fixation and devitillinazation (e.g., Sarstedt polystyrene tubes).
6. Aqueous solution interferes with the efficiency of the subsequent devitillinzation step. This has likely occurred if the devitillinization solution is cloudy and less than 80% of embryos have moved from the interphase to the bottom of the tube. Care should be taken to minimize uptake of the lower aqueous phase when drawing up embryos from the fixative into the pipet tip. Quite often, if this occurs, the phases will separate in the tip, and the lower aqueous phase can be returned to the scintillation vial. If transfer of aqueous solution has already occurred, the devitillinization step can be repeated as necessary by removing as much heptane and methanol as possible, replacing with fresh heptane and methanol and shaking again.
7. The extent of proteinase K digestion is a very important consideration. In general, proteinase K digestion enhances probe accessibility and hence the strength of the signal. However, overdigestion results in poor embryo morphology and ruptured embryos. Also, when double-labeling for proteins, proteinase K digestion can destroy the epitope. This can be remedied by lowering the working concentration of proteinase K as required. In fact, some in situ probes work very well with little or no proteinase K digestion. Newly prepared proteinase K stock solutions should be tested at several dilutions and/or digestion times. Prepare a 20 mg/mL

stock of proteinase K by dissolving in sterile water and storing at –20°C in 20–50 µL aliquots. Repeated freeze–thaw cycles appear to increase the activity of the enzyme.

8. The antibodies used here come lyophilized. For uniformity and convenience, we resuspend the powders in 50% glycerol and then aliquot and store at –70°C. One of the aliquots can be kept at –20°C for convenience (does not freeze and is relatively stable).

9. Antibodies described here have been chosen with usefulness in double-labeling in mind. The primary antibodies are whole IgGs raised in different hosts. Similarly, secondary antibodies are selected so that they are unlikely to crossreact with the other primary and secondary antibodies. Jackson laboratories, from which the recommended secondary antibodies were obtained, provide information, suggestions, and many products that make choosing and obtaining the appropriate antibodies relatively easy. Secondary antibodies most suitable for multiple-labeling are designated "ML." These are generally comprised of the $F(ab')_2$ portion of IgG antibodies that recognize both heavy and light (H and L) chains of their target antibodies. ML antibodies are also preabsorbed against multiple host sera. For this reason, and because the antibodies contain light-sensitive molecules, we do not bother to preabsorb them against embryos. However, if background is obtained, this may help. Cyanine-conjugated secondaries were chosen because of their strong emission spectra and resistance to photo-bleaching. The latter is particularly important with the high energy lasers used for LSCM.

Acknowledgments

We would like to thank Andrew J. Simmonds for his assistance with the imaginal disk staining and confocal microscopy. The *en* 4D9 antibody was a gift from T. Kornberg. This technology was developed with support from the National Cancer Institute of Canada (NCIC).

References

1. Tautz, D. and Pfeiffle, C. (1989) A non-radioactive in situ hybridization method for the localization of specific RNAs in *Drosophila* embryos reveals translational control of the segmentation gene *hunchback*. *Chromosoma* **98,** 81–85.
2. Harlow, E. and Lane, D. (1988) *Antibodies: A Laboratory Manual*, Cold Spring Harbor Laboratory Press, Cold Spring Harbor, NY, pp. 396–399.
3. Paddock, S. W., Langeland, A., DeVries, P. J., and Carroll, S. B. (1993) Three-colour immunofluoresence imaging of *Drosophila* embryos by laser scanning confocal microscopy. *BioTechniques* **14,** 42–47.
4. Hughes, S. C. and Krause, H. M. (1998) Double labeling with FISH in *Drosophila* whole-mount embryos. *BioTechniques* **24,** 530–532.
5. Lehmann, R. and Tautz, D. (1994) *In situ* hybridization to RNA, in *Methods in Cell Biology*, Vol **44,** (L. S. B. Goldstein and E. A. Fyrberg, eds.), Academic Press, San Diego, pp. 575–598.
6. Patel, N. H., Martin-Blanco, E., Coleman, K. G., Poole, S. J., Ellis, M. C., Kornberg, T. B., and Goodman, C. S. (1989) Expression of *engrailed* proteins in arthropods, annelids, and chordates. *Cell* **58,** 955–968.

6

Confocal Microscopy of Plant Cells

Carol L. Wymer, Alison F. Beven, Kurt Boudonck, and Clive W. Lloyd

1. Introduction

The increasing availability of confocal microscopy has begun a revolution in plant biology in which microscopy has again become a powerful tool for understanding structure and function. Examples of applications include: three-dimensional (3D) reconstruction of the interphase microtubule array in large vacuolated epidermal cells *(1)*; measuring cytoplasmic free calcium changes in whole maize coleoptile segments in response to phototropic and gravitropic stimuli *(2)*; and studying symplastic phloem connections in intact *Arabidopsis* roots *(3)*. The major reason for this revolution is the ability to collect clear images in three dimensions due to the lack of image degradation caused by out-of-focus light. Plant cells can attain very large sizes (hundreds of micrometers, in some cases) and are very thick. Thus the ability of the confocal microscope to obtain optical sections of tissues from which 3D reconstructions can be made surpasses the limitations of conventional "wide-field" microscopic techniques where microtome sectioning is often required and cells must be viewed as flat, two-dimensional objects. Furthermore, the reduction in out-of-focus flare increases depth discrimination.

The purpose of this chapter is twofold: to discuss some of the unique features of plant cells that should be borne in mind during sample preparation for confocal microscopy and to provide some example protocols as start-off points for new applications. The discussion is directed mainly toward fluorescence imaging because it is the major application of our laboratories and virtually all biological confocal microscopy is fluorescence. These protocols are based on those used in conventional epifluorescence microscopy with the key point being that in preparing tissue for confocal microscopy it is essential to retain the tissue's 3D structure.

From: *Methods in Molecular Biology, vol. 122: Confocal Microscopy Methods and Protocols*
Edited by: S. Paddock Humana Press Inc., Totowa, NJ

1.1. Unique Features of Plant Cells

1.1.1. The Cell Wall

The cell wall is a composite material composed primarily of polysaccharides and proteins. Cellulose microfibrils, distributed in highly ordered patterns, are the main component of the cell wall. These are interlinked by hemicelluloses, pectins, and proteins. The result is an extracellular matrix that can be 100–200 µm thick in primary cell walls and can withstand 0.3 MPa of pressure. It is estimated that the pore size of this matrix would seriously restrict molecules larger than 5 nm *(4)*, and so provides a serious barrier for antibody or nucleic acid probes. This penetration problem is generally dealt with in one of two ways. The first is to section the material, thus bypassing the need for the probe to penetrate the cell wall. The second is to partially degrade the cell wall with specific enzymes for cellulose, hemicellulose, or pectin. Enzymatic digestion can be very effective, but it is important to keep the treatment time short to minimize the loss of cell morphology and cell separation. In addition, these enzyme preparations are often contaminated with proteases. The goal of enzymatic digestion is to loosen the cell wall enough to allow probes to penetrate, without sacrificing structure.

In addition to polysaccharide and protein, cell walls may also contain waxy substances such as cutin (this is especially true for epidermal cells) or lignin. Although such compounds provide the cell with an excellent first line of defense against pathogen attack and water loss, they also make penetration of fixatives and probes difficult. In the case of tissues and whole organ segments, vacuum infiltration can help to get the fixative throughout tissue more completely and so improve fixation. Extraction of the tissue with an nonionic detergent such as 0.1% Tween-20 or 0.5% Triton X-100 for 10-15 min can also improve penetration.

One final consideration with respect to the cell wall is its autofluorescence. The occurrence of autofluorescence in primary cell walls is highly variable, but autofluorescence of secondary cell walls (as in xylem thickenings) is quite common. This fluorescence is relatively broadband, and hence difficult to remove with filtration. Although the source of the fluorescence is not certain, it is believed to be caused by phenolic compounds.

1.1.2. Vacuole

The vacuole has a number of functions including maintaining cell shape, providing the force necessary for cell expansion, and acting as a repository for select molecules. The size of the vacuole is greatly dependent upon the developmental stage of the cell. Rapidly dividing cells contain a number of small vacuoles so that the ratio of the vacuole to the cytoplasm volume is relatively low. Once cells stop dividing, the small vacuoles coalesce into a single, large central vacuole. Most mature, terminally differentiated cells have a very high vacuole to cytoplasm volume ratio. The presence of the vacuole, then, has several implications for the microscopist. For the microscopist interested in preserving structure in fixed specimens, it may be necessary to include an osmoticum in the fixative to maintain osmotic balance during the fixation process or to embed the tissue to prevent it becoming flattened. For the microscopist trying to study living cells on the microscope, efforts must to be made to prevent the cell from drying which can cause changes in turgor and thus changes in cell behavior. Methods for maintaining living cells on the microscope stage have been discussed previously *(5)* and so are discussed only briefly below.

A particular consideration when studying living cells is the potential for the sequestration of dyes by the vacuole; this is especially important when using dyes for quantification *(6)*. A number of workers have noted this when using calcium-measuring dyes such as fura-2 *(7)*. To combat sequestration, dyes can be bound to high molecular weight dextrans to prevent their transport across the tonoplast. Alternately, ratiometric dyes can be utilized. Ratiometric dyes experience a shift in either excitation or emission properties upon binding an ion. In this way, the ratio of bound to unbound dye can be measured making accumulation of dye by the vacuole much less important. Regardless of the method of quantification used, it is critical that proper control measurements are made.

1.1.3. Chloroplasts

Although it is easy to remember the role of chlorophyll role in absorbing light for photosynthesis (peaks at 430, 454, 595, and 643 nm for chlorophyll *b*), it is often forgotten that chlorophyll fluoresces in the red. This fluorescence should be borne in mind when choosing an appropriate fluorochrome for fluorescence applications, particularly since because the excitation and emission peaks of chlorophyll are similar to the rhodamine-type dyes (excitation at 541-570 nm; emission at 570–595 nm). Partly for this reason, fluorochromes such as fluorescein isothiocyanate (FITC), which are excited in the blue (around 488 nm) and emit in the green (around 520 nm), are popular for plant applications. Chloroplasts still fluoresce when excited with blue light, but the fluorescence is red and can be distinguished from the green light emitted by the fluorochrome. In some cases, conflicts between fluorochromes and chlorophyll can be avoided by using dark-grown (etiolated) tissue or tissue lacking chloroplasts such as roots or the white sectors of variegated tissue. For situations that require the use of red-emitting fluorochromes in tissues containing chloroplasts, it is sometimes possible to use secondary filters such as a 575 nm long pass filter to remove chlorophyll fluorescence but retain the rhodamine fluorescence.

1.2. Data Acquisition and Presentation

As previously mentioned, the real advantage of confocal microscopy, as opposed to standard epifluorescence, is the increased resolution of cellular components due to the removal of out-of-focus light. In this way, fine detail can be resolved, even in a thick plant cell. The result is an optical section with a very narrow depth of field. This means that to image the whole of an organelle or network, numerous optical sections are required. Once such a series of images has been gathered, the images can be viewed individually or, by means of some calculations, collectively as projections. Using computer software programs, one can flip "up" and "down" through the series of sections manually or through animation. This gives the effect of focusing through the structure as if it was still on the microscope stage. Although this provides the most detail, it is not possible to view the structure as a whole (**Fig. 1A–C**). To view the structure as a whole, projections can be calculated. For a projected image, each pixel in the image is calculated as the sum or the maximum of the pixel values encountered along a line of sight through the data stack. This could be compared to having pictures on several overhead transparencies (optical sections) and overlaying all of them (**Fig. 1D–F**). A projection provides information about the structure as a whole, but results in a flattened representation of the structure. To obtain a more 3D view of the structure, projections can be calculated at incremental angles to produce a rotation. In this way the image appears to have been spun about a chosen axis (**Fig. 1D,F**). These two methods of presentation provide different types of information and the method used obviously depends on individual requirements. For instance, to sample the cortical microtubules beneath the convex outer wall of an epidermal cell three to five optical sections taken at 0.5– or 1–μm intervals can be gathered and projected to provide a full (but flattened) view of this microtubule array (**Fig. 1E**) *(8)*. However, to look at the microtubules as they move out of this plane and turn around the side walls, it is necessary to gather many more optical sections and then make rotations (**Fig. 1D,F**) *(8)*. In addition, a pair of projections calculated at a suitable angle of separation (approximately 6°) can be viewed as a stereo-pair. Stereo-pairs are a typical way of presenting 3D information for publication, but this is not as effective as animating the structure on a viewing screen. Authors are beginning to take advantage of Internet capabilities by including Internet sites where readers can view animated data. It is anticipated that this will be an increasingly popular way for authors to display 3D images.

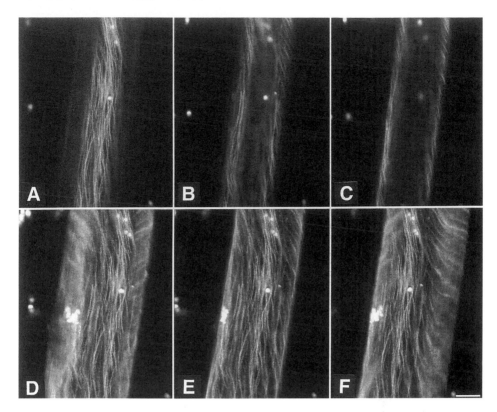

Fig. 1. Images collected with the confocal microscope can be displayed as individual images or as projections. All of the images are of a pea (*Pisum sativum*) epidermal cell and are part of the same data set. (**A–C**) Three successive optical sections taken at 1.4-μm intervals. (**D**) Rotation to show the cortical microtubule array under the left-hand wall of the cell. (**E**) A simple projection of the cell, with no rotation, to show the microtubules under the outer epidermal wall. (**F**) Rotation of the cell to show the microtubules under the right-hand wall. Data courtesy of Ming Yuan. Bar = 20 μm.

2. Materials

Unless otherwise stated, reagents can be obtained from Sigma (St. Louis, MO, USA).

1. Fixatives: The choice of fixatives varies widely, but they almost always contain formaldehyde as the chemical crosslinker. Formaldehyde is usually used at 4% (w/v), but it is often easier to prepare an 8% (w/v) solution and mix it with 2× buffers. For 8% formaldehyde, weigh 0.8 g of paraformaldehyde into a small vial or beaker containing a small stir-bar. Paraformaldehyde should be weighed out in a fume hood. Add 10 mL of water and cover loosely with a lid or foil. Place the container in a beaker of water and heat to 65°C in a fume hood. After about

10–15 min at 65°C, add one or two drops of 1*N* NaOH. The solution should clear instantly.
   a. Fixative 1: 4% (v/v) Formaldehyde freshly prepared from paraformaldehyde
      0.1% (v/v) Glutaraldehyde
      0.25*M* Mannitol
      50 m*M* Sodium phosphate
   b. Fixative 2: 4% (v/v) Formaldehyde freshly prepared from paraformaldehyde
      0.1% (v/v) Glutaraldehyde (preferably EM grade)
      2.75% (w/v) Sucrose
      0.1% (v/v) Nonidet P-40 (NP-40) in PEM 50:5:5
   c. Fixative 3: 4% (v/v) Formaldehyde
      0.1% (v/v) Glutaraldehyde (preferably EM-grade)
      50 m*M* PIPES (pH 6.9)
      1 m*M* $MgSO_4$
      5 m*M* EGTA
      1% (v/v) Glycerol
   d. Fixative 4: Freshly prepared 4% (v/v) formaldehyde in PEM 50:5:5
      0.2% (v/v) NP-40
   e. Fixative 5: Freshly prepared 4% (v/v) formaldehyde in PEM 50:5:5
      0.1% (v/v) Glutaraldehyde
      10% (v/v) DMSO
      0.1% (v/v) Tween-20
2. Enzymatic solutions: (*see* **Note 1**)
   a. Enzyme solution 1: 0.5% (w/v) Cellulase RC (Seishin Pharmaceuticals, Tokyo)
      0.05% (w/v) Pectolyase Y-23 (Seishin Pharmaceuticals, Tokyo)
      0.5% (v/v) Triton X-100
      0.25*M* Mannitol
      1 µg/mL Protease inhibitors (such as chymostatin, and/or pepstatin) in PEM 50:1:1 buffer
   b. Enzyme solution 2: 0.5% (w/v) Cellulase R-10 (Yakult Pharmaceuticals, Tokyo)
      0.05% (w/v) Pectolyase Y-23 (Seishin Pharmaceuticals, Tokyo)
      0.1% (v/v) Triton X-100 in PEM 50:5:5
   c. Enzyme solution 3: 0.1% (w/v) Cellulase RC (Seishin Pharmaceuticals, Tokyo)
      0.01% (w/v) Pectolyase Y-23 (Seishin Pharmaceuticals, Tokyo) in PEM 50:1:1
   d. Enzyme solution 4: 0.05% (w/v) Cellulase RS (Yakult Pharmaceuticals, Tokyo)
      0.025% (w/v) Pectolyase Y-23 (Seishin Pharmaceuticals, Tokyo)
      1% (w/v) Driselase (Sigma) in PEM 50:5:5
(*Note:* Driselase comes from the manufacturer attached to starch grains. Prepare a stock solution at 2% (w/v) and centrifuge it to remove the starch. Aliquots of this stock can be stored for many months at –20°C.)
3. Poly-L-lysine: To attach cells to slides, poly-L-lysine with a molecular weight of >250,000 can be used. It is prepared at 1 mg/ml in water and stored at –20°C. Once thawed, aliquots can be kept at 4°C for up to 2 weeks. Microscope slides

should not be used directly from the manufacturer; instead they should be washed in detergent, rinsed with copious amounts of distilled water, and dried in an oven. Apply the poly-L-lysine with a cotton swab to the warm slides.

4. γ-Aminopropyl triethoxysilane (APTES): APTES (Sigma) is a bonding agent that sticks tissue down very firmly and seems to give a lower background for antibody and *in situ* hybridization procedures. The microscope slides are first washed in detergent and placed in a freshly prepared 2% (v/v) solution of APTES in acetone for 10 s (longer incubations can result in increased background). The slides are then briefly treated in acetone and allowed to air dry. Slides prepared in this way should be used within 1 month. Just before use, the slides are placed in 2.5% (v/v) glutaraldehyde (reagent grade) in phosphate-buffered saline (PBS) (pH 7.4) for at least 30 min. Finally, the slides are rinsed in distilled water and air dried.

5. Subbing slides:
   a. Wash slides in 95% ethanol and dry.
   b. Dip in subbing solution.
      Subbing solution: Heat 500 mL of water and add 2 g of gelatin. After gelatin dissolves, add 0.2 g of chrom alum (chromic potassium sulphate) while stirring. Let mixture cool, then add azide to 0.02% (w/v). Filter solution and store in refrigerator. Filter each time before use.
   c. Dry slides at 60°C.

6. Blocking solution: 3% (w/v) bovine serum albumin (BSA), 0.2% (v/v) NP-40, in PEM 50:5:5.

7. Phosphate buffered saline: PBS can be prepared according to many different recipes based on the buffer's intended use. The following recipes are for PBS at two different pHs. PBS should be stored at 4°C, although it can be stored at room temperature if azide is added.
      PBS, pH 7.4: 8 mM Na$_2$HPO$_4$, 1.47 mM KH$_2$PO$_4$, 137 mM NaCl, 2.68 mM KCl, adjust pH to 7.4.
      PBS, pH 8.0: 4 mM Na$_2$HPO$_4$, 2 mM KH$_2$PO$_4$, 140 mM NaCl, 3 mM KCl, adjust pH to 8.0.

8. Tris-buffered saline (TBS): 25 mM Tris-HCl (pH 7.4), 140 mM NaCl, 3 mM KCl.

9. Physiological buffer (PB): 100 mM potassium acetate, 20 mM KCl, 20 mM HEPES, pH 7.4 with KOH, 1 mM MgCl$_2$, 1 mM ATP (disodium salt), 1% (v/v) thiodiglycol, 2 μg/mL aprotinin, 0.5 mM phenylmethylsulfonyl fluoride.

10. PEM buffers: All of the following protocols utilize a buffer composed of PIPES (pH 6.9), EGTA, and MgSO$_4$ which is referred to as "PEM." Because each of the buffers has been developed by different individuals for different purposes, the buffer components are used in different proportions. The buffers will be referred to as "PEM 50:1:1" and "PEM 50:5:5" based on the mM concentration of their components. PEM buffers should be stored at 4°C.

11. Saline sodium citrate (SSC): This buffer is used at several different concentrations in the following protocols. It is often useful to prepare a 20× stock and prepare the dilutions as needed.
      1× SSC: 150 mM NaCl, 15 mM sodium citrate.

12. Transcription mix: 500 μ*M* CTP (sodium salt; Pharmacia), 500 μ*M* GTP (sodium salt; Pharmacia), 250 μ*M* BrUTP (sodium salt; Sigma), 125 μ*M* MgCl$_2$, 100 U/mL RNA Guard (Pharmacia), in physiological buffer.

13. Antibodies: Commercially available antibodies can generally be used at dilutions of 1:100 to 1:400, although it is best to check the manufacturer's recommendations. To prevent nonspecific binding, antibodies are usually diluted in 3% (w/v) BSA. In cases where antibody penetration is a problem, better results can be obtained by using an F(ab')$_2$ fragment instead of the whole IgG. Most antibody incubations are not carried out at 37°C as is common for animal applications. Instead, incubations are carried out at room temperature, or 4°C in the case of overnight incubations. To prevent the antibodies from evaporating, the incubations are carried out in humid chambers: boxes containing wet paper towel. Finally, it is advisable to keep the slides out of the light during secondary antibody incubation to minimize fading of the fluorochrome.

14. Probe mixture for *in situ* hybridization:
    ~200 ng/μL Digoxigenin or biotin-labeled RNA probe
    ~1000 ng/μL Unlabeled RNA transcribed from a plasmid containing an unrelated insert
    50% (v/v) Deionized formamide (reagent grade; Fisons)
    10% (w/v) Dextran sulfate (sodium salt; Sigma)
    300 m*M* NaCl
    10 m*M* PIPES (pH 8.0)
    1 m*M* EDTA

15. Fluorescent mounting media: There are a number of commercially available mounting media including Citifluor (Agar Scientific, UK) and Vectashield (Vector Laboratories, Peterborough, UK). Alternatively, a recipe for mounting medium is given below. Glycerol-based media that contain an antifading agent such as *p*-phenylenediamine-HCl are most frequently used. Best results are obtained by leaving the slides overnight to allow the anti-fading agent time to penetrate. In addition, a chromatin binding dye such as DAPI (4',6-diamidino-2-phenylindole) or Hoechst 33258 (both at 1 μg/mL) can be added to the medium immediately before use. Store the mounting medium at 4°C and add 1 mg/mL of *p*-phenylenediamine dihydrochloride immediately before use.
    Mounting medium: 0.1*M* Tris-HCl (pH 9.0), 50% (v/v) glycerol.

3. Methods

The following protocols have been selected because they represent a wide range of plant material (suspension cultured cells to whole roots) and because they illustrate several types of procedures (indirect immunofluorescence and *in situ* hybridization). They are intended as start-off points for similar applications. Additional methods for staining a variety of plant cells are given in **ref. 9**.

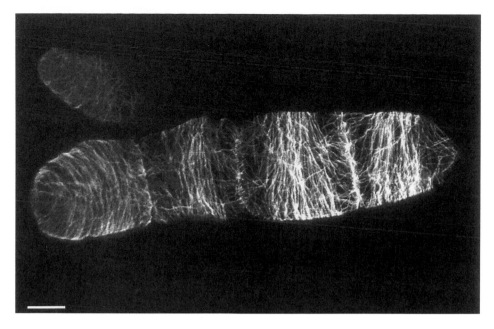

Fig. 2. Immunolocalization of cortical microtubules in the tobacco (*Nicotiana tabacum*) suspension culture line BY-2. The cells were taken from a 5-day-old culture which was rapidly dividing and the microtubules were labeled with YOL 1/34 (Sera-Lab, Crawley Down, UK). Projection of nine optical sections taken at 0.5-μm intervals. Bar = 20 μm.

3.1. Protocol 1: Suspension Cultured Cells

This protocol describes the immunolocalization of microtubules in the tobacco (*Nicotiana tabacum*) suspension culture line BY-2 (Bright Yellow-2) (**Fig. 2**). The BY-2 cell line has been used extensively for studying the microtubule cytoskeleton *(10)*. The following protocol has been slightly modified from **ref. 11**. The same protocol has also been used for examining microtubules in carrot suspension cultures.

1. Transfer a sample of cells to a test tube.
2. Gently pellet by centrifuging at 100 rpm for 10–15 min.
3. Remove excess medium and add at least 2× the cell volume of Fixative 1.
4. Fix cells for 1 h.
5. Pellet cells as before and remove the fixative.
6. Gently rinse the cells 3× by resuspending in PBS (pH 7.4) and then pelleting. At this point cells can be stored in PBS for several weeks without loss of antigenicity.

7. Place a couple of drops onto poly-L-lysine-coated slides and allow the cells to settle for 10–15 min.
8. Wick away the unbound cells with a piece of paper towel and extract the cells for 4–10 min using Enzyme Solution 1 to allow antibodies to penetrate.
9. Rinse for 10 min in PBS (pH 7.4) by dipping the slides into a beaker of PBS. When necessary, any of the following steps can be carried out overnight at 4°C.
10. Wick away excess PBS and block nonspecific antibody labeling by treating with 3% (w/v) BSA in PEM 50:1:1 for 15 min.
11. Treat with an anti-tubulin antibody (such as YOL 1/34 from Sera-Lab, Crawley Down, UK or anti-β-tubulin from Amersham) diluted in 3% (w/v) BSA for at least 45 min at room temperature in a humid chamber.
12. Rinse in fresh PBS for 15 min.
13. Wick away excess PBS and treat with an appropriate secondary antibody diluted in 3% (w/v) BSA for at least 45 min. at room temperature.
14. Rinse in fresh PBS for 15 min.
15. Remove excess PBS and add fluorescent mounting medium containing an antifade agent.

3.2. Protocol 2: Epidermal Peels

Epidermal cells have long been studied by plant physiologists because of their exposed nature and because of their unique features. For example, special cell divisions within the epidermis give rise to guard cells, trichomes, and root hairs. To aid experimentation, especially microscopy, it was discovered that the epidermis of some organs of some species could be peeled away from the organ and the epidermal cells would still be viable. Many studies of guard cell behavior have been carried out this way, e.g., **ref. *12***. Generally, peels consist of more than the epidermis; they also contain two or three layers of cortical cells. It is critical to examine the peels under the microscope before using them because cell viability after peeling is highly variable. The following protocol is for the immunolocalization of microtubules in epidermal cells of pea (*Pisum sativum*) epicotyls (**Fig. 3**). There is great difficulty in penetrating the cell walls of these epidermal cells because of their waxy cuticle, so a brief treatment with ethanol has been added. This protocol has also been used successfully for thin, paradermal slices obtained by hand-sectioning pea stems.

1. With a flexible, double-edged razor blade, cut the stem at the base of the region of interest. This provides a large piece of stem above the region of interest to hold onto. Lightly score the stem at the top of the area of interest with the razor blade. This will provide an end for the peel, otherwise the whole segment will be peeled. Grasp the epidermis at the cut end with a pair of fine forceps and peel the epidermis away from the rest of the stem.
2. Place the peels (epidermal side up) on the sticky side of adhesive tape (a hydrophobic type such as Cellux) that has been secured to a microscope slide with double-sided tape (**Fig. 3**). The adhesive tape is used because the double-sided tape does not adhere

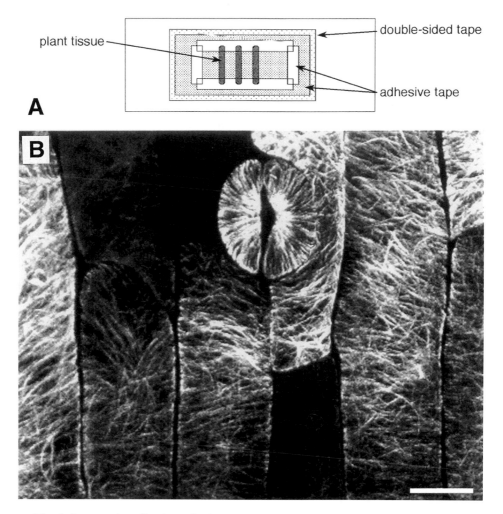

Fig. 3. Immunolocalization of microtubules in pea (*Pisum sativum*) epidermal cells. (**A**) Configuration of the tape used to secure the pieces of tissue. Double-sided tape is placed on the microscope slide and a piece of waterproof adhesive tape is placed on top of this with the sticky side up. The tissue is placed on the adhesive tape and secured with the same type of tape, placed sticky side down. (**B**) Cortical microtubules in the epidermal cells. The image is a projections of 11 optical sections taken at 0.5-µm intervals. Bar = 10 µm.

well once it is wet. Place a small piece of adhesive tape, sticky side down, over one end of the peels to be sure that they stay affixed to the slides during processing. By adding additional pieces of tape around the sections, a hydrophobic ring is formed.

3. Cover with buffer (50 mM potassium glutamate, 0.5 mM magnesium chloride, and 2.75% sucrose) while preparing all of the slides.
4. Treat with 95% ethanol for 20 s.
5. Remove the ethanol and rinse briefly with buffer.
6. Cover the peels with a large drop of Fixative 2 and fix for at least 2 h. During this and subsequent incubations, the slides should be placed in a humid box to prevent drying and to contain the aldehyde fumes.
7. Remove the fixative and rinse by dipping the slide in a beaker of PBS (pH 7.4) for 15 min.
8. Extract for 10 min in Enzyme Solution 2.
   Note: The duration of the extraction and the choice of detergent is highly dependent on the thickness of the cell wall and the cuticle.
9. Rinse as before in PBS for 15 min.
10. Incubate in anti-tubulin antibody [diluted in 3% (w/v) BSA] overnight at 4°C. Inclusion of detergent [0.1% (v/v) NP-40] can be used to increase antibody penetration.
11. Rinse in PBS for 30 min.
12. Incubate in secondary antibody (diluted as for the primary antibody) for 6 h.
13. Rinse in PBS for 30 min.
14. Mount in fluorescent mounting medium containing an antifade agent.

3.3. Protocol 3: Immunofluorescence of Tissue Sections

This protocol describes immunolocalization in plant root sections cut on a vibratome. The vibratome uses a razor blade that vibrates to and fro across the tissue as the section is cut. Typical sections from pea (*Pisum sativum*) roots are 30–40 µm thick and contain about two layers of cells. Vibratome sectioning is most successful using dense, rigid tissue such as root tips. Much less success is achieved using highly vacuolated tissue such as leaf tissue. Also, species with very small roots, e.g., wheat (*Triticum*), are more difficult to section than those of species with larger roots, e.g., pea (*Pisum sativum*) and maize (*Zea mays*). The use of sections, rather than other preparative methods such as squashes, has the advantage that both good 3D structure and positional information are preserved (**Fig. 4**).

1. Excise the apical 3–4 mm of a freshly germinated pea root and fix in 4% (v/v) formaldehyde in PEM 50:5:5 for 1 h at room temperature.
2. Wash the roots in PEM 50:5:5 for 15 min.
3. Cut 30–40 µm sections on the vibratome.
4. Allow the sections to air dry onto APTES-coated multiwell slides.
5. Digest the cell wall by incubating in 2% (w/v) Cellulase R-10 (Yakult Pharmaceutical, Tokyo) in TBS for 1 h.
6. Wash the slides 2× for 10 min each in TBS.
7. Apply the primary antibody in 3% (w/v) BSA (prepared in TBS) and incubate for approximately 45 min at room temperature.

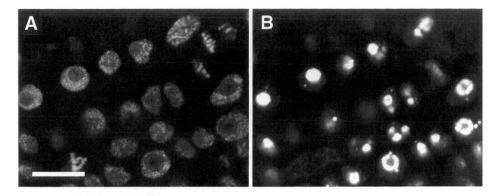

Fig. 4. Distribution of the nuclear protein fibrillarin in tissue sections of pea (*Pisum sativum*) roots. (**A**) This image is a single confocal section showing DAPI-labelling of the chromatin. (**B**) A single confocal section of the same area as in (A) showing the distribution of fibrillarin detected with anti-fibrillarin antibody (72B9; the generous gift from E. Tam and B. Ochs of Scripps Research Institute, La Jolla, CA, USA). The contrast has been adjusted in (B) to be able to see the fainter labeling in mitotic cells which has resulted in over-saturation of the interphase cells. Bar = 10 µm. (Figure reprinted with permission from **ref. *13*.**)

8. Wash slides 3× for 10 min each in TBS.
9. Incubate in secondary antibody diluted in 3% (w/v) BSA (prepared in TBS) for approximately 45 min at room temperature.
10. Wash again in TBS: 3× each for 10 min.
11. Treat sections with DAPI (1 µg/mL) for 5 min to stain chromatin.
12. Wash briefly in TBS and mount in Vectashield.

3.4. Protocol 4: In Vitro Transcription in Tissue Sections

The following protocol describes the localization of transcription sites in root tip sections. Because this procedure uses unfixed material, several factors must be carefully controlled to ensure optimal detection of transcription. The quality of the starting material is crucial. Only healthy, recently germinated, well growing roots should be used. The sectioning procedure, from the excision of the root tip and cutting of sections on a vibratome, to the permeabilization step (which allows substrates to pass into the cells) should be as quick as possible, ideally no longer than 5 min. The solutions used are designed to match, as nearly as possible, the ionic conditions inside the cell to ensure the RNA polymerases can continue to transcribe during the procedure. The sections are incubated in a transcription mix that contains a modified nucleotide triphosphate (BrUTP) and under these conditions, any engaged polymerases will continue to transcribe and incorporate the modified nucleotide. This modified base

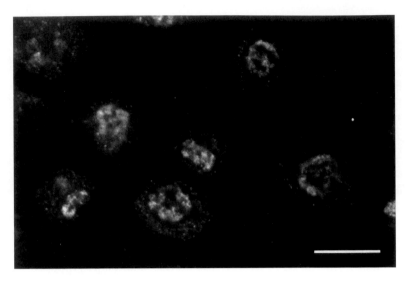

Fig. 5. In vitro transcription in pea (*Pisum sativum*) root sections. A single optical section of pea root tissue showing BrUTP incorporation by RNA polymerases I and II. The labeling comprises many closely packed small foci, which in the nucleolus form a more continuous region. Bar = 10 μm.

can subsequently be detected by a specific antibody. The following protocol was modified for plant systems from an original method by Hozak *et al.* *(14)*. It has been used with good success in pea (*Pisum sativum*) *(15)* (**Fig. 5**), maize (*Zea mays*), rye (*Secale*), and wheat (*Triticum*).

1. Excise the apical 3–4 mm of a freshly germinated root tip and mount it on a vibratome. Cut sections (40–50 μm) in physiological buffer (PB) containing 1 *M* hexylene glycol (2-methyl-2,4-pentanediol) and transfer to tissue handling device *(16)*.
2. Incubate in 0.05% (v/v) Triton X-100 in PB for 1 min.
   Note: Shorter Triton treatments give better nuclear transcription, whereas longer treatments result in predominantly nucleolar transcription.
3. Wash in PB alone—three washes over ~30 s.
4. Incubate in Transcription Mix for 2–10 min depending on the degree of incorporation required.
5. Wash again in PB alone—three washes over ~30 s.
6. Fix in 4% (v/v) formaldehyde prepared in PEM 50:5:5 for 1 h.
7. Wash in TBS for 10 min followed by distilled H$_2$O for 10 min.
8. Remove from tissue handling device and allow to air dry onto APTES-coated multiwell slides.
9. Treat sections with 2% (w/v) Cellulase R-10 (Yakult Pharmaceutical, Tokyo) in TBS for 1 h.
10. Wash in TBS—three changes over 15 min.

11. Incubate for 1 h with primary antibody: mouse anti-BrDU (Boehringer Mannheim) diluted 1:20 with 3% (w/v) BSA in TBS.
12. Wash thoroughly with TBS.
13. Incubate for 1 h with fluorescent secondary antibody: Cy3-conjugated donkey anti-mouse (Jackson ImmunoResearch Laboratories, West Grove, PA, USA) diluted 1:100 with 3% (w/v) BSA in PEM 50:5:5.
14. Mount in Vectashield or other antifade mountant.

3.5. Protocol 5: Tissue In Situ Hybridization with RNA-Digoxygenin or RNA–Biotin Probes

The following protocol describes *in situ* localization in root tissue sections using RNA probes labeled with either digoxygenin or biotin [modified from a method by Highett *et al.* *(17)*]. This procedure can also be used with a minor modification to localize DNA probes, but in this case the probes need to be heat denatured prior to the *in situ* hybridization. In adapting this method to different species, consideration of the enzymatic digestion step could be important for enabling optimal penetration of the probes. In all cases, we recommend using a probe size of no more than 100 base pairs, again, to ensure good tissue penetration. The stringency of the procedure can be controlled by varying both the temperature and the salt concentration of the washing step. Higher temperature, combined with a low-concentration salt solution, gives the highest stringency. This procedure has been used with success on pea (*Pisum sativum*) *(18)* (**Fig. 6**), maize (*Zea mays*), rye (*Secale*), wheat (*Triticum*), and onion (*Allium cepa*).

1. Fix roots in 4% (v/v) formaldehyde and 0.1% (v/v) glutaraldehyde [or 4% (v/v) formaldehyde alone] in PEM 50:5:5 for 1 h at room temperature. Addition of glutaraldehyde yields better tissue preservation, but can make penetration of the probe more difficult.
2. Wash in PEM 50:5:5.
3. Cut 30–40 µm sections on the vibratome.
4. Allow sections to dry onto APTES-coated multiwell slides.
5. Treat sections with a freshly prepared solution of $NaBH_4$ (1 mg/mL) in PBS (pH 8.0) for 15 min. Aspirate off and repeat three more times. This step is necessary only when the fixative contains glutaraldehyde. PBS is used in this step instead of TBS because $NaBH_4$ reacts with Tris.
6. Incubate in 2% (w/v) Cellulase R-10 (Yakult Pharmaceuticals, Tokyo) in TBS for 1 h at room temperature.
7. Wash in TBS for 5 min and then in 0.1× SSC for 5 min.
8. Incubate in 0.1× SSC at 98°C for 5 min. and then transfer immediately into ice-cold 0.1× SSC. This step is necessary only when detecting double-stranded DNA.
9. Remove surface liquid from the slide wells and add 10 µL of ice-cold Probe Mixture. Place slides in a humid chamber at 37°C overnight.

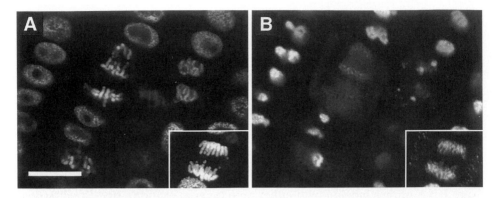

Fig. 6. Determination of the distribution of the external transcribed spacer (ETS) probe in pea (*Pisum sativum*) root sections using in situ hybridization. (**A**) This image is a single confocal section showing DAPI-labelling of the chromatin. (**B**) A single confocal section of the same area as in (A) showing the distribution of nascent and newly completed pre-rRNA transcripts detected with an ETS probe *(18)*. The contrast has been adjusted in (B) to make it possible to use the fainter labeling in mitotic cells which has resulted in over-saturation of the interphase cells. The inserts show a projection of a whole cell that is further down in the same 3D data set. Bar = 10 μm. (Figure reprinted with permission from **ref. *13*.**)

10. Wash in 0.1× SSC at 50°C for 1.5 h with three changes.
11. Antibody detection—Remove excess wash buffer and apply antibodies diluted in 3% (w/v) BSA prepared in TBS. Use dilutions recommended by manufacturer and incubate at room temperature for approximately 45 min each. Wash thoroughly with TBS between antibody incubations.

    Antibody detection schemes:

    For digoxigenin-labeled probes:

    1° Mouse anti-digoxin (anti-digoxin crossreacts with digoxigenin; Sigma)

    2° Cy3-conjugated anti-mouse (Jackson ImmunoResearch Labs., West Grove, PA, USA)

    For biotin-labeled probes:

    Extravidin Cy3-conjugate (Sigma)
12. Counterstain with DAPI (1 μg/μL) for 3 min.
13. Rinse in TBS and mount in Vectashield (Vector Laboratories, Peterborough, UK) or other antifade mountant.

3.6. Protocol 6: Microtubules and Chromosomes in Root Tip Squashes

In this procedure microtubules are immunolocalized in root tip cells. The great advantages of these cells are: (1) they have a high proportion of well-characterized mitotic figures; and (2) these small, nonvacuolated cells can be

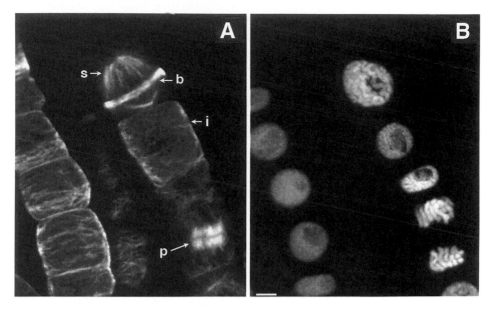

Fig. 7. Microtubule arrays in the meristem of onion (*Allium*). Eight-day-old *Allium* root tips were fixed, squashed, and labeled using an anti-tubulin antibody (YOL 1/34, Sera-Labs, Crawley Down, UK). All types of microtubule arrays are illustrated in this field of cells: i, interphase cortical array; b, pre-prophase band; s, mitotic spindle; p, phragmoplast. Bar = 20 μm.

air dried without destroying their 3D structure as can happen with vacuolated cells, such as suspension culture cells. This procedure was originally developed by Wick *et al.* *(19)* and has been modified by a number of laboratories. For the best results, use root tips from freshly germinated seedlings that are rapidly growing. This procedure works well with a number of dicot and monocot species (**Fig. 7**).

1. Excise the root tip from the seedling.
2. Fix roots at room temperature for 45–60 min in Fixative 3.
3. Remove the fixative and rinse 3× in PEM 50:1:1 buffer. Roots can be stored for days to weeks at 4°C in buffer.
4. Place roots in Enzyme Solution 3 for 10 min.
5. Wash 2× in PEM 50:1:1 buffer.
6. Place root tips onto subbed slides, put a coverslip on top, and press down hard (squash).
7. Gently remove the coverslips and dry slides at 4°C overnight. Alternatively, one can adsorb the cells to poly-L-lysine-coated slides and process the same day. The cells do not have to be dried if poly-L-lysine-coated slides are used.
8. Place slides in a humid box and extract for 10 min with 0.1% (v/v) Triton X-100 in PEM 50:1:1.

9. Block any free aldehyde groups with 3% (w/v) BSA for 15 min at room temperature.
10. Remove excess blocking buffer and add anti-tubulin antibody [diluted in 3% (w/v) BSA] place for 45 min at room temperature.
11. Rinse primary antibody from the slide using a gentle stream of PBS (pH 7.4) in a wash bottle. Place slides in a beaker of PBS for 15 min.
12. Remove excess PBS from slides and absorb secondary antibody [diluted in 3% (w/v) BSA] for 45 min. Protect from light.
13. Wash slides as before.
14. Add a few drops of fluorescent mounting medium containing a chromatin binding dye and secure the coverslips in place.

3.7. Protocol 7: Whole Mount Immunofluorescence in Arabidopsis Roots

This protocol describes the immunofluorescence localization of proteins in whole-mount *Arabidopsis thaliana* roots. The combination of whole-mount immunofluorescence and confocal microscopy allows the localization of highly abundant proteins in all root tissues without the need for sectioning. The following protocol is now routinely used to localize the U2B" splicing protein (detected by 4G3 antibody) (**Fig. 8**).

1. Place three day old seedlings in Fixative 4 for 1 h.
2. Wash the seedlings 3× for 5 min each in PEM 50:5:5 containing 0.2% (v/v) NP-40, then resuspend them in PEM 50:5:5 alone.
3. Excise the roots and transfer them to APTES-treated multiwell slides and let them dry down.
4. Digest the roots in Enzyme Solution 4 for 10 min.
5. Wash the roots 6× in PEM 50/5/5 containing 0.2% (v/v) NP-40 and allow to air dry afterwards.
6. Wash in Blocking Solution for 90 min.
7. Incubate in primary antibody for 2 h at 37°C or overnight at 4°C: mouse 4G3 (Euro-diagnostica B.V., The Netherlands) diluted 1:10 with 3% (w/v) BSA in PEM 50:5:5.
8. Wash in Blocking Solution for 1 h.
9. Incubate in secondary antibody for 2 h at 37°C: Cy3-conjugated donkey anti-mouse (Jackson ImmunoResearch Laboratories, West Grove, PA, USA) diluted 1:100 with 3% (w/v) BSA in PEM 50:5:5.
10. Wash in PEM 50:5:5 containing 0.2% (v/v) NP-40 for 2 days.
11. Counterstain with DAPI if required and mount in Vectashield.

3.8. Protocol 8: Whole Mount In Situ Hybridization in Arabidopsis Roots

The following protocol describes the *in situ* localization of relatively highly abundant RNAs in whole-mount *Arabidopsis* roots. It is based on protocols from **refs. *20–22***, although it has been slightly simplified and adapted for *in*

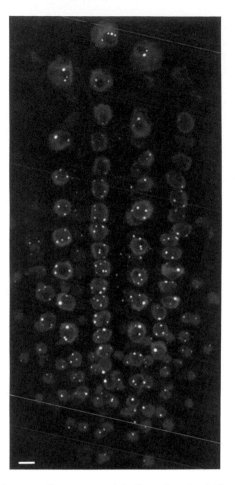

Fig. 8. Whole mount immunofluorescence labeling of an *Arabidopsis thaliana* root with the 4G3 antibody. The 4G3 antibody labels the snRNP-specific protein U2B". It is interesting to note that the label is confined to the nucleus and is excluded from the nucleolus. The round bright spots, often associated with the nucleolus, are coiled bodies. Bar = 5 μm.

situ hybridization. For the RNA probe used to develop this procedure, the images obtained appear to be nearly identical to those obtained by immunolocalization (*see* **Fig. 8**).

1. Fix 3-day-old seedlings for 1 h in Fixative 5.
2. Wash 2× for 3 min each in methanol followed by 4× for 5 min each in ethanol at –20°C.
3. Wash for 30 min in 1:1 ethanol/xylene.
4. Wash for 3 min in ethanol followed by a wash for 3 min in methanol.
5. Wash 2× for 5 min each in PEM 50:5:5 containing 0.1% (v/v) Tween, then resuspend in PEM 50:5:5 alone.

6. Excise the roots and transfer them to APTES-treated multiwell slides and let them dry down.
7. Digest the roots for 10 min in Enzyme Solution 4.
8. Wash the roots 6× in PEM 50:5:5 containing 0.1% (v/v) Tween 20 and allow to air dry.
9. Treat with 40 µg/mL of proteinase K [in PEM 50/5/5 containing 0.1% (v/v) Tween-20] for 10 min at 37°C.
10. Neutralize the proteinase K by treating for 5 min with 0.2% (w/v) glycine prepared in PEM 50:5:5 containing 0.1% (v/v) Tween-20.
11. Wash the roots a few times in PEM 50:5:5 and allow them to air dry.
12. Treat the roots 4× in 1 mg/mL of $NaBH_4$ in PEM 50:5:5 for 15 min each.
13. Wash the roots in PEM 50:5:5 and give them a final wash in 0.1× SSC.
14. Add Probe Mixture and incubate at 37°C for 24 h.
15. Wash in 0.1× SSC for 1 h at 50°C.
16. Wash 2× in PEM 50:5:5 containing 0.1% (v/v) Tween-20 and allow to dry.
17. Wash in Blocking Solution for 90 min.
18. Incubate in primary antibody for 2 h at 37°C or overnight at 4°C: mouse anti-digoxin diluted 1:5000 with 3% (w/v) BSA in PEM 50:5:5.
19. Wash in Blocking Solution for 1 h.
20. Incubate in secondary antibody for 2 h. at 37°C: Cy3-conjugated donkey anti-mouse (Jackson ImmunoResearch Laboratories, West Grove, PA, USA) diluted 1:100 with 3% (w/v) BSA in PEM 50:5:5.
21. Wash in PEM 50:5:5 containing 0.1% (v/v) Tween-20 for 2 days.
22. Mount in Vectashield.

3.9. Protocol 9: Studying Living Cells

The ability to study living cells has great advantages for understanding active processes. We therefore provide some information for those who would like to pursue such avenues. However, methods of visualizing fluorescent probes in living cells are largely dependent on the cell type and fluorescent probe being used; so no one protocol is preferred. Instead we will discuss the imaging of living cells in terms of being a four-step process and present alternative ways to perform each step. The use of fluorescent probes for studying living cells has also been discussed by Oparka and Read *(23)*.

3.9.1. Step 1: Immobilization

If visualization is to be successful, the first step is to immobilize the cell or tissue in a way that keeps the cells alive. The challenge of effective immobilization is to keep cells in the same place for repetitive measurements while maintaining the cells in a healthy state. The most popular way to do this is to embed the sample in agarose. Low-melting-point agarose is often used to minimize the potential for cell damage or induction of heat-shock responses. When using this method for microinjection, a balance must be struck between increasing the agarose percentage to keep the cells in place during microinjection while decreasing the percentage for ease of needle movement. This technique has been used successfully for protoplasts *(24)*, pollen tubes *(25,26)*, stamen hairs *(27)*, and root hairs *(7,28)*. Another method of immobilization is the use of adhesive tape. Adhesive tape (a hydrophobic type such as Cellux) is affixed to a microscope slide with its sticky side up by means of double-sided tape. The tissue to be examined is then simply stuck on top. The tissue can be injected "dry" without anything covering it or it can be covered with inert oil such as Voltalef PCTFE Oil (Atochem, Pierre-Bénite, France). The oil prevents drying of the tissue and improves the optical properties of the system. This method is somewhat limited to dry or only slightly moist material; however, with extra precautions, the tissue can be covered with aqueous solutions *(29)*. Staiger *et al. (30)* and Yuan *et al. (31)* have used this method for stamen hairs and pea epidermal strips, respectively. A similar approach is to use a water-insoluble, biologically inert adhesive such as dental adhesive or prosthetic adhesive (Secure B401, Factor II, Lakeside, AZ, USA).

3.9.2. Step 2: Keeping Cells Alive and Healthy

The easiest way to assess the health of the cells is by carefully examining them with the microscope. The outlines of the organelles should be distinct and it may be possible to see cytoplasmic streaming. When cells become unhealthy, there are often changes in the position and shape of the organelles, particularly the nucleus and cytoplasmic strands, and Brownian motion can sometimes be observed. To keep the cells healthy on the microscope stage they must be kept at an appropriate turgor pressure. For applications utilizing pressure microinjection, turgor may need to be decreased slightly to inject solution from the needle, but care must be taken that this does not induce plasmolysis. Turgor is maintained by keeping the cells covered with buffer or embedded in a gel at an appropriate osmotic strength. The strength of the required osmoticum can be determined by placing cells in buffers containing a range of sorbitol concentrations and using one that does not induce plasmolysis, and that still allows pressure injection. (Plasmolysis is most easily observed in cell corners where the protoplasm retracts from the cell wall.) Even when the cells have been covered with buffer, evaporation can readily occur and will increase the concentration of the bathing solution. In the shortterm (10–15 min), evaporation is usually not a problem, but for long-term experiments (>20 min) it may be better to keep the specimen in a humid box between measurements or to use some type of perfusion system.

Although maintaining turgor is the most important and most obvious consideration for keeping cells healthy, there are also other considerations including oxygen deprivation, temperature, and light damage. Oxygen deprivation is not a problem for open systems where oxygen can diffuse into the bathing solution and then into the cells, but it can be a problem in a closed chamber. Simple perfusion systems can provide aeration, however. On a large scale, temperature control is rarely a problem on the open microscope stage when room temperature is well controlled. Although some lasers produce a large amount of heat, this is easily compensated for by good ventilation or air conditioning. However, temperature should be a considered when embedding the cells in agarose or other gels. Plant cells can be severely damaged by high-energy laser light and this is due, at least in part, to large local increases in the temperature of laser-irradiated areas. When actively growing pollen tubes are irradiated at the tip with high doses of UV laser light, their growth is retarded and they can burst (Grant Calder, *personal communication*). The rule of thumb is to use the lowest power that still allows for a clear signal. The important thing to remember is that cell viability is extremely important and that it should be monitored throughout the experiment.

3.9.3. Step 3: Introduction of Fluorescent Probes

The main objective of this step is to introduce the probe into as many cells as possible while maintaining their viability. This area has been covered in other reviews *(5,23)* so we will confine our discussion to the two methods that have been most successful: direct uptake of permeant probes and microinjection. Uptake procedures have the advantages of being able to label many cells simultaneously and that the procedures are straightforward, requiring no special equipment. The experimenter is limited, however, to cell types that can directly take up the probes and to commercially available probes that permeate plant cells. Antibodies and other proteins cannot enter cells this way. Microinjection can deliver antibodies, proteins, and nucleic acids into specific cells. This specificity, however, comes at the cost of reduced sample sizes due to the laborious nature of microinjection. In addition, the apparatus needed for microinjection is costly and microinjection is a skill that requires much practice to attain.

3.9.3.1. DIRECT UPTAKE OF PERMEANT PROBES

Some probes, due to their lipophilic nature, are readily cell permeant. Procedures for loading such probes into the cell simply require incubating the cells in an appropriate concentration of the dye. Other probes are not cell permeant because of carboxyl groups on the molecules at physiological pH. These molecules can be made membrane permeant by masking their charge with acetoxymethyl groups. This method relies on intracellular esterases to hydrolyze the probe releasing the free dye. There can be problems with this method if hydrolysis occurs outside of the cell *(32)* or if there is incomplete hydrolysis within the cell *(33)*. Protons can also be used to mask the charge of carboxyl groups on dyes. In this case, the cells are incubated in the dye at low pH (around 4.5) causing the dye to be protonated and membrane permeant. Once the dye is in the cytoplasm (pH 7), the protons dissociate and the dye becomes trapped within the cell. This method is often referred to as acid-loading. It has been used with some success *(28,34)*, but not all cells can survive the pH stress *(35)* and there can also be problems with dyes sticking to the cell wall.

3.9.3.2. MICROINJECTION OF PROBES

There are two types of microinjection: iontophoretic injection and pressure injection. In iontophoretic injection an electrical potential is induced between the bathing solution within the cell and the needle solution by means of a wire immersed in each. A weak current passes through the cell carried by the probe being introduced. Thus, this method requires that the probe be charged and relatively small. This procedure will not work for most proteins. In pressure injection, the microinjection needle is attached to an air supply or syringelike apparatus such that hydraulic pressure is used to force probes into the cell. There are few limitations to the type of probe injected, so probes can range from dyes to proteins to whole virus particles. With pressure microinjection, the volume of the cell is altered (increased) so injection solutions should be kept as small and concentrated as possible. Also, for this reason, the dye-to-protein ratio should be as high as possible while still maintaining probe functionality. In addition, successful injection requires that turgor be exceeded by the injection pressure, so the control of cell turgor is very important (as mentioned previously).

3.9.3.3. EXAMPLE PROCEDURE: MICROINJECTION OF LABELED TUBULIN

What follows is an example of one application, pressure microinjection of rhodamine-labeled tubulin into pea stem epidermal cells, which illustrates some of the points discussed previously and provides some specific details (**Fig. 9**). Pieces of the epidermis are peeled from the stem and placed, epidermal side up, onto tape. The ends of the peels are secured with tape to prevent the peels from lifting off and floating away. Additional pieces of tape are applied parallel to the peels, overlapping the pieces of tape securing the tissue (**Fig. 3**). In this way a hydrophobic ring is formed around the peels in which buffer can be added to keep the peels wet. This buffer consists of 50 mM potassium glutamate, 0.5 mM magnesium chloride, and 2.75% (w/v) sucrose. Sucrose is added to the buffer to regulate cell turgor. When not on the microscope stage, the slide containing the tissue is kept in a Petri dish with moist paper in order to prevent evaporation. Before starting injections, the tissue is given about 15 min to recover. Using a water immersion objective lens (Plan-NEOFLUAR

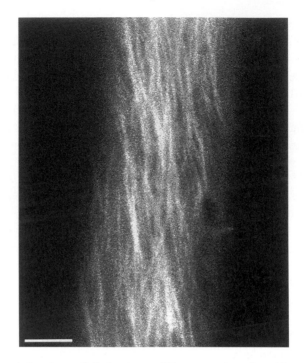

Fig. 9. Microinjection of living pea (*Pisum sativum*) epidermal cells with rhodamine-tubulin. Projection image of a living pea epidermal cell injected with 1.5 mg/mL of rhodamine-labeled tubulin. Eight optical sections were taken at 0.3-μm intervals. Bar = 5 μm.

63/1.2), the peel is examined and an area is chosen in which there are many cells that are streaming. We have found that tubulin incorporation is correlated with cytoplasmic streaming. Once an area for injection has been chosen, the water immersion objective lens is replaced with a long working distance objective lens (Nikon Plan 40X/0.35 SLWD).

Tubulin is purified from pig brains and labeled with tetramethyl-rhodamine according to standard procedures [details in *(29)*]. Because tubulin can self-assemble at room temperature, it is kept on ice. Similarly, the injection needles are kept on ice to reduce the possibility of tubulin assembly in the needle tip. It is generally best to keep purified proteins on ice while injecting them, although chilling of the needles is not usually necessary. The size and shape of the injection needles are very important. The shape is designed to be sharp and the size to be small, but large enough not to become clogged (in this case, approximately 750 nm outer diameter). One of the most frustrating parts of microinjection of plant cells is the propensity for needles to become clogged by compounds in the cell wall. The injection of viscous solutions such as proteins also requires a larger diameter than is needed for injecting aqueous solutions such as dyes. Once the microinjection needles are fitted to their holder, they are tip-filled by placing the needle into a 2-µL drop of tubulin in a Petri dish and applying a slight suction to the needle. One advantage of tip-filling is that if the protein goes up the needle easily, it is more likely to be able to come back out of the needle easily. After several hours of injecting, we often observe that it becomes more difficult to get tubulin into the needle. At this point we start using a fresh drop of tubulin because this is an indication that the tubulin is aggregating or becoming assembly incompetent.

Once the needle is filled, it is positioned for injection. The geometry of the cells or tissue should be borne in mind; use a shallow angle of attack that will allow the needle to enter the cytoplasm rather than a steep angle that would enter the vacuole. In the case of the epidermal peel, which is largely flat, the angle of attack is not very important, but for a large cylindrical cell, such as a stamen hair or trichome, it can make a large difference. Similarly, there may be areas within the cell that make better targets for microinjection; e.g., some cells have pockets of cytoplasm at their corners. Once the needle is positioned, it is poked into the cell using movement in the axis of the needle. It is often necessary to apply a small amount of back-pressure to the needle when entering the cell to decrease the chance that turgor will push cytoplasm up the needle. A small amount of tubulin is then injected into the cell. After allowing 1 or 2 min for tubulin assembly, the needle is slowly removed using movement in the axis of the needle. Removal of the needle is the most dangerous part of the injection process. If the needle is removed too quickly, the cell can be damaged and protoplasm leaks out. Pea epidermal cells seem to reseal fairly well such that the needle can be removed within 2 min with little, if any, loss of cytoplasm. Each cell type and tissue has slightly different requirements, but once proficiency is attained with one cell type, others seem easier.

3.9.4. Step 4: Imaging

Other chapters in this book deal with the details of different imaging techniques so discussion here is limited. The major consideration with confocal imaging of living cells is avoiding photodamage. Laser intensity should be kept to a minimum as should scan duration. We have been able to follow microtubule reorientation for more than hour while scanning at 3 min intervals by using the fast-photon counting setting and 10% laser power (from an argon/krypton laser) on a Bio-Rad MRC-600 confocal microscope. With careful handling, living plant cells can be used very effectively on the confocal microscope.

4. Notes

1. A brief treatment with cell wall degrading enzymes is often used to increase antibody penetration into the tissue. The purity and specificity of the enzymes

will influence the efficiency of this step and the degree of nonspecific tissue damage. Detergents may also be added to extract the tissue.

Acknowledgments

This work was supported by a Biotechnology and Biological Research Council Linked Research Grant (to R. Warn, P. Shaw, and C. Lloyd), a Biotechnology and Biological Research Council grant in aid to the John Innes Centre, a John Innes Foundation studentship (to K. Boudonck), and an EU Grant (to K. Boudonck).

References

1. Flanders, D. J., Rawlins, D. J., Shaw, P. J., and Lloyd, C. W. (1989) Computer-aided 3-D reconstruction of interphase microtubules in epidermal cells of *Datura stramonium* reveals principles of array assembly. *Development* **106,** 531–541.
2. Gehring, C. A., Williams, D. A., Cody, S. H., and Parish, R. W. (1990) Phototropism and geotropism in maize coleoptiles are spatially correlated with increases in cytosolic free calcium. *Nature* **345,** 528–530.
3. Oparka, K. J., Prior, D. A. M., and Wright, K. M. (1995) Symplastic communication between primary and developing lateral roots of *Arabidopsis thaliana. J. Exp. Bot.* **283,** 187–197.
4. Carpita, N., Sabularse, D., Montezinos, D., and Delmer, D. (1979) Determination of the pore size of cell walls of living plant cells. *Science* **205,** 1144–1147.
5. Fricker, M., Tester, M., and Gilroy, S. (1994) Fluorescence and luminescence techniques to probe ion activities in living plant cells, in *Fluorescent and Luminescent Probes for Biological Activity; A Practical Guide to Technology for Quantitative Real-Time Analysis* (W. T. Mason, ed.), Academic Press, London, pp. 360–377.
6. Bush, D. S. and Jones, R. L. (1990) Measuring intracellular Ca^{2+} levels in plant cells using the fluorescent probes, indo-1 and fura-2. *Plant Physiol.* **93,** 841–845.
7. Clarkson, D. T., Brownlee, C., and Ayling, S. M. (1988) Cytoplasmic calcium measurements in intact higher plant cells: Results from fluorescence ratio imaging of fura-2. *J. Cell Sci.* **91,** 71–80.
8. Yuan, M., Warn, R. M., Shaw, P. J., and Lloyd, C. W. (1995) Dynamic microtubules under the radial and outer tangential walls of microinjected pea epidermal cells observed by computer reconstruction. *Plant J.* **7,** 17–23.
9. Goodbody, K. C. and Lloyd, C. W. (1994) Immunofluorescence techniques for analysis of the cytoskeleton, in *Plant Cell Biology: A Practical Approach* (N. Harris, and K. J. Oparka, eds.), Oxford University Press, Oxford, pp. 221–243.
10. Nagata, T., Nemoto, Y., and Hasezawa, S. (1992) Tobacco BY-2 cell line as the "HeLa" cell in the cell biology of higher plants. *Int. Rev. Cytol.* **132,** 1–30.
11. Kuss-Wymer, C. L. and Cyr, R. J. (1992) Tobacco protoplasts differentiate into elongate cells without net microtubule depolymerization. *Protoplasma* **168,** 64–72.
12. Gilroy, S., Fricker, M. D., Read, N. D., and Trewavas, A. J. (1991) Role of calcium in signal transduction of *Commelina* guard cells. *Plant Cell* **3,** 333–344.

13. Beven, A. F., Lee, R., Razaz, M., Leader, D. J., Brown, J. W. S., and Shaw, P. J. (1996) The organization of ribosomal RNA processing correlates with the distribution of nucleolar snRNAs. *J. Cell Sci.* **109**, 1241–1251.
14. Hozak, P., Hassan, A. B., Jackson, D. A., and Cook, P. R. (1993) Visualization of replication factories attached to a nucleoskeleton. *Cell* **73**, 361–373.
15. Thompson, W. F., Beven, A. F., Wells, B., and Shaw, P. J. (1997) Sites of rDNA transcription are widely dispersed through the nucleolus in *Pisum sativum* and can comprise single genes. *Plant J.* **12**, 571–581.
16. Wells, B. (1985) Low temperature box and tissue handling device for embedding biological tissue for immunostaining in electron microscopy. *Micron Microscop. Acta* **16**, 49–53.
17. Highett, M. I., Rawlins, D. J., and Shaw, P. J. (1993) Different patterns of rDNA distribution in *Pisum sativum* nucleoli correlate with different levels of nucleolar activity. *J. Cell Sci.* **104**, 843–852.
18. Shaw, P. J., Highett, M. I., Beven, A. F., and Jordan, E. G. (1995) The nucleolar architecture of polymerase I transcription and processing. *EMBO J.* **14**, 2896–2906.
19. Wick, S. M., Seagull, R. W., Osborn, M., Weber, K., and Gunning, B. E. S. (1981) Immunofluorescence microscopy of organized microtubule arrays in structurally stabilized meristematic plant cells. *J. Cell Biol.* **89**, 685–690.
20. Ludevid, D., Höfte, H., Himelbrau, E., and Chrispeels, M. J. (1992) The expression pattern of the tonoplast intrinsic protein gamma-TIP in *Arabidopsis thaliana* is correlated with cell enlargement. *Plant Physiol.* **100**, 1633–1639.
21. Bauwens, S., Katsanis, K., Van Montagu, M., Van Oostveldt, P., and Engler, G. (1994) Procedure for whole mount fluorescence *in situ* hybridization of interphase nuclei on *Arabidopsis thaliana*. *Plant J.* **6**, 123–131.
22. de Almeida Engler, J., Van Montagu, M., and Engler, G. (1994) Hybridization *in situ* of whole-mount messenger RNA in plants. *Plant Mol. Biol. Rep.* **12**, 321–331.
23. Oparka, K. J. and Read, N. D. (1994) The use of fluorescent probes for studies of living plant cells, in *Plant Cell Biology: A Practical Approach* (N. Harris, and K. J. Oparka, eds.), Oxford University Press, Oxford, pp. 27–50.
24. Gilroy, S. and Jones, R. L. (1994) Perception of gibberellin and abscisic acid at the external face of the plasma membrane of barley (*Hordeum vulgare* L.) aleurone protoplasts. *Plant Physiol.* **104**, 1185–1192.
25. Miller, D. D., Callaham, D. A., Gross, D. J., and Hepler, P. K. (1992) Free Ca^{2+} gradient in growing pollen tubes of *Lilium*. *J. Cell Sci.* **101**, 7–12.
26. Pierson, E. S., Miller, D. D., Callaham, D. A., Shipley, A. M., Rivers, B. A., Cresti, M., and Hepler, P. K. (1994) Pollen tube growth is coupled to the extracellular calcium ion flux and the intracellular calcium gradient: effect of BAPTA-type buffers and hypertonic media. *Plant Cell* **6**, 1815–1828.
27. Zhang, D., Wadsworth, P., and Hepler, P. K. (1990) Microtubule dynamics in living dividing plant cells: confocal imaging of microinjected fluorescent brain tubulin. *Proc. Natl. Acad. Sci. USA* **87**, 8820–8824.

28. Wymer, C. L., Bibikova, T. N., and Gilroy, S. (1997) Cytoplasmic free calcium distributions during the development of root hairs of *Arabidopsis thaliana*. *Plant J.* **12,** 427–439.

29. Wymer, C. L., Shaw, P. J., Warn, R. M., and Lloyd, C. W. (1997) Microinjection of fluorescent tubulin into plant cells provides a representative picture of the cortical microtubule array. *Plant J.* **12,** 229–234.

30. Staiger, C. J., Yuan, M., Valenta, R., Shaw, P. J., Warn, R. M., and Lloyd, C. W. (1994) Microinjected profilin affects cytoplasmic streaming in plant cells by rapidly depolymerizing actin microfilaments. *Curr. Biol.* **4,** 215–219.

31. Yuan, M., Shaw, P. J., Warn, R. M., and Lloyd, C. W. (1994) Dynamic reorientation of cortical microtubules, from transverse to longitudinal, in living plant cells. *Proc. Natl. Acad. Sci. USA* **91,** 6050–6053.

32. Cork, R. J. (1986) Problems with the application of quin-2-AM to measuring cytoplasmic free calcium in plant cells. *Plant Cell Environ.* **9,** 157–161.

33. Brownlee, C. and Wood, J. W. (1986) A gradient of cytoplasmic free calcium in growing rhizoid cells of *Fucus serratus*. *Nature* **320,** 624–626.

34. Bush, D. S. and Jones, R. L. (1987) Measurement of cytoplasmic calcium in aleurone protoplasts using indo-1 and fura-2. *Cell Calcium* **8,** 455–472.

35. Elliott, D. C. and Petkoff, H. S. (1990) Measurement of cytoplasmic free calcium in plant protoplasts. *Plant Sci.* **67,** 125–131.

7

Preparation of Yeast Cells for Confocal Microscopy

Audrey L. Atkin

1. Introduction

Confocal scanning microscopy has been successfully used for immunofluo-rescence work in yeast. The major axis of a *Saccharomyces cerevisiae* haploid cell is approx 4 µm. Optical sections of approximately 2× 0.4 µm thickness from fluorescently-labeled yeast cells can be obtained using the laser scanning confocal microscope *(1)*. This means that it is possible to look at optical sections corresponding to about one tenth of a yeast cell. For example, confocal scanning microscopy has been used to examine the distribution of actin in fixed cells prepared from reverting protoplasts, and to show that a monoclonal anti-body raised to rat liver nuclear proteins recognized two protein components of the yeast nuclear pore complex, p95 and p110 *(2,3)*.

Confocal scanning microscopy is especially useful for determining the cellu-lar localization of proteins that may localize to more than one structure or com-partment within the yeast cell. For instance, confocal scanning microscopy was used to show that Kap95p accumulated both at the nuclear envelope and inside the nucleus, and to look for potential nuclear localization sequences in eIF4E, a cap binding protein that cycles between the nucleus and the cytoplasm *(4,5)*.

The distribution of more than one macromolecule can be determined by simultaneously labeling cells with different fluorochromes *(6)*. The images are collected separately for each fluorochrome and then merged to determine if there is overlap in the subcellular distribution of macromolecules of interest *(6)*. This approach was used to show that epitope tagged allele of *UPF1* is found primarily in the cytoplasm *(7)*.

This chapter describes a method for the staining of fixed *S. cerevisiae* cells with fluorescently labeled antibodies (**Fig. 1**) and 4',6-diamidino-2-phenylindale (DAPI) for confocal microscopy (*7*; similar to a procedure

From: *Methods in Molecular Biology, vol. 122: Confocal Microscopy Methods and Protocols*
Edited by: S. Paddock © Humana Press Inc., Totowa, NJ

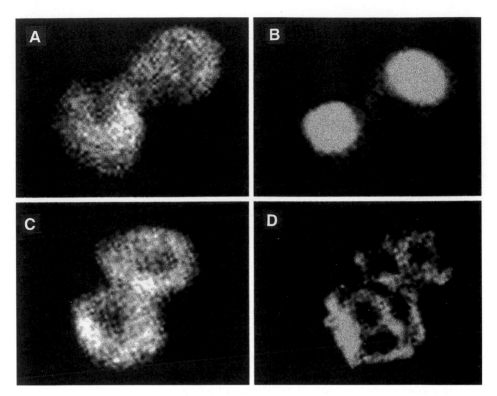

Fig. 1. Single optical sections of double-labeled yeast cells. The UPF1-3EP protein (**A,C**) is localized in the cytoplasmic compartment because it does not co-localize with the SEN1::βgal protein in the nucleus (**B**) or the KAR2 protein in the endoplasmic reticulum (**D**), but it does co-localize with the L1 protein, of the large ribosomal subunit, in the cytoplasm (results not shown). *See* **ref. 7** for more details.

described in **ref. 8**). This method can be adopted for other dyes for chromatin such as ToPro™ or propidium iodide have excitation and emission spectra that are better suited to the lasers supplied with confocal imaging systems (*see* Chapter 1).

2. Materials

1. 1% Polyethyleneimine: Mix 20 μL of 50% aqueous polyethyleneimine (Sigma P-3143) with 980 μL of dH₂O. Store at 4°C in the dark (*see* **Note 1**).
2. Glass slides
3. 37% Formaldehyde solution: (Sigma F-1635) prewarmed to 37°C
4. Phosphate-buffered saline (PBS): Potassium phosphate dibasic (7.0 g), potassium phosphate monobasic (1.4 g), sodium chloride (8.8 g), made up to 1 L with dH₂O (distilled water), and adjusted to pH 7.2. Sterilize by autoclaving. Store at room temperature.

5. Lyticase solution: Dissolve 1900 U of lyticase (Sigma L-4025) in 200 µL of PBS. Prepare immediately before use.
6. 1M Potassium phosphate (pH 6.5) stock solution. Combine 342.5 mL of 1M potassium phosphate monobasic, and 157.5 mL of 1M potassium phosphate dibasic. Sterilize by autoclaving. Store at room temperature.
7. Wash buffer 1 [1.2M Sorbitol, 0.1M potassium phosphate (pH 6.5), 1% 2-mercaptoethanol]: Mix 60 mL of 4M sorbitol and 20 mL of 1M potassium phosphate (pH 6.5) with 118 mL of dH$_2$O. Sterilize by autoclaving. Cool to room temperature. Add 2 mL of 2-mercaptoethanol. Store at room temperature.
8. Wash buffer 2 [1.2M sorbitol, 0.1M potassium phosphate (pH 6.5)]. Mix 60 mL of 4M sorbitol, and 20 mL of 1M potassium phosphate (pH 6.5) with 120 mL of dH$_2$O. Sterilize by autoclaving. Store at room temperature.
9. 10% Igepal CA-630 stock solution (*see* **Note 2**): Add 10 mL of Igepal CA-630 (Sigma I-3021) to 90 mL of dH$_2$O. Mix well (*see* **Note 3**). Store at room temperature.
10. Blocking buffer: Dissolve bovine serum albumin (0.1 g) in 10 mL of PBS. Add 50 µL of 10% Igepal CA-630. Store at –20°C in 1-mL aliquots.
11. Primary antibodies: Dilute the primary antibodies in PBS to 2× the desired final concentration immediately before use. Remove unwanted particles by centrifugation at 10,000g for 10 min at 4°C.
12. 0.05% Igepal CA-630/PBS: Add 5 mL of 10% Igepal CA-630 to 995 mL of PBS. Mix well. Store at room temperature.
13. Labeled secondary antibodies: Dilute the secondary antibodies in PBS to 2× the desired final concentration immediately before use. Centrifuge at 10,000g for 10 min at 4°C to remove unwanted particles.
14. 1 µg/mL DAPI solution: Prepare a 1 mg/mL DAPI stock solution by dissolving DAPI (1 mg; Sigma D-9542) in 1 mL of PBS. Prepare the 1 µg/mL working DAPI solution by diluting 1 µL of the DAPI stock solution in 999 µL of PBS. Store the stock and working DAPI solutions at 4°C.
15. 0.5M Carbonate buffer (pH 8.0): Add sodium bicarbonate (0.42 g) to 9 mL of dH$_2$O. Adjust the pH to 8.0 using 1N NaOH. Bring the volume up to 10 mL.
16. Glass coverslips, 18 × 18 mm, No. 1 thickness
17. Mounting medium: Dissolve p-phenylenediamine (30 mg; Sigma P-6001) in 4 mL of dH$_2$O. Add 6 mL of glycerol and 1 mL of 0.5M carbonate buffer (pH 8.0). Filter through a 0.2-µm filter to remove any undissolved chemical. Store in 0.5-mL aliquots at –70°C in the dark (*see* **Note 4**).

3. Method
3.1. Preparation of Slides

1. Prepare one slide for each sample.
2. Etch a 3/8-inch circle in the center of each slide with a diamond pen. Thoroughly rinse the slides with dH$_2$O and dry with a Kimwipe to remove the glass chips and any dust.
3. Outline the circles with a black permanent marker. Pipet 30 µL of 1% polyethyleneimine into each circle (*see* **Note 5**). Let the slide sit for approximately

5 min. Aspirate off the remainder of the solution. Rinse the slides thoroughly with dH$_2$O. Shake the slide dry and then place in a moist chamber (*see* **Note 6**).

3.2. Growth, Harvesting and Fixation of Yeast Cells

1. Grow cells to an OD$_{600}$ of 0.4–0.6 in 10 mL of the appropriate medium (*see* **Note 7**). This can be done in two steps. First grow an overnight culture to saturation in 3 mL of medium. Second, inoculate 10 mL of fresh medium with different amounts of the overnight culture. Inoculation of 10 mL of selective medium with 5 µL, 10 µL, 50 µL, and 100 µL of the overnight culture the afternoon of the second day will generally ensure that at least one of the cultures will be at an OD$_{600}$ of 0.4–0.6 the following day.
2. Add 1.4 mL of 37% formaldehyde per 10 mL-culture. Incubate for 5 min at 37°C with gentle shaking (*see* **Notes 8** and **9**).
3. Let the culture sit at room temperature for 1 h (*see* **Note 8**).
4. Pellet the yeast cells by centrifugation for 3 min at 3000g. Discard the supernatant. Be careful not to disturb the yeast cell pellet.
5. Wash the yeast cells three times in wash buffer 1. This is done by resuspending the yeast cells in 3 mL of wash buffer 1, pelleting the yeast cells by centrifugation at 3000g for 3 min, and discarding the supernatant.
6. Resuspend the yeast cells in 0.5 mL of wash buffer 1.

3.3. Spheroplasting of the Yeast Cells

1. Add 50 µL of freshly prepared lyticase solution to the yeast cell slurry.
2. Incubate for 9 min at 30°C with gentle shaking (*see* **Note 10**).
3. Pellet the cells immediately after the incubation by centrifugation for 3 min at 3000g.
4. Wash the cells two times with 3 mL of wash buffer 2.
5. Resuspend the cells in wash buffer 2 at approximately 5.0 A_{600}/mL (*see* **Note 11**).

3.4. Primary Antibody Incubations

1. Pipet 60 µL of cells onto each circle on the prepared slides in the moist chamber.
2. Let the cells settle for 30 min.
3. Rinse the slides twice by dipping the slides into coplin jars filled with PBS. These rinses will remove any cells that did not adhere to the slide.
4. Check the cells for density using a light microscope.
5. Immerse the slides into a coplin jar containing methanol that had been prechilled to –20°C, for 6 min (*see* **Note 12**).
6. Immediately transfer the slides to a coplin jar containing acetone that had been prechilled to –20°C, for 30 s (*see* **Note 12**).
7. Tilt the slides to drain on a paper towel. Allow the slides to air dry.
8. Place the slides back in the moist chamber.
9. Block the cells with 30 µL of blocking buffer. Do not let the cells dry up after this step.
10. Incubate 15 min at room temperature.

11. Add 30 μL of PBS containing the primary antibodies to the cells (*see* **Notes 13–19**).
12. Incubate overnight at room temperature (*see* **Note 14**).

3.5. Washes

1. Immerse the slides in a coplin jar containing 0.05% Igepal CA-630/PBS (*see* **Note 15**).
2. Immediately transfer the slides to a second coplin jar containing 0.05% Igepal CA-630/PBS. Incubate for 5 min.
3. Wash in PBS by transferring the slides to a coplin jar filled with PBS. Incubate for 5 min. Repeat this step a total of three times.
4. Drain the slides on a paper towel to remove excess PBS. Immediately place the slides in the moist chamber.

3.6. Secondary Antibody Incubations

1. Add 30 μL of blocking buffer to the cells on each slide.
2. Incubate for 5–10 min at room temperature.
3. Add 30 μL of PBS containing the secondary antibodies (*see* **Notes 16–19**).
4. Incubate for 1 h at room temperature in the dark (*see* **Note 20**).

3.7. Washes

1. Immerse the slides in a coplin jar containing 0.05% Igepal CA-630/PBS.
2. Immediately transfer the slides to a second coplin jar with 0.05% Igepal CA-630/PBS. Let sit for 5 min in the dark.
3. Wash three times in PBS for 5 min in the dark.
4. Drain the slides on a paper towel. Place the slides in the moist chamber.

3.8. DAPI Staining

1. Add 30 μL of a 1 μg/mL DAPI solution to the cells.
2. Incubate for 5 min at room temperature in the dark.
3. Rinse two times in PBS.
4. Drain off the excess liquid. Allow the slides to air dry in the dark.

3.9. Mounting

1. Place one drop of mounting medium on a 18 mm × 18 mm coverslip (*see* **Note 21**).
2. Carefully place the coverslip over the circle. Avoid trapping air bubbles under the coverslip. Cover the slide with filter paper and press gently on the coverslip to spread the mounting medium evenly. Be careful not to move the coverslip. Wick any excess mounting medium away with a Kimwipe.
3. Incubate the slides for 1 h at room temperature.
4. Place a drop of clear acrylic nail polish at each corner of the coverslip. Allow the nail polish to harden. Seal coverslip to the slide with additional clear nail polish.
5. The slides can be stored at –20°C in light tight boxes.
6. View the slides within 24–48 h (*see* **Note 22**).
7. Optical sections of 0.4 μm can be obtained by using a 60× objective lens, numerical aperture 1.4, with the pinhole closed (1 mm) and a laser emission of 0.1%

transmittance. Select laser excitation wavelength and emission filters appropriate for the fluorochrome and DAPI stain.

4. Notes

1. Phenylenediamine is very viscous and light sensitive. It can be pipetted using a tip that has had the end cut off.
2. Igepal CA-630 is chemically identical to Nonidet P-40. Nonidet P-40 is no longer commercially available.
3. Igepal CA-630 may require heat to dissolve in water.
4. The mounting medium is light sensitive. It is colorless and clear when freshly made. Discard this solution when it develops an orange color.
5. The spheroplasted yeast cells must be attached to the slides for efficient staining and washing. The slides can be prepared by coating the surface of the slide with polyethyleneimine. Polyethyleneimine prepares the surface of the slide so that the spheroplasted yeast cells will adhere to the slide. Polylysine is also commonly used for this purpose. A protocol for preparing the surface of slides with polylysine can be found in **ref. 8**.
6. The use of a moist chamber slows evaporation during the longer incubation steps. This is important to ensure that the cells stay moist during the staining process. A moist chamber can be prepared by lining a shallow, flat-bottomed container that has a tight-fitting lid with a sheet of 3 MM chromatography paper saturated with dH_2O.
7. The choice of yeast strain and growth conditions can affect the results obtained with immunofluorescence methods (8). Ideally diploid cells harvested in the exponential growth phase should be used. Diploid cells are usually larger than haploid cells and visualization of subcellular structures is easier in larger cells. Effective removal of the cell wall is essential for access of antibodies to the corresponding antigens within the cell. The yeast cell wall is easiest to remove from cells harvested during exponential growth. However, the growth conditions must be optimized for cellular components that are present transiently. Standard protocols for the growth and maintenance of yeast are available (9–11).
8. The fixation protocol presented in this chapter has worked well for a number of antigens (7). Cells are rapidly fixed by adding the fixative to the culture medium. This ensures that cellular structures are fixed before they can be potentially disturbed by harvesting. Several subcellular structures including actin in budding cells, as well as vacuolar and mitochondrial morphology, have been shown to be disturbed during the process of centrifugation (8). Optimal fixation times can vary between 30 minutes and 3 h. The best length of time for fixation must be determined for each antigen.
9. Optimal fixation protocols depend on the antigen and the location of the antigen. The goal of fixation is to preserve the ultrastructure of the cell as much as possible without destroying the antigenic determinants recognized by the antibody. Strong fixation generally gives better structural preservation, but leads to weaker antibody labeling. This is because the fixation protocol can inactivate an antigen,

the antibody may not be able to gain access to the antigen, the antigen can be extracted by the fixation procedure, or the 3° structure of the antigen may not be recognized by the antibody. Commercial formaldehyde is not always the best fixative for all antigens because these solutions are stabilized by the addition of 10–15% methanol, which is also a fixative. Paraformaldehyde has been successfully used as a fixative for yeast cells and a protocol is available *(12)*. Additional fixation protocols have been described *(13)*.

10. Spheroplasting helps to permeablize the yeast cells to reagents such as antibodies and DAPI. The time that it takes to spheroplast a yeast cell varies with the yeast strain and the growth conditions. Incubation times for spheroplasting must be optimized for each new yeast strain and each change in growth conditions. This is done by incubating the cells in the presence of lyticase for different times and monitoring cell wall removal by a dilution lysis assay *(14)*. The dilution assay is done by mixing an aliquot of cells (~0.4 OD unit cells) with 1 mL of water. Mix a second aliquot of cells with 1 mL of PBS. Determine the OD_{600} of the two samples. Complete spheroplasting results in a greater than 10-fold drop in the OD_{600}.

11. After spheroplasting, the cells can be stored at 4°C for up to 16 h. Longer times at 4°C adversely affect the quality of the cells *(8)*.

12. Methanol and acetone are inorganic solvents. Immersion of the fixed, spheroplasted cells attached to slides in methanol followed by acetone helps to lyse the yeast cells. This step can be omitted for staining with most antibodies. However, this step is essential for staining with some antibodies, such as the antibodies that recognize Mas2p, a mitochondrial matrix protein (Atkin, *unpublished data*). In some cases methanol and/or acetone can destroy the antigenicity of some antigens because they are also weak fixatives. When working with a new antigen or antibody preparation, try staining the cells with antibody after methanol/acetone treatment and in the absence of methanol/acetone treatment.

13. Determine the concentration of the primary (monoclonal or polyclonal, affinity-purified) antibodies needed to produce a suitable signal. The working concentrations can vary between 1:10 to 1:10,000. Test each new lot of primary antibody at 1:10, 1:100, 1:500, 1:1000, and 1:10,000 dilutions to determine which concentration gives the best signal relative to background.

14. Optimize primary antibody incubation time. Times can vary from 1 h to overnight.

15. The presence of the detergent Igepal CA-630 in the PBS helps to reduce nonspecific staining by the primary and secondary antibodies.

16. Secondary antibodies are directed against the IgGs of the species in which the first antibodies have been made. Many animals have had yeast infections, so the secondary antibodies should be affinity-purified. Affinity-purified, fluorochrome-labeled secondary antibodies can be purchased from Boehringer Mannheim Biochemicals. The working concentration of the secondary antibody needs to be determined for each new lot of antibody. The working concentrations can vary between 1:10 to 1:10,000. Optimal antibody concentrations will vary

according to the nature of the sample and the assay conditions. Test each new lot of secondary antibodies at 1:10, 1:100, 1:500, 1:1000, and 1:10,000 dilutions to determine which concentration gives the best signal relative to background.

17. The choice of fluorochrome conjugated to the secondary antibody is limited by the filter sets that are available for the microscope that will be used. Fluorescein, rhodamine, and Texas Red™-labeled secondary antibodies are available. Fluorescein emits a green light that is easy to detect; however, it is prone to rapid photobleaching. Rhodamine emits a red light. It is not as prone to photobleaching. However, it yields a higher background than fluorescein because it is hydrophobic. Secondary antibodies labeled with Texas Red are becoming more readily available. Texas Red also emits a red light. It is more resistant than fluorescein and rhodamine to photobleaching. Fluorescein-labeled secondaries can be combined with either rhodamine or Texas Red conjugates in double labeling experiments.

18. Controls are essential to help identify the source of any background problems. Stain cells with secondary antibody alone, and with no antibodies at all. If possible, a control of the immunofluorescence in a strain that has a null mutation in the gene encoding the protein of interest should be determined. Nonspecific background can be reduced by blocking in 5% serum derived from the same species as the labeled antibody, 3% bovine serum albumin (BSA) and 10% nonfat dry milk, diluting the antibody in blocking solution, adding 0.2% Tween 20 to all buffers and wash solutions, reducing the incubation time of the primary antibody or the labeled secondary antibody, and increasing the number and the duration of the washes.

19. Cells can be double labeled for comparing the intracellular localization of two cellular constituents within the same cell. Use primary antibodies from different types of animals or different immunoglobulin classes, and secondary antibodies for each primary antibody with different fluorochrome labels. Care must be taken to ensure that no cross-reaction is possible between the labeled detection reagents. Successful double labeling requires that each primary and secondary antibody combination produce intense staining with a clean background.

20. A dark chamber can be prepared by lining a small box with black paper.

21. Mounting medium includes specific antifade reagents. These reagents are antioxidants that minimize the concentration of free oxygen radicals *(15)*. Free oxygen radicals attack the unexcited fluorochromes and damage them. p-Phenylenediamine is the antifade reagent in the mounting medium. Other commonly used mounting media are described elsewhere *(16)*.

22. Formaldehyde fixation is not permanent *(17)*. The crosslinks slowly reverse in aqueous buffers. The yeast cells should be stained and observed within a short time of fixation.

References

1. White, J. G., Amos, W. B., and Fordham, M. (1987) An evaluation of confocal versus conventional imaging of biological structures by fluorescence light microscopy. *J. Cell Biol.* **105,** 41–48.

2. Kobori, H., Yamada, N., Taki, A., and Osumi, M. (1989) Actin is associated with the formation of the cell wall in reverting protoplasts of the fission yeast *Schizosaccharomyces pombe. J. Cell Sci.* **94,** 635–646.
3. Aris, J. P. and Blobel, G. (1989) Yeast nuclear envelope proteins cross react with an antibody against mammalian pore complex proteins. *J. Cell Biol.* **108,** 2059–2067.
4. Iovine, M. K. and Wente, S. R. (1997) A nuclear export signal in Kap95p is required for both recycling the import factor and interaction with the nucleoporin GLFG repeat regions of Nup116p and Nup100p. *J. Cell Biol.* **137,** 797–811.
5. Ptushkina, M., Vasilescu, S., Fierro-Monti, I., Rohde, M., and McCarthy, J. E. G. (1996). Intracellular targeting and mRNA interactions of the eukaryotic translation initiation factor eIF4E in the yeast *Saccharomyces cerevisiae. Biochim. Biophys. Acta* **1308,** 142–150.
6. Paddock, S. W. (1994) To boldly glow. Applications of laser scanning confocal microscopy in developmental biology. *BioEssays* **16,** 357–365.
7. Atkin, A. L., Altamura, N., Leeds, P., and Culbertson, M. R. (1995) The majority of yeast UPF1 co-localizes with polyribosomes in the cytoplasm. *Mol. Biol. Cell* **6,** 611–625.
8. Pringle, J. R., Adams, A. E. M., Drubin, D. G., and Haarer, B. K. (1991) Immunofluorescence methods for yeast. *Methods in Enzymol.* **194,** 565–602.
9. Rose, M. D., Winston, F. and Hieter, P. (1990) Methods in Yeast Genetics, Cold Spring Harbor Laboratory Press, Cold Spring Harbor, NY.
10. Sherman, F. (1991) Getting started with yeast. *Methods in Enzymol.* **194,** 3–21.
11. Ausubel, F. M., Brent, R., Kingston, R. E., Moore, D. D,. Seideman, J. G., Smith, J. A., and Struhl, K., eds. (1993) *Current Protocols in Molecular Biology*, Green Publishing Associates and Wiley-Interscience, New York.
12. Harlow, E. and Lane, D. (1988) *Antibodies: A Laboratory Manual.* Cold Spring Harbor Laboratory Press, Cold Spring Harbor, NY.
13. Osborn, M. (1994) Immunofluorescence microscopy of cultured cells, in *Cell Biology: A Laboratory Handbook,* vol. 2 (J. E. Celis, ed.), Academic Press, San Diego, pp. 347–354.
14. Franzusoff, A., Rothblatt, J., and Schenkman, R. (1991) Analysis of polypeptide transit through yeast secretory pathway. *Methods Enzymol.* **194,** 662–674.
15. Johnson, G. D., Davidson, R. S., McNamee, K. C., Russell, G., Goodwin, D., and Holborow, E. J. (1982) Fading of immunofluorescence during microscopy: a study of the phenomenon and its remedy. *J. Immunol. Methods* **55,** 231–242.
16. Pawley, J. B. and Centonze, V. E. (1994) Practical laser-scanning confocal light microscopy: obtaining optimal performance from your instrument, in *Cell Biology: A Laboratory Handbook,* vol. 2 (J. E. Celis, ed.), Academic Press, San Diego, pp. 44–64.
17. Reinsch, S. and Stelzer, E. H. K. (1994). Confocal microscopy of polarized MDCK epithelial cells, in *Cell Biology: A Laboratory Handbook,* vol. 2 (J. E. Celis, ed.), Academic Press, San Diego, pp. 89–95.

8

Confocal Methods for *Caenorhabditis elegans*

Sarah L. Crittenden and Judith Kimble

1. Introduction

The use of antibodies to visualize the distribution and subcellular localization of gene products powerfully complements genetic and molecular analysis of gene function in *Caenorhabditis elegans*. Double and triple staining protocols are particularly useful for several reasons. First, colocalization of proteins either within tissues or at a subcellular level can be examined. Second, costaining with stage-specific or tissue-specific markers can define the timing and tissue specificity of antigen expression. For these types of studies it is useful to be able to collect data from multiple fluorescence wavelengths simultaneously. A confocal microscope equipped with a krypton/argon laser can simultaneously detect up to three different antigens. Using a confocal microscope it is also possible to collect a series of optical sections through a sample that allows observation of changes in distribution of the antigen in different focal planes of the tissue or cell.

In this chapter we outline procedures for fixing and staining *C. elegans* adult tissues, embryos, and larvae. Because embryos have an impermeable chitinous eggshell and larvae and adults have an impermeable collagenous cuticle, it is important to permeabilize animals without destroying morphology or antigenicity. We routinely use two different procedures depending on which tissue and which stage we want to stain. Whole mount freeze cracking *(1–3)* has been used for years and is a good starting point; it is easy and it works well with most antibodies and with embryos, larvae, and adults. In addition the fixation can be varied easily to suit different antigens. Another useful method is to extrude tissues, e.g., gonads and intestines, from the carcass *(4,5)*. The extruded tissues are then easily fixed and permeabilized. Tissues remaining in the carcass usually do not stain well. This procedure is most useful for L4 larvae and

From: *Methods in Molecular Biology, vol. 122: Confocal Microscopy Methods and Protocols*
Edited by: S. Paddock © Humana Press Inc., Totowa, NJ

adults, although it is possible to extrude tissues from younger larvae, too. In addition, we discuss DNA binding dyes that are useful with the confocal microscope and procedures for obtaining images from live worms expressing green fluorescent protein (GFP) reporter constructs.

2. Materials

1. Subbing solution for slides:
   a. Bring 200 mL of distilled water to 60°C.
   b. Add 0.4 g of gelatin
   c. Cool to 40°C.
   d. Add 0.04 g of chrome alum
   e. Add 1 mM sodium azide
   f. Add polylysine to 1 mg/mL
      Put the subbing solution in a coplin jar and store it at 4°C. Soak clean slides in subbing solution for 5 min to 1 h, air dry and store at room temperature. Subbed slides can be used for weeks. Several batches of slides can be subbed in the same subbing solution. When slides become less sticky, it is time to make new solution.
2. 10× TBS (1 L): 250 mM Tris (pH 8.0), 1.5M NaCl, (30 g of Tris-base, 80.2 g NaCl). Add distilled water to 800 mL, pH to 8.0 (approx. 15 mL HCl). Bring volume to 1 L with distilled water.
3. 5% Bovine serum albumin (BSA) (makes 25 1-mL aliquots): 1.25 g BSA (FractionV), 2.5 mL of 10× TBS (pH 8.0), 22.5 mL of distilled water.
   Mix well, make 1-mL aliquots in 15-mL tubes and freeze at –20°C.
4. TBSB (enough for one or two staining experiments): Add 9 mL of 1× TBS to 1 mL of 5% BSA.
5. Acetone powder (modified from Harlow and Lane, **ref. 6**)
   a. Homogenize worms in Dounce homogenizer - use about 1g of worms/mL of M9.
   b. Set on ice for 5 min.
   c. Add 4 mL of –20 acetone/mL of worm suspension. Mix vigorously.
   d. Set on ice for 30 min with occasional vigorous mixing.
   e. Centrifuge at 10,000g for 10 min.
   f. Resuspend pellet with fresh –20 acetone.
   g. Mix vigorously on ice for 10 min.
   h. Centrifuge at 10,000g for 10 min.
   i. Spread pellet on clean filter paper and allow to dry at room temperature.
   j. When dry, break up chunks in a mortar and pestle, then transfer powder to a microfuge tube and store at 4°C.
6. M9 Buffer (1 L): KH$_2$PO$_4$ (3.0 g), Na$_2$HPO$_4$ (6.0 g), NaCl (0.5 g), NH$_4$Cl (1.0 g). Bring to 1 L with distilled water
7. 10× PBS (1 L): 80 g of NaCl, 2.0 g of Kcl, 14.4 g of Na$_2$HPO$_4$, 2.4 g of KH$_2$PO$_4$. Add distilled water to 800 mL (pH to 7.2), bring volume to 1 L.
8. 5% Paraformaldehyde (5 mL).
   a. Add 0.25 g of paraformaldehyde to 4.2 mL of distilled water.
   b. Add 2 µL of 4N NaOH.

    c. Heat at 65°C until dissolved.

    d. Add 0.5 mL 10× PBS, Bring volume to 5 mL with distilled water.

    e. Filter.

    f. Store at 4°C for no more than 1 week.

  9. 0.25 *M* levamisole (5 mL): Dissolve 0.3 g of levamisole in 5 mL of M9. Aliquot and store at –20°C. For 0.25 m*M* levamisole, add 1 µL of 0.25*M* levamisole to 1 mL of M9. Store at room temperature.

10. Fluorescently labeled secondary antibodies (Jackson Immunoresearch, 872 West Baltimore Pike, P.O. Box 9, West Grove, PA 19390; 800/367-5296).

11. Propidium iodide (Molecular Probes; Catalog No. P-3566; 4849 Pitchford Ave., Eugene, OR 97402; 541/465-8300).

12. To-Pro™-3 (Molecular Probes; Catalog No. for sampler kit N-7566; 4849 Pitchford Ave., Eugene, OR 97402; 541/465-8300). Sampler kit is a good value and you can try six different dyes.

13. DAPI (4',6-diamidino-2-phenylindole) (Sigma; Catalog No. D-1388, P.O. Box 14508, St. Louis, MO. 63178; 800/325-3010).

14. Mab030 (Chemicon; Catalog No. MAB030; 28835 Single Oak Dr., Temecula, CA 92590; 800/437-7500).

15. Anti-β tubulin (Amersham; Catalog No. N357; Life Sciences Division, 2636 S. Clearbrook Dr., Arlington Heights, IL 60005; 800/323-9750).

16. Anti-phosphohistone H3 (Upstate biotechnology; Catalog No. 06-570; 199 Saranac Ave. Lake Placid, NY 12946; 800/233-3991).

17. anti-actin (monoclonal C4; ICN; Catalog No. 69-100-1; ICN Plaza, 3300 Hyland Ave., Costa Mesa, CA 92626; 714/545-0113).

18. anti-GFP (Clontech; Catalog No. 8363; 1020 East Meadow Circle, Palo Alto, CA 94303; 800/662-2566).

19. *C. elegans* website: http://eatworms.swmed.edu.

3. Methods

3.1. Fixation Methods

3.1.1. Whole-Mount Freeze Cracking Method (see **Notes 1–6**)

1. Using a pick, place animals into 6 µL of M9 on a subbed slide. Cut the animals open with a 25-gauge syringe needle if early embryos or extruded germ lines or intestines are to be stained; to try whole larvae, leave the animals intact.

2. Add 2 µL of 5% paraformaldehyde (*see* **Note 7**).

3. Set an 18 mm ×18 mm coverslip carefully on top of the animals. Use a needle to apply gentle pressure several times over each animal or region of the slide. The animals will flatten; usually a few burst. This procedure aids in opening the eggshell or cuticle (*see* **Note 8**).

4. Put the slide on a metal plate on top of dry ice for at least 10 min.

5. Pop the coverslip off with a razor blade and immerse the slide immediately in 100% cold methanol for 5 min followed by 100% cold acetone for 5 min (*see* **Note 7**).

6. Air dry the slide for 5 min. This helps the animals adhere to the slide better.
7. Gently drop 200 µL of TBS containing 0.5% BSA (TBSB) onto the animals and incubate for 30 min at room temperature.
8. Follow the antibody incubation procedure (**Subheading 3.2.**).

3.1.2. Tissue Extrusion Method (see **Note 9**)

1. Using a pick, place 10–15 animals into 5 µL of M9 containing 0.25 m*M* levamisole on a subbed slide. Using a 25-gauge syringe needle, cut off the heads or tails of the animals, allowing the gonad and intestine to extrude from the animal (*see* **Note 10**).
2. Gently drop 100 µL of 1.0% paraformaldehyde in PBS onto the cut animals. Incubate for 10 min at room temperature in a humidified chamber.
3. Remove the paraformaldehyde and add 50 µL of TBSB containing 0.1% Triton X100 (TBSBTx) for 5 min at room temperature.
4. Remove the TBSBTx and wash two times with 200 µL of TBSB.
5. Incubate samples in 200 µL of TBSB for approximately 30 min at room temperature.
6. Follow the antibody incubation procedure (**Subheading 3.2.**).

3.2. Antibody Incubation Procedure (see **Notes 11–15**)

1. Incubate worms with 30–50 µL of primary antibody overnight at 4°C or for several hours at room temperature in a humidified chamber.
2. Wash by gently covering the worms with 200 µL of TBSB three times for 15 min each at room temperature (*see* **Note 16**).
3. Dilute secondary antibodies to the recommended concentration in TBSB. Use 100 µL per slide. If desired, add the DNA stain DAPI to 0.5 µg/mL. To reduce nonspecific background, preabsorb with worm acetone powder (*see* **Note 17**).
4. Incubate the worms with the secondary antibody mix for 1–2 h at room temperature.
5. Wash worms as in **step 2**.
6. After removing the last wash, add 8 µL of mounting medium (Vectashield, Vector Laboratories) and wipe off excess moisture with a tissue. Then put an 18 mm × 18 mm coverslip over the worms and seal with nail polish.
7. After the nail polish is dry, worms can be viewed through the confocal microscope (*see* **Note 18**).

3.3. DNA Binding Dyes

3.3.1. DAPI

DAPI staining is useful for finding worms by epifluorescence and gives the best detail if you want to look at nuclear morpholgy. DAPI images can be obtained on confocals that have UV capability.

1. Add 0.5 µg/mL of DAPI to the secondary antibody mix and proceed according to the antibody staining procedure (**Subheading 3.2.**).

3.3.2. To-Pro-3

To-Pro-3 emits in the far-red range and can be used with a krypton/argon laser. Staining with To-Pro-3 does not work as consistently or with as much detail as DAPI; however, for marking nuclei it works well (*see* **Notes 19** and **20**).

1. After the secondary antibody incubation and one wash, add a 1:5000 to 1:10,000 dilution of a 1 m*M* stock to 200 µL of TBSB per slide.
2. Incubate at room temperature for 10 min.
3. Wash three more times as with TBSB and mount the slides as described in the antibody staining procedures (**Subheading 3.2.**).

3.3.3. Propidium Iodide

Propidium iodide emits in the red/yellow range and can be used with a krypton/argon laser. The quality of DNA staining is similar to that achieved with To-Pro-3.

1. Use the same procedure as for To-Pro-3 except put 2.5 µ*M* propidium iodide in TBSB.

3.4. Detection of GFP in Living Worms

GFP is routinely used as a reporter for looking at the localization of gene products in living worms. It can be readily viewed in living or fixed worms using the fluorescein isothiocyanate (FITC) filter sets on the confocal microscope.

1. Using a pick, place worms onto an agarose pad of 4% agarose in dH$_2$O (*7*).
2. Add 8 µL of M9 containing 0.25 mM levamisole to immobilize the worms (*see* **Note 21**).
3. Gently put coverslip over the worms.
4. Worms are ready to view.
5. Set up the confocal to view FITC emission. Keep the laser power as low as possible to avoid damage to the worms. While viewing, watch for movement and bleaching (*see* **Notes 22–25**).

4. Notes

1. Worms generally do not stick well to slides, so start with plenty of animals and be gentle when doing washes. Use >10 adults or >40 larvae. Some people find it helpful to monitor the worms under a dissecting microscope during the washes.
2. A humidified chamber can be made from a plastic container with a wet paper towel taped to the lid.
3. Larvae and adults can either be picked from plates or washed off with M9. Young embryos (1–50 cells) are easily obtained by cutting open gravid hermaphrodites. Older embryos can be obtained by adding M9 to a plate, washing the adults and larvae off, then scraping the remaining embryos off with a pasteur pipet into

additional M9. The worms or embryos are then pelleted by centrifuging for 1–2 min at 1000 rpm in a microfuge. To remove *E. coli*, more M9 is added and the worms are pelleted again. Six to eight microliters of concentrated worms can then be dropped onto a slide.

4. When staining larvae, it helps to stage the animals so that they are similar in size. This way the amount of pressure can be adjusted for the size of the worms being fixed. For example, if there are many large adults (or larvae), it is difficult to permeabilize L1s without completely squashing the adults. It becomes increasingly more difficult to effectively permeabilize the worms as they get older.

5. For tissues that can be extruded from the cuticle, such as germ lines and intestines, the morphology is generally better using a non-freeze-crack method (*see* extrusion method, **Subheading 3.1.2.**).

6. Another commonly used procedure for whole mounts is the reduction/oxidation method of Finney and Ruvkun *(8)*. This method is described in detail in Miller and Shakes *(9)*.

7. Variations in fixation protocols. Paraformaldehyde fixation improves morphology; however; it can interfere with antibody binding and can be omitted. Concentrations of between 1% and 5% paraformaldehyde and incubation times of 30 s to 30 min at room temperature in a humidified chamber are commonly used *(3,10,11)*; adjust time and concentration for ideal staining. Acetone incubation can also interfere with some antibody reactions and can be omitted or shortened. For some antigens, incubation in cold N, N-dimethyl formamide works better than MeOH or acetone *(12,* SLC and Voula Kodoyianni, *unpublished observations)*. Finally, for some antigens, rehydration through a series of increasingly aqueous solutions of MeOH *(11)*, EtOH *(9)* and acetone *(13)* have been used.

8. Instead of putting pressure on the coverslip, a Kimwipe can be used to wick excess liquid from under the coverslip until the worms flatten.

9. This fix works well for at least some membrane proteins *(5,14)*, but not for the cytoskeletal proteins actin and tubulin. For cytoskeleton, try fixing first in 100% MeOH for 5 min at room temperature followed by 1% paraformaldehyde for 25 min at room temperature *(5)*. In addition, other protocols have been used for fixing extruded tissues *(4,15)* and/or cytoskeleton *(4,15,16)*.

10. Levamisole causes the animals to contract, which results in their germ lines and intestines being extruded efficiently.

11. Common background problems are intestine autofluorescence in the DAPI and fluorescein channels, dim nuclear stain, some cuticle fluorescence. If the background is high, determine whether it is due to the primary or secondary antibody. Try diluting the antibodies further; affinity purify the primary antibody, and preabsorb primary and/or secondary antibodies with worm or bacterial acetone powder or with fixed worms. Null mutants should be used so that the specific antibody will not be depleted.

12. Do not let the worms dry after they have been fixed and rehydrated; this tends to give a nonspecific haze to nuclei and cytoplasms.

13. If the morphology looks poor, try to fix the worms more quickly. Alternatively, contaminated solutions can distort the morphology. DAPI-stained DNA should look well defined and crisp; if it doesn't, be suspicious. Using different fixes or making small changes in concentration of fixative or time of fixation can make a big difference in the quality of staining.

14. Use an antibody that is known to work to test for good morphology, permeability, and fixation. Some useful control antibodies are anti-DNA monoclonal mAb 030 *(5,10)*, anti-actin clone C4 *(4,10)*, anti-β-tubulin *(5,10,15,16)*; anti-phosphohistone H3.

15. Several approaches can be used to test for specificity of staining:
    a. Stain a null mutant and look for lack of staining. If a null mutant is not available, it is possible to eliminate the antigen (Voula Kodoyianni, *personal communication*) in embryos from animals that have been injected with antisense RNA *(17,18)*.
    b. Compete with proteins that contain the antigen used to raise the antibodies.
    c. Compare patterns obtained with antibodies to different regions of the protein.

16. Alternatively, the slides can be immersed in a coplin jar if the worms are well attached to the slide.

17. To preabsorb with acetone powder:
    a. Add approximately 1 mg of worm acetone powder/200 μL secondary antibody solution.
    b. Incubate the secondary mix at 4°C for 15 min to 1 h.
    c. Centrifuge the mix at 10,000 rpm for 5 min in a microfuge to pellet the acetone powder.
    d. Use the supernatant.

18. Viewing worms on the confocal.
    a. For most samples we use a 63× (NA 1.4) oil immersion lens (**Fig. 1**). In some cases, e.g. if an image of a whole worm is required, it is useful to use a 25× oil immersion lens (**Fig. 2**).
    b. For single focal plane images, we use the standard triple labeling methods provided with the confocal. Use the lowest laser power that gives a reasonable signal. We usually collect double- and triple-labeled images simultaneously. If there is significant bleedthrough from one channel to another, images can be taken sequentially using a single laser line. In addition, if one fluor is brighter than another images can be taken sequentially using different laser powers.
    c. Z-series are useful for recording data in all focal planes. We usually use 8–15 1-μm sections depending on how flat the sample is. This has been useful for looking at the difference in morphology between the surface and the middle of the germ line *(5)* and for looking at staining patterns in different focal planes of embryos and larvae.
    d. Projection of Z-series has been useful to show the complete number of nuclei (e.g., Kodoyianni *et al.*, in preparation), P granules *(19)*, or distal tip cell processes (e.g., Gao *et al.*, in preparation) in a single image.

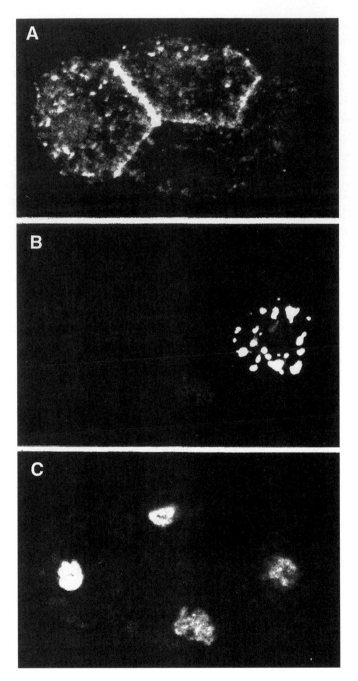

Fig. 1. Four-cell embryo stained with (**A**) anti-GLP-1 antibodies, (**B**) anti-P-granule antibodies, and (**C**) Yo-Pro™-1. Embryos were prepared by freeze-cracking followed by incubation in methanol, then acetone.

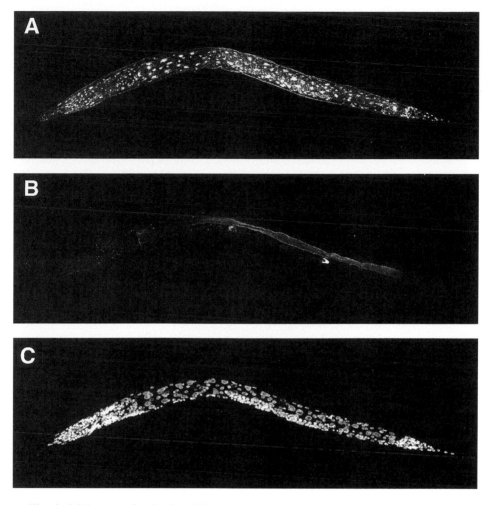

Fig. 2. L3 larva stained with (**A**) anti-LAG-1 antibodies, (**B**) anti-GFP antibodies, and (**C**) To-Pro-3. Larvae were prepared by freeze-cracking followed by incubation in DMF. This animal carries a GFP reporter that is expressed in the two distal tip cells. LAG-1 is a transcription factor expressed in nuclei (Kodoyianni et al., in preparation).

19. Molecular Probes also has other DNA binding dyes. They do not appear to be as clean as To-Pro-3 and propidium iodide; however if To-Pro-3 or propidium iodide do not work well on your sample or if you want a dye that emits in another channel, it may be worth working out conditions for the other dyes.
20. To-Pro-3 is often dimmer than antibody staining. To obtain a good image, try using the far red line of the laser alone and increase the laser power. You can then take sequential images if you have a double- or triple-stained sample. This way the other channels will not be bleached by the high laser power.

21. Levamisole will make the animals contract and look shorter and fatter than usual.
22. It is possible to see very dim (i.e., not visible in the intact animal) GFP fluorescence in tissues extruded from the animal (e.g., gonad, intestine or pharynx; Gao *et al.*, in preparation).
23. GFP fluorescence is detectable in fixed worms; however, it is often dimmer than in live worms. If it is difficult to see, anti-GFP antibody works well. Use a fluorescein-labeled secondary antibody so that both endogenous and antibody-stained GFP are seen in the same channel.
24. Z-series (Gao *et al.*, in preparation) and four-dimensional data sets (Sam Henderson, *personal communication*) of GFP fluorescence can be obtained.
25. With a transmission detector it is possible to take simultaneous Nomarski and fluorescent images. Because Nomarski filters severely cut back the laser power (e.g., we use 100% laser power for Nomarski images) it is often useful to collect an image of GFP first and then to collect the transmitted Nomarski image. The Nomarski image usually is not great.

References

1. Strome, S. and Wood, W. B. (1982) Immunofluorescence visualization of germ-line-specific cytoplasmic granules in embryos, larvae, and adults of *Caenorhabditis elegans*. *Proc. Natl. Acad. Sci. USA* **79,** 1558–1562.
2. Albertson, D. G. (1984) Formation of the first cleavage spindle in nematode embryos. *Dev. Biol.* **101,** 61–72.
3. Bowerman, B., Draper, B. W., Mello, C. C., and Priess, J. F. (1993) The maternal gene *skn-1* encodes a protein that is distributed unequally in early *C. elegans* embryos. *Cell* **74,** 443–452.
4. Strome, S. (1986) Fluorescence visualization of the distribution of microfilaments in gonads and early embryos of the nematode *Caenorhabditis elegans*. *J. Cell Biol.* **103,** 2241–2252.
5. Crittenden, S. L., Troemel, E. R., Evans, T. C., and Kimble, J. (1994) GLP-1 is localized to the mitotic region of the *C. elegans* germ line. *Development* **120,** 2901–2911.
6. Harlow, E. and Lane, D. (1988) *Antibodies: A Laboratory Manual.* Cold Spring Harbor Laboratory Press, Cold Spring Harbor, New York.
7. Sulston, J. and Hodgkin, J. (1988) Methods, in *The Nematode Caenorhabditis elegans* (W. B. Wood, ed.), Cold Spring Harbor Laboratory Press, Cold Spring Harbor, NY, pp. 587–606.
8. Finney, M. and Ruvkun, G. (1990) The *unc-86* gene product couples cell lineage and cell identity in *C. elegans*. *Cell* **63,** 895–905.
9. Miller, D. M. and Shakes, D. C. (1995) Immunofluorescence microscopy, in *Caenorhabditis elegans: Modern Biological Analysis of an Organism* (H. F. Epstein and D. C. Shakes, eds.). Academic Press, San Diego, pp. 365–394.
10. Evans, T. C., Crittenden, S. L., Kodoyianni, V., and Kimble, J. (1994) Translational control of maternal *glp-1* mRNA establishes an asymmetry in the *C. elegans* embryo. *Cell* **77,** 183–194.

11. Lin, R., Thompson, S., and Priess, J. R. (1995) *pop-1* encodes an HMG box protein required for the specification of a mesoderm precursor in early *C. elegans* embryos. *Cell 83,* 599–609.

12. Lin, R., Hill, R. J., and Priess J. R. (1998) *POP-1* and anterior–posterior fate decisions in *C. elegans* embryos. *Cell* **92,** 229–239.

13. Goldstein, B. and Hird, S. N. (1996) Specification of the anteroposterior axis in *Caenorhabditis elegans. Development* **122,** 1467–1474.

14. Henderson, S. T., Gao, D., Lambie, E. J., and Kimble, J. (1994) *lag-2* may encode a signaling ligand for the GLP-1 and LIN-12 receptors of *C. elegans. Development* **120,** 2913–2924.

15. Francis, R., Barton, M. K., Kimble, J., and Schedl, T. (1995) *gld-1*, a tumor suppressor gene required for oocyte development in *Caenorhabditis elegans. Genetics* **139,** 579–606.

16. Waddle, J. A., Cooper, J. A., and Waterston, R. H. (1994) Transient localized accumulation of actin in *Caenorhabditis elegans* blastomeres with oriented asymmetric divisions. *Development* **120,** 2317–2328.

17. Guo, S. and Kemphues, K. J. (1995) *par-1*, a gene required for establishing polarity in *C. elegans* embryos, encodes a putative Ser/Thr kinase that is asymmetrically distributed. *Cell* **81,** 611–620.

18. Fire, A., Xu, S., Montgomery, M., Kostas, S. A., Driver, S. E., and Mello, C. C. (1998) Potent and specific genetic interference by double-stranded RNA in *Caenorhabditis elegans. Nature* **391,** 806–811.

19. Crittenden, S. L., Rudel, D., Binder, J., Evans, T. C., and Kimble, J. (1997) Genes required for GLP-1 asymmetry in the early *Caenorhabditis elegans* embryo. *Dev. Biol.* **181,** 36–46.

9

Imaging Sea Urchin Fertilization

Jon M. Holy

1. Introduction

Imaging animal fertilization presents a number of unique challenges that arise from the unusual nature of the cells involved. Eggs are among the largest cells produced by animals, and sperm are frequently the smallest. Observation of the initial steps of fertilization, including sperm–egg binding and fusion, can be challenging owing to the size discrepancy between these two cell types and, in many species, the rapid time course over which they occur. Subsequent events in fertilization, including the remarkable activities of the microtubular cytoskeleton during the first cell cycle, are usually obscured by the mass of the egg cytoplasm, and only the most obvious features, such as pronuclear formation, are apparent by routine microscopic methods. The application of confocal microscopy methods to the demands of imaging studies in fertilization has proven to be an extremely valuable approach, as the morphological features of most types of animal eggs are extremely well suited to the specific advantages this technology offers. Confocal microscopy is a powerful and flexible tool, and has been skillfully used and developed by a number of researchers not only to study the structural features of fertilization at high resolution, but also to examine dynamic events in living gametes, including imaging signaling events that occur during and after gamete fusion.

Fertilization and subsequent developmental events have now been studied in a wide variety of species by confocal microscopy. The gametes and embryos of each species have their own special characteristics, but three features detrimental to microscopic studies are exhibited by most types of animal eggs: their large size, the presence of extensive yolk or other type of cytoplasmic inclusions, and the presence of specialized extracellular coats of material that not only can obscure cytoplasmic detail, but can significantly interfere with fixation and immunolabeling efforts. A brief summary of a number of studies

From: *Methods in Molecular Biology, vol. 122: Confocal Microscopy Methods and Protocols*
Edited by: S. Paddock © Humana Press Inc., Totowa, NJ

Table 1
Examples of Fertilization-Related Confocal Microscope Studies

Species	Study focus	References
Human	Chromatin organization	*(1)*
	Sperm penetration	*(2)*
	Cytoskeletal organization	*(3,4)*
Rhesus monkey	Cytoskeletal organization	*(5)*
Pig	Cortical granule exocytosis	*(6)*
Cow	Centrosome activity	*(7)*
	Immunocontraception	*(8)*
Rodents	ER organization	*(9,10)*
	Calcium dynamics	*(11)*
	pH measurements	*(12)*
	Cytoskeletal dynamics	*(13)*
	Cortical granules	*(14)*
	Fertilin expression	*(15)*
Xenopus	Cytoskeletal organization	*(16–25)*
	Yolk platelet pH	*(26)*
Sea urchin	Protein kinase C activation	*(27)*
	Cortical granule exocytosis	*(28)*
	Calcium dynamics	*(29)*
	ER organization	*(30)*
	Cytoskeletal organization	*(31,32)*
Starfish	Calcium dynamics	*(33)*
	ER organization	*(34)*
Ascidian	ER organization	*(35)*
Drosophila	Sperm nonequivalence	*(36)*
	Cytoskeletal organization	*(37)*
Silkmoth	Cytoskeletal organization	*(38)*

employing a variety of animal species is listed in **Table 1** to illustrate the types of fertilization studies that have been undertaken using confocal microscopy, and as a launching point toward obtaining further information regarding specimen preparation approaches for specific species. In this chapter, methods to prepare sea urchin eggs and embryos for immunochemical confocal studies of cytoskeletal organization are described.

2. Materials
2.1. Equipment and Supplies

1. A dozen 50-mL beakers, a half dozen 300- and 600-mL glass beakers, and a few 1- and 2-L beakers.

2. Disposable pasteur pipets and bulbs.
3. Microscope slides; 22 mm^2, 1.5 thickness coverslips.
4. Plastic six-well flat-bottom culture dishes (Falcon No. 3046 or equivalent).
5. Half dozen 60-mm to 90-mm diameter bacteriological plastic disposable Petri plates.
6. Two large shallow pans.
7. Inexpensive inverted or upright microscope with 10× or 20× objective lens.
8. 80 μm and 65 μm mesh Nitex nylon cloths.

2.2. Reagents and Solutions

1. Instant Ocean AFSW (artificial sea water); make up following manufacturer's instructions.
2. CaFSW (calcium-free artificial sea water): 16 mM Tris-HCl, 2.5 mM EGTA, 488 mM NaCl, 10 mM KCl, 29 mM MgSO$_4$, 27 mM MgCl$_2$. Add 0.8 g of NaHCO$_3$/4 L, pH to 8.1–8.3 with NaOH.
3. 500 mL of 0.5M KCl.
4. PBST (phosphate-buffered saline with Tween-20): 140 mM NaCl, 13 mM KCl, 2 mM KH$_2$PO$_4$, 8 mM Na$_2$HPO$_4$, 5 mL/L of a 20% stock solution of Tween-20 in distilled H$_2$O.
5. 1 mg/mL poly-L-lysine (mol wt >300,000) in water, aliquoted in 1-mL volumes and stored in freezer (*see* **Note 1**).
6. Anti-β-tubulin monoclonal antibody E7 (Developmental Studies Hybridoma Bank, Iowa City, IA). This antibody reacts with tubulin from a wide variety of species from yeast to humans, and works extremely well with fertilized sea urchin eggs and sea urchin sperm. Cells or culture supernatant can be purchased from DSHB. From a flask of cells grown to exhaustion, centrifuge the culture medium (2000g for 10 min) to remove cells, add sodium azide to 0.05%, and aliquot in desired volumes and store at –80C. For immunofluorescence microscopy, culture supernatant can usually be used at a 1:10 dilution in PBST.
7. Goat anti-mouse polyvalent (anti-IgG and -IgM) antibody conjugated to fluorescein isothiocyanate (FITC; Jackson Immunoresearch, Malvern, PA). Antibody is reconstituted according to manufacturer's directions, and aliquoted into 25-μL volumes in 200-μL microcentrifuge tubes and stored at –80°C.
8. Hoechst 33258 (Sigma Chemical Co., St. Louis, MO). A stock solution of 5 mg/mL is made in water and stored in the dark at 4°C. Hoechst strongly binds DNA and is a suspected mutagen; wear gloves when handling.
9. Mounting medium: in a 15-mL disposable centrifuge tube, combine 8 mL of glycerol, 2 mL of 0.5M CAPS buffer (pH 9.0), and 0.02 g of DABCO (1,4-diazabicyclo[2.2.2]octane) and 0.01 g p-phenylenediamine (all from Sigma; phenylenediamine is a suspected carcinogen and it is best to wear gloves when handling). Wrap in aluminum foil and rotate end-over-end for 1–2 h at room temperature, or until phenylenediamine is dissolved. Solution should be a faint tan color. Aliquot into 1-mL microcentrifuge tubes and store at –80°C (this medium will darken with exposure to air, which is accelerated by temperatures above –10°C; it usable for a number of weeks if stored at –10°C, but is usable for

many months when stored at −80°C. It can be used until quite a bit of discoloring has occurred). PBS or PBST may be used instead of CAPS, but make sure pH is not below 7.4; FITC fluorescence is quenched at an acidic pH.

2.3. Animals

Sea urchins used for these studies were obtained from Marinus, Inc. (Long Beach, CA). Two species were used: *Lytechinus pictus*, which are gravid from approximately April through September, and *Strongylocentrotus purpuratus*, which are gravid from approx. November through March. Gravid *L. pictus* females usually yield a few (1 to 10) mL of eggs when induced to spawn, whereas over 50 mL of eggs can be obtained from ripe *S. purpuratus* females. *L. pictus* usually ship well, but in peak season, *S. purpuratus* may begin to spawn during shipment; therefore, one should be ready to begin a number of experiments on the day of *S. purpuratus* arrival.

3. Methods
3.1. Spawning

1. Prepare for spawning by setting up a workstation containing disposable glass pasteur pipets and bulbs; a number of 50-mL, 300-mL, and 600-mL glass beakers; a small scissors; a blunt forceps, an ice bucket filled with crushed ice; 0.5*M* KCl; and two large shallow waterproof trays (large photographic developing trays work well). Place 1–2 inches of water in the first tray, and add small amounts of ice until the desired temperature is reached (about 12°C for *S. purpuratus*, and 18°C for *L. pictus*; small amounts of ice should be added periodically to maintain the temperature in the desired range. A piece of styrofoam can be placed under the tray of water to reduce the warming rate of the water bath). Cover the bottom of the second tray with 1–2 inches of crushed ice. Fill a number of 50-mL beakers to the brim with AFSW and place in the first water bath tray. Place a few opened 60 to 90-mm diameter plastic petri dishes on the ice in the second tray.
2. Transport urchins in a small container filled with chilled AFSW (the same temperature as the holding aquarium) to the workstation. Open the coelomic cavity by cutting the mouth around the peristomial membrane with a scissors, and discard the mouthparts. Examine gonads to determine the sex of the animal and to assess the reproductive status; ripe ovaries are tan-to-yellow to orange in *L. pictus*, and a brighter orange in *S. purpuratus*. The testes in both species are more tan, and exude white semen when nicked. Ripe gonads should fill a large percentage of the coelomic cavity, whereas poorly developed gonads resemble thin shreds of brown-to-tan material lying along the wall of the coelomic cavity.
3. Gently deliver a few milliliters of 0.5*M* KCl into the coelomic cavity (a wash bottle is convenient), and immediately place females upside down (opened coelomic cavity facing up) on the top of the 50-mL beakers filled with AFSW. Place males upside down on petri plates over ice. Allow spawning to continue for approximately 10 min (until the flow of eggs becomes slow), then remove females

from the beakers of AFSW. If pigment has been released into the AFSW (frequently a problem with *S. purpuratus*), immediately aspirate to just above the level of the eggs and replace with fresh AFSW. Check for thick puddles of sperm underneath males; collect and transfer the thick semen with a pasteur pipette to 1-mL centrifuge tubes, and place on ice (*see* **Note 2**).

3.2. Fertilization

1. Strip jelly coats from eggs by passing eggs over a Nitex screen (80 μ*M* diameter for *L. pictus*, and 65 μ*M* for *S. purpuratus*). To do this, gently transfer eggs to pour this suspension over the Nitex cloth into a second empty 600-mL beaker; repeat this step two to three times, collecting the filtrate from the last nitex passage into a clean 600-mL beaker. Let eggs settle (about 10 min), remove AFSW by aspiration, and resuspend in fresh AFSW. Stripping the jelly coats allows for a greater degree of developmental synchrony among a population of eggs by allowing sperm rapid access to the egg surfaces at the time of addition.

2. While the eggs are settling in the final AFSW wash in step 1, prepare 100 mL of a 2× fertilization envelope stripping solution consisting of 0.01% trypsin and either 12 or 20 m*M* dithiothreitol (DTT) (12 m*M* for *L. pictus*, and 20 m*M* for *S. purpuratus*) in CaFSW at the desired temperature (*see* **Note 3**). This solution should be used within approx. 30 min or so of preparation. Also at this time, make a 1:1000 dilution of sperm by adding about 10 μL of semen to 10 mL of AFSW. Mix to obtain a homogeneous suspension.

3. Gently pull up 4 to 5 mL (about two Pasteur pipetfuls) of settled dejellied and washed eggs (collected as they rest in a layer on the bottom of the 600-mL beaker) to 100 mL of fresh AFSW in a 300-mL beaker. Add one half Pasteur pipetful (about 1 mL) of diluted sperm, immediately gently mix by passing eggs back and forth once or twice between another beaker, and start a timer. Remove small aliquots of eggs to check under a microscope at low magnification. The fertilization envelope should begin to rise from one point on each egg within about 1 min. At 2 min from the addition of sperm, fertilization envelopes should be rising from 95% or more of the eggs. At no later than the 2-min mark, add an equal volume of 2× stripping solution and gently mix by passing the egg suspension back and forth between two beakers once or twice. Let the egg beaker rest in the ice bath without disturbance for 5–10 min.

4. At no later than the 10-min mark (earlier, if the eggs settle more quickly) aspirate off as much of the sea water as possible (at least 70–80%), and add fresh CaFSW to fill the beaker. Exposure of eggs to the full-strength stripping solution for longer than 10 minutes frequently results in the formation of small membrane "blebs" as the first cell cycle progresses. Let eggs settle for another 5–10 min, and repeat the wash with fresh CaFSW. At this point, fertilized eggs can be cultured by one of two methods. Relatively low numbers of eggs can be cultured in Petri dishes in a water bath or refrigerated incubator at the appropriate temperature (adjust concentration so that only about 50% of the surface of the dish is covered with eggs), or larger numbers can be cultured in 1-L beakers with stir-

ring. A simple stirring device can be constructed from a clock motor attached to a Plexiglas plate, a glass rod to act as a shaft, and a Teflon paddle attached to the glass rod.

3.3. Fixation and Immunolabeling

1. Calculate the number of samples required, and fill a corresponding number of wells in six-well dishes with freezer temperature (–10°C) absolute methanol. Keep dishes in the freezer until used.
2. Two minutes before each desired timepoint, remove zygotes from the culture with a Pasteur pipet and puddle on polylysine-coated glass coverslips (22 mm^2, 1.5 thickness). After 2 min, pick up each coverslip with forceps and lower straight down into a well filled with cold methanol. Tip the coverslip during this process as little as possible, i.e., do not let the puddle run off coverslip, and do not "slide" the coverslip into methanol, which can result in mechanical damage to the fragile zygotes. Replace the dish in the freezer, and fix cells for a minimum of 5–10 min (*see* **Note 4**). Cells can be stored in methanol in the freezer for extended periods of time (sea urchin embryos stored for 4 years in absolute methanol at –80°C have been successfully labeled with anti-tubulin and anti-lamin antibodies).
3. To rehydrate cells, remove coverslips from the methanol dishes, briefly drain, and place in similar six-well dishes filled with PBST. After 5 min, transfer coverslips to wells containing 0.5% Carnation nonfat powdered milk in PBST ("blocking" solution), and incubate for 30–60 min at 37°C.
4. Remove coverslips from blocking solution, briefly drain, and place coverslips cell-side-up on a flat sheet of parafilm in a humid container. Apply 100–200 µL of primary antibody at an appropriate dilution and incubate 4–5 h at room temperature (*see* **Note 5**). The parafilm can be labeled with a "Sharpie" indelible marker to identify coverslips.
5. Remove and save antibody, if desired, and wash coverslips in three changes of PBST over 15–20 min. Serially transferring the coverslips through PBST-filled wells in six-well dishes works well for this purpose. During this washing step, rinse residual antibody off of the parafilm and dry with Kimwipes.
6. Replace coverslips cell-side-up on parafilm and apply 100–200 µL secondary antibody at a 1:50 dilution in PBST. If 5 mL or less of secondary antibody is prepared, clear the solution before use by centrifuging in a microcentrifuge (10,000g for 3 min), and avoid using solution from the bottom of the microcentrifuge tube. If larger volumes are used, it is usually more convenient to clear the solution by Millipore filtration (0.22 µm pore size) with a disposable syringe filter unit. Failure to clear the secondary antibody frequently results in contamination of coverslips with finely particulate fluorescent background material. Label cells for 1 h at 37°C.
7. Remove secondary antibody from the surface of the coverslip and save, if desired (with the addition of sodium azide to 0.05%, diluted secondary antibody is usually good for 3–4 wk when stored at 4°C), and apply 100–200 µL of PBST containing 5 µg/mL Hoechst 33258. Incubate 5–10 min at room temperature.

8. Briefly drain coverslips and wash in PBST as in step 5.
9. Mount coverslips on glass microscope slides, using phenylenediamine/DABCO/ CAPS mounting medium. Remove an aliquot of mounting medium from the freezer and allow it to come to room temperature. Place one drop in the middle of a clean microscope slide, and also place a few small broken pieces of a no. 1.5 coverslip around the mounting medium to support the coverslip and reduce egg deformation. Pick up the coverslip from the last PBST wash with a forceps and carefully wipe the back dry with a Kimwipe. Invert coverslip (cell side down) over the drop of mounting medium and glass pieces. Seal the edges of the coverslip with nail polish, and store slides in a –10°C or –80°C freezer. Slides may be stable for many weeks in the freezer, but should be photographed as soon as possible (immediately is best, but within 1–3 days is usually acceptable) because some redistribution of fluorochrome does occur with time (in particular, nuclear fluorescence frequently appears with extended storage).

3.4. Cytoskeletal Experiments

The first cell cycle in fertilized sea urchin eggs takes approximately 1.5 h to complete, depending on the species and culture temperature. During this time, the microtubule cytoskeleton undergoes a series of remarkable structural reorganizations, which attend its functions in bringing the pronuclei together and in forming the mitotic spindle (**Table 2**). The dynamic behavior of microtubules during the first cell cycle is readily amenable to disruption by a wide variety of pharmacological and physical treatments. For an illustration of a typical cytoskeletal disruption experiment, the treatment of *L. pictus* zygotes with the carbamate herbicide CIPC (chlorpropham) is described (for further details, *see* **ref. 39**).

Prepare a 100 m*M* stock solution of CIPC [N-(3-chlorophenyl)carbamate] (Sigma) in dimethyl sulfoxide (DMSO). This solution appears to retain its activity for weeks, but not months, when stored at room temperature. Add 1, 10, or 100 µL of CIPC stock solution/100 mL of egg culture just before adding sperm, to obtain a 1 µ*M*, 10 µ*M*, and 100 µ*M* concentration, respectively. Fertilize and culture the eggs as described previously, with CIPC added to all AFSW and CaFSW solutions. Fix and immunolabel as described, and compare microtubule patterns between CIPC-treated embryos and control embryos (treated with equivalent amounts of DMSO only). A range of cytoskeletal perturbations at each timepoint should be apparent, with little or no effects noticeable at 1 µ*M*, significant alterations occurring at 10 µ*M*, and gross disruption of cytoskeletal organization and cleavage patterns resulting from treatment with 100 µ*M* CIPC (**Fig. 1**).

4. Notes

1. The polylysine is used to make polylysine-coated coverslips in the following way. Transfer a few hundred 22 mm² 1.5 thickness glass coverslips to a beaker

**Table 2
Developmental and Cytoskeletal Landmarks, First Cell Cycle,
L. pictus, 18°C**

0–5 min:	Sperm incorporation
5–10 min:	Beginning of sperm aster formation; sperm-associated centrosomal material only faintly detectable
10–20 min:	Sperm aster greatly increases in size; pronuclei approach each other; maternal centrosomal material recruited around nuclei (especially sperm nucleus)
25–35 min:	Pronuclei fuse; centrosomal material completely surrounds the zygote nucleus; bipolarization of microtubule organizing centers occurs at the end of this period
45–55 min:	Chromatin in zygote nucleus appears more homogeneous; microtubules form "streak"; centrosomal material becomes elongated in plane of "streak"
55–65 min:	Chromosomes begin to condense; microtubules begin to form bipolar spindle apparatus; centrosomal material surrounds nucleus, with concentrations at the spindle poles
65–75 min:	Prophase progresses to prometaphase; centrosomal material completely separated and localized to each of the spindle poles
75–80 min:	Metaphase
80–85 min:	Anaphase
85–95 min:	Telophase and cytokinesis

containing hot water and standard laboratory detergent, such as Alconox. Let sit for 5–10 min, with periodic agitation. Decant detergent, and rinse extremely well with warm tap water (at least 10 changes), followed by a final rinse with distilled water. Decant distilled water and add 95% ethanol to completely immerse coverslips, and cover with parafilm to store. For coating, remove desired number of coverslips and place on paper towels. Rub the ethanol off of the exposed surface of the coverslip with a Kimwipe until the coverslip "squeaks." Let dry completely, and add a 10- to 15-µL puddle of 1 mg/mL polylysine to one half of the coverslips. Invert the other one half of the coverslips over the polylysine-coated coverslips to form pairs of

---

Fig. 1. (*facing page*) First cell cycle mitosis in CIPC-treated zygotes (**A–D**) and Triton X-100/glycerol extracted specimens (**E,F**). (**A–D**) Confocal microscopy (CF); (**E,F**) routine epifluorescence microscopy (EPI). (**A**) Prophase; the incipient spindle poles have failed to become aligned in a straight line with the main mass of chromatin (i.e., have not become perfectly apposed). (**B**) Metaphase; note the clublike appearance of the spindle, and the truncated astral microtubules. (**C**) Anaphase; astral microtubules are again truncated, and the chromosomes are arrayed over an unusually broad plane. (**D**) Telophase; microtubules are relatively dense, but fail to penetrate the most peripheral areas of the cortex. Cytokinesis in embryos treated with 10 µM or higher CIPC frequently fails to be

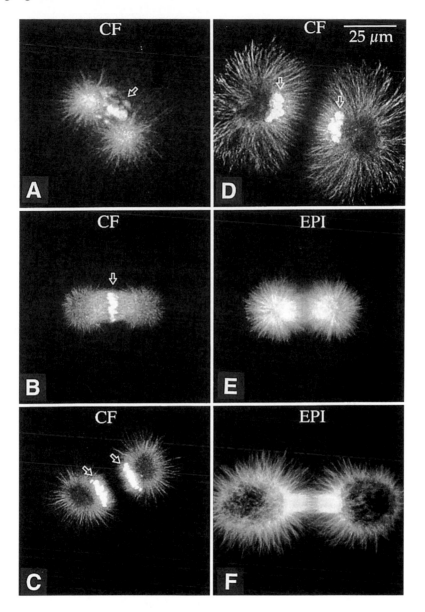

completed. (**E,F**) Control (non-CIPC-treated) zygotes subjected to Triton X-100/glycerol extraction prior to fixation in cold methanol. Routine epifluorescence microscopy. (**E**) Metaphase; (**F**) anaphase. Note the distortion of the microtubule cytoskeleton in the anaphase spindle (compare with **Fig. 2F**). The metaphase spindle in this example is very well preserved for extracted specimens; frequently, detectable physical distortion occurs in this stage, as well as all others, following extraction. (**A–D**) Double labeled with E7 anti-β-tubulin antibody and Hoechst; arrows indicate chromatin.

coverslip "sandwiches" (cleaned surface to cleaned surface). Let sit for 5 min, and slide each pair apart, and place wet-side-up on paper towels. Examine the wet surfaces of each coverslip: if the residual polylysine immediately beads up into tight droplets, discard coverslip (this indicates surface was not sufficiently clean, and eggs will not stick well). Save those coverslips on which the polylysine remains largely spread out (some contraction of the puddle is acceptable).

2. Some researchers prefer to use a syringe and inject 0.5M KCl into the coelomic cavity through the peristomial membrane. However, removing the mouth allows the status of the gonads to be visually determined; in addition, animals appear to generally release more gametes with this approach. On the other hand, the more traumatic method of mouth removal also seems to result in the release of more pigment along with the eggs (especially in *S. purpuratus*), which can be inhibitory to fertilization if not washed away.

3. There are a number of alternate methods to strip the fertilization membrane from the fertilized egg, including the use of urea or aminotriazole (ATA) instead of trypsin/DTT. The ATA method *(40)* is very convenient, and has the advantage that close attention does not have to be paid to the time of addition of stripping solution. However, the quality of fixation and immunolabeling is very sensitive to the presence of hyalin, which is secreted onto the surface of the egg. The trypsin/DTT method, followed by culturing in CaFSW, appears to be the most effective in removing this material, and usually results in the best images after processing is complete.

4. Alternate approaches toward preparing sea urchin embryos for microscopy include the use of formaldehyde-based fixatives, and extracting eggs and embryos prior to fixation. Direct methanol or formaldehyde fixations usually result in excellent morphological preservation, but the large size of the eggs results in a "muddy" image of immunolabeled cytoskeletal elements when viewed by routine epifluorescence microscopy (compare **Fig. 2 A, C, E** with **B, D, F**). To circumvent this problem, extraction methods have been developed to visualize the cytoskeleton more clearly by routine epifluorescence microscopy *(41)*. However, extraction can produce significant artifacts (compare **Figs. 1F** and **2F**), and the most faithful preservation of morphological relationships is obtained by direct fixation in either methanol or paraformaldehyde. Methanol fixation is usually better for immunocytochemistry, and produces excellent images when coupled with confocal microscopy.

5. Primary antibodies are diluted so that a 1-h incubation at 37°C produces acceptable labeling results (strong specific labeling with low background). The following incubation times and temperatures yield roughly comparable labeling intensities: 1 h at 37°C, 3–5 h at room temperature, or overnight (about 16 h) at 4°C. Subjectively, the best results are obtained from a 3- to 5-h incubation at room temperature. The intensity of immunolabeling can be markedly increased by incubating overnight at room temperature. Many solutions have been used prior to the primary antibody to "block" nonspecific protein binding sites, including BSA (bovine serum albumin) and 10–20% normal goat serum. However, a solution of 0.5% nonfat milk is at least as effective for all antibodies so far tested in this laboratory, and is extremely simple and economical to use.

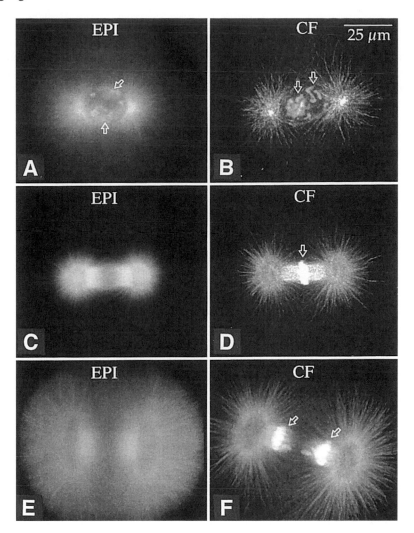

Fig. 2. First cell cycle mitosis by epifluorescence microscopy (EPI) and confocal microscopy (CF). (**A,B**) Prophase. (**C,D**) Metaphase. (**E,F**) Anaphase. All specimens were fixed by immersion in cold methanol and labeled with E7 anti-β-tubulin antibody. All images except for (**C**) and (**E**) are double exposures, showing both tubulin and Hoechst staining; *arrows* indicate Hoechst-stained chromatin. Note the marked increase in clarity and detail in the CF images.

References

1. Zalensky, A. O., Allen, M. J., Kobayashi, A., Zalenskaya, I. A., Balhorn, R., and Bradbury, E. M. (1995) Well-defined genome architecture in the human sperm nucleus. *Chromosoma* **103,** 577–590.

2. Blerkom, J. V., Davis, P. W., and Merriam, J. (1994) A retrospective analysis of unfertilized and presumed parthenogenetically activated human oocytes demonstrates a high frequency of sperm penetration. *Hum. Reprod.* **9,** 2381–2388.

3. Blerkom, J. V., Davis, P., Merriam, J., and Sinclair, J. (1995) Nuclear and cytoplasmic dynamics of sperm penetration, pronuclear formation, and microtubule organization during fertilization and early preimplantation development in the human. *Hum. Reprod. Update* **1,** 429–461.

4. Simerly, C., Wu, G., Zoran, S., Ord, T., Rawlins, R., Jones, J., Navara, C., Gerrity, M., Rinehart, J., Binor, Z., Asch, R., and Schatten, G. (1995) The paternal inheritance of the centrosome, the cell's microtubule organizing center, in humans and the implications for infertility. *Nature Med.* **1,** 47–53.

5. Wu, G.-J., Simerly, C., Zoran, S. S., Funte, L. R., and Schatten, G. (1996) Microtubule and chromatin dynamics during fertilization and early development in rhesus monkeys and regulation by intracellular calcium ions. *Biol. Reprod.* **55,** 260–270.

6. Kim, N. H., Funahashi, H., Abeydeera, L. R., Moon, S. J., Prather, R. S., and Day, B. N. (1996) Effects of oviductal fluid on sperm penetration and cortical granule expcytosis during fertilization of pig oocytes in vitro. *J. Reprod. Fertil.* **107,** 79–86.

7. Navara, C. S., First, N. L., and Schatten, G. (1996) Phenotypic variations among paternal centrosomes are expressed within the zygote as disparate microtubule lengths and sperm aster organization: correlations between centrosome activity and development success. *Proc. Natl. Acad. Sci. USA* **93,** 5384–5388.

8. Coonrod, S. A., Westhusin, M. E., and Naz, R. K. (1994) Monoclonal antibody to human fertilization antigen-1 (FA-1) inhibits bovine fertilization in vitro: application in immunocontraception. *Biol. Reprod.* **51,** 14–23.

9. Shiraishi, K., Okada, A., Shirakawa, H., Nakanishi, S., Mikoshiba, K., and Miyazaki, S. (1995) Developmental changes in the distribution of the endoplasmic reticulum and inositol 1,4,5-triphosphate receptors and the spatial pattern of Ca^{2+} release during maturation of hamster oocytes. *Dev. Biol.* **170,** 594–606.

10. Mehlman, L. M., Terasaki, M., Jaffe, L. A., and Kline, D. (1995) Reorganization of the endoplasmic reticulum during meiotic maturation of the mouse oocyte. *Dev. Biol.* **170,** 607–615.

11. Ayabe, T., Kopf, G. S., and Schultz, R. M. (1995) Regulation of mouse egg activation: presence of ryanodine receptors and effects of microinjected ryanodine and cyclic ADP ribose on uninseminated and inseminated eggs. *Development* **121,** 2233–2244.

12. House, C. R. (1994) Confocal ratio-imaging of intracellular pH in unfertilized mouse oocytes. *Zygote* **1,** 37–45.

13. Simerly, C. R., Hecht, N. B., Goldberg, E., and Schatten, G. (1993) Tracing the incorporation of the sperm tail in the mouse zygote and early embryo using an anti-testicular alpha-tubulin antibody. *Dev. Biol.* **158,** 536–548.

14. Ducibella, T., Duffy, P., and Buetow, J. (1994) Quantification and localization of cortical granules during oogenesis in the mouse. *Biol. Reprod.* **50,** 467–473.

15. Carroll, D. J., Dikegoros, E., Koppel, D. E., and Cowan, A. E. (1995) Surface expression of the pre-beta subunit of fertilin is regulated at a post-translational level in guinea pig spermatids. *Dev. Biol.* **168,** 429–437.

16. Gard, D. L. (1993) Confocal immunofluorescence microscopy of microtubules in amphibian oocytes and eggs. *Methods Cell Biol.* **38,** 241–264.

17. Gard, D. L. (1991) Organization, nucleation, and acetylation of microtubules in *Xenopus laevis* oocytes: a study by confocal immunofluorescence microscopy. *Dev. Biol.* **143,** 346–362.

18. Gard, D. L. (1992) Microtubule organization during maturation of *Xenopus* oocytes: assembly and rotation of the meiotic spindles. *Dev. Biol.* **151,** 516–530.

19. Gard, D. L. (1993) Ectopic spindle assembly during maturation of *Xenopus* oocytes: evidence for functional polarization of the oocyte. *Dev. Biol.* **159,** 298–310.

20. Gard, D. L. (1994) Gamma-tubulin is asymmetrically distributed in the cortex of *Xenopus* oocytes. *Dev. Biol.* **161,** 131–140.

21. Gard, D. L., Cha, B. J., and Schroeder, M. M. (1995) Confocal immunofluorescence microscopy of microtubules, microtubule-associated proteins, and microtubule-organizing centers during amphibian oogenesis and early development. *Curr. Top. Dev. Biol.* **31,** 383–431.

22. Gard, D. L., Cha, B. J., and Roeder, A. D. (1995) F-actin is required for spindle anchoring and rotation in *Xenopus* oocytes; a re-examination of the effects of cytochalasin B on oocyte maturation. *Zygote* **3,** 17–26.

23. Gard, D. L., Affleck, D., and Error, B. M. (1995) Microtubule organization, acetylation, and nucleation in *Xenopus laevis* oocytes: II. A developmental transition in microtubule organization during early diplotene. *Dev. Biol.* **168,** 189–201.

24. Roeder, A. D. and Gard, D. L. (1994) Confocal microscopy of F-actin distribution in *Xenopus* oocytes. *Zygote* **2,** 111–124.

25. Schroeder, M. M. and Gard, D. L. (1992) Organization and regulation of cortical microtubules during the first cell cycle of *Xenopus* eggs. *Development* **114,** 699–709.

26. Fagotto, F. and Maxfield, F. R. (1994) Changes in yolk platelent pH during *Xenopus laevis* development correlate with yolk utilization. A quantitative confocal microscopy study. *J. Cell Sci.* **107,** 3325–3337.

27. Olds, J. L., Favit, A., Nelson, T., Ascoli, G., Gerstein, A., Cameron, M., Cameron, L., Lester, D. S., Rakow, T., and DeBarry, J. (1995) Imaging protein kinase C activation in living sea urchin eggs after fertilization. *Dev. Biol.* **172,** 675–682.

28. Terasaki, M. (1995) Visualization of exocytosis during sea urchin egg fertilization using confocal microscopy. *J. Cell Sci.* **108,** 2293–2300.

29. Stricker, S. A., Centonze, V. E., Paddock, S. W., and Schatten, G. (1992) Confocal microscopy of fertilization-induced calcium dynamics in sea urchin eggs. *Dev. Biol.* **149,** 370–380.

30. Terasaki, M. and Jaffe, L. A. (1991) Organization of the sea urchin egg endoplasmic reticulum and its reorganization at fertilization. *J. Cell Biol.* **114,** 929–940.

31. Holy, J. and Schatten, G. (1991) Spindle pole centrosomes of sea urchin embryos are partially composed of material recruited from maternal stores. *Dev. Biol.* **147,** 343–353.

32. Holy, J. and Schatten, G. (1997) Recruitment of maternal material during assembly of the zygote centrosome in fertilized sea urchin eggs. *Cell Tiss. Res.,* **289,** 285–297.

33. Stricker, S. A. (1995) Time-lapse confocal imaging of calcium dynamics in starfish embryos. *Dev. Biol.* **170,** 496–518.
34. Jaffe, L. A. and Terasaki, M. (1994) Structural changes in the endoplasmic reticulum of starfish oocytes during meiotic maturation and fertilization. *Dev. Biol.* **164,** 579–587.
35. Speksnijder, J. E., Terasaki, M., Hage, W. J., Jaffe, L. F., and Sardet, C. (1993) Polarity and reorganization of the endoplasmic reticulum during fertilization and ooplasmic segregation in the ascidian egg. *J. Cell Biol.* **120,** 1337–1346.
36. Snook, R. R., Markow, T. A., and Karr, T. L. (1994) Functional nonequivalence of sperm in Drosophila. *Proc. Natl. Acad. Sci. USA* **91,** 11,222–11,226.
37. Theurkauf, W. E., Smiley, S., Wong, M. L. and Alberts, B. M. (1992) Reorganization of the cytoskeleton during *Drosophila* oogenesis: implications for axis specification and intercellular transport. *Development* **115,** 923–936.
38. Watson, C. A., Sauman, I. and Berry, S. J. (1993) Actin is a major structural and functional element of the egg cortex of giant silkmoths during oogenesis. *Dev. Biol.* **155,** 315–323.
39. Holy, J. (1997) Chlorpropham [N-(3-chlorophenyl)carbamate] disrupts microtubule organization, cell division, and early development of sea urchin embryos. *J. Toxicol. Env. Health* **54,** 319–333.
40. Showman, R. M. and Foerder, C. A. (1979) Removal of the fertilization membrane of sea urchin embryos employing aminotriazole. *Exp. Cell Res.* **120,** 253–255.
41. Balczon, R. and Schatten, G. (1983) Microtubule-containing detergent-extracted cytoskeletons in sea urchin eggs from fertilization through cell division: antitubulin immunofluorescence microscopy. *Cell Motil. Cytoskel.* **3,** 213–226.

10

Imaging Immunolabeled *Drosophila* Embryos by Confocal Microscopy

James A. Langeland

1. Introduction

Monoclonal and polyclonal antisera raised against recombinant proteins are highly sensitive probes that reveal cellular and subcellular protein localization in developing embryos. For *Drosophila* researchers, the ease of generating such antisera *(1)* and the number and widespread availability of existing antibodies make immunofluroescence of embryos an indispensable technique. The use of fluorochrome-conjugated secondary and/or tertiary antibodies on *Drosophila* embryos and detection by confocal microscopy offers two, critical advantages over enzyme-mediated detection methods, such as alkaline phosphatase and horseradish peroxidase. First, the sensitivity achieved with confocal microscopy may be difficult to match with enzyme-mediated detection especially when imaging cells deep within later stage embryos. Second, immunofluorescence and confocal detection allow the selective and simultaneous labeling with up to three different primary antisera. The following protocol contains instructions for the fixation, labeling, and detection of *Drosophila* embryos with one, two, and three different primary antisera.

Considerable forethought must go into designing a multiple labeling experiment. The critical considerations are: (1) the number of primary antibodies to be used, (2) the host species for each of the primary antibodies, (3) the availability of appropriately specific and appropriately conjugated secondary and/or tertiary antisera, and (4) the order and timing of antibody incubation. In general, each of the primary antibodies should be raised in a separate species. Primary antibodies can be labeled directly with fluorescently tagged secondary antibodies, or detected via a biotin-labeled secondary antibody and a

From: *Methods in Molecular Biology, vol. 122: Confocal Microscopy Methods and Protocols*
Edited by: S. Paddock © Humana Press Inc., Totowa, NJ

strepavidin-conjugated fluorochrome. Secondary and tertiary reagents must be non-cross-reactive and able to distinguish between different reagents used in previous incubation.

Fluorochromes that may be used in the terminal labeling step include fluorescein (excites at 498 nm), rhodamine (excites at 568 nm depending on the exact subtype), and Cy5 (excites at 647 nm).

2. Materials

2.1. Embryo Fixation

1. Embryo wash buffer: 0.4% NaCl, 0.3% Triton X-100
2. Clorox
3. Fix buffer: 100 mM PIPES (pH. 6.9) 2 mM EGTA, 1 mM MgSO4
4. Heptane
5. 37% Formaldehyde
6. Methanol
7. Methanol/EGTA: 90% methanol, 0.05M EGTA
8. Ethanol

2.2. Immunolabeling

1. Phosphate-buffered saline (PBS): 13.7 mM NaCl, 0.27 mM KCl, 0.43 mM Na$_2$HPO$_4$·7H$_2$O, 0.14 mM KH$_2$PO$_4$
2. PBS-Triton: PBS with 0.1% Triton X-100
3. Blocking solution: PBS; 0.1% Triton X-100, 1% BSA. Make fresh and store at 4°C
4. Primary antisera: Raised against the protein of interest in any one of several hosts *(1)*
5. Secondary and/or tertiary antisera: Commercially manufactured antisera conjugated with specific fluorochromes
6. Mounting medium: 50 mM Tris-HCl (pH 8.8), 10% glycerol, 0.5 mg/ml *p*-phenylenediamine

3. Methods

3.1. Embryo Fixation

1. Embryos from the appropriate wild-type or mutant stock are collected on agar-molasses caps for approximately at 25°C *(2)* and *see* **Note 2**.
2. Wash the collected embryos in embryo wash buffer.
3. Dechorionate embryos in 50% Clorox for 2 min (*see* **Note 3**).
4. Thoroughly rinse in embryo wash buffer.
5. Transfer the embryos to a 15-mL screwtop tube containing 4 mL of fix buffer.
6. Add 5 mL of heptane and 1 mL of 37% formaldehyde, and shake the tube by inversion for 20–25 min.
7. Devitellinize the embryos by removing the fix buffer and adding 5 mL of 100% methanol to the embryos in heptane and shaking vigorously for 2–3 min.

8. The devitellinized embryos sink to the bottom of the tube and are removed to a 1.5-mL snap-top tube with a Pasteur pipet.
9. Fixed, devittelinized embryos are rinsed 2× in 90% methanol/0.05*M* EGTA.
10. For long-term storage, embryos are dehydrated in five or six washes in 100% ethanol and stored at –20°C. For immediate use, or to recover from long-term storage, embryos are rehydrated in two changes of PBS-Triton (PBT) and then placed in PBT at 4°C for blocking.

3.2. Single Immunofluorescent Labeling of Embryos

1. Primary antisera should initially be used at a variety of concentrations to determine the optimal concentration for detection by the specific confocal imaging system (*see* **Note 4**). Typical final concentration ranges are from 0.5 to 2 ug/mL.
2. Fixed, blocked embryos are incubated in a 1.5-mL snap-top tube with primary antibodies in a volume of 200 μL overnight at 4°C.
3. Wash the embryos 10 times in PBT over a period of 1 h.
4. Incubate with the secondary antibody in a 1-mL volume for 2 h at 4°C with rocking.
5. Embryos are again washed 10 times in PBT over a period of 1 h.
6. After the addition of fluorescent conjugates, the embryos are protected from light as much as possible (i.e., wrapped in aluminum foil). If needed, incubation with tertiary antibodies takes place again in a volume of 1 mL for 1 h at 4°C with rocking.
7. After a final series of 10 washes, the embryos were equilibrated in mounting medium and then mounted on slides in this same solution (*see* **Notes 5** and **6**).

3.3. Double and Triple Immunofluorescent Labeling

A sample labeling scheme using antibodies against three *Drosophila* segmentation proteins is shown in **Table 1**, and the resulting single- and triple-labeled images are shown in **Fig. 1**.

1. For multiple labeling, embryos are incubated in a 1.5-mL centrifuge tube with primary antibodies in a volume of 200 μL overnight at 4°C.
2. Embryos are then washed 10 times in PBT over a period of 1 h.
3. Incubation with secondary antibodies takes place in a 1-mL volume for 2 h at 4°C with rocking.
4. Embryos are again washed 10 times in PBT over a period of 1 h.
5. Incubate with tertiary antibodies in a volume of 1 mL for 1 h at 4°C with rocking.
6. Again, after the addition of fluorescent conjugates, the embryos are protected from light as much as possible (i.e., wrapped in aluminum foil).
7. After a final series of washes, the embryos are equilibrated in mounting medium and then mounted on slides in this same solution.

3.4. Confocal Microscopy

Fluorescent images were collected using a Bio-Rad MRC600 Laser Scanning Confocal Microscope equipped with a krypton/argon laser as a light source

Table 1
A Sample Labeling Scheme for the Simultaneous Detection of Three Different *Drosophila* Segmentation Proteins

Primary antibody	α-Hairy	α-Krüppel	α-Giant
Host species for primary antibody	Mouse	Rat	Rabbit
Secondary antibody	Cy5 α-Mouse IgG	α-rat IgG	Biotinylated α-rabbit IgG
Host species for secondary antibody	Donkey	Goat	Donkey
Tertiary antibody or conjugate or conjugate	None	Rhodamine α-goat IgG	Fluorescein streptavidin
Host species for tertiary antibody	Donkey	N/A	

The primary antisera are each raised in a different host species. Secondary antibodies are directed specifically against the primary host species to avoid cross-hybridization. In the case of the mouse a hairy antibody, a Cy5 conjugated secondary antibody is used in a two-step detection. For both Rat a Krüppel and Rabbit a Krüppel, a three-step detection was used.

(*see* **Note 7**). This light source has three major lines that emit at 488 nm (for detection of fluorescein), 568 nm (for detection of rhodamine), and 647 nm (for detection of Cy5). Two or three single optical planes are Kalman averaged and collected at a resolution of 8 bits (**Fig. 1**). The images are assigned specific colors and then merged into a single two or three color image using methods described previously *(3,4)* and in Chapter 21.

4. Notes

1. Embryos may be stored for several hours at 4°C prior to fixation, as development is essentially arrested at this temperature.
2. The timing of embryo collection depends upon the desired developmental stages. For example, if the desired antigen(s) is expressed during hours 4–8 of development, then embryos should be collected for 4 h, and then aged for 4 h prior to fixation.
3. Dechorionation (removal of the chorion shell) and devitellinization (removal of the vitelline membrane) are essential to allow probes to penetrate the embryos.
4. During the initial stages of staining embryos with novel antibody probes a series of antibody concentrations should be tested on embryos to fine tune the protocol to the confocal imaging system.
5. Coverslips should be sealed with clear nail polish, and slides should be clean and dry before viewing. Make sure to select the correct coverslip thickness for the objective lens.

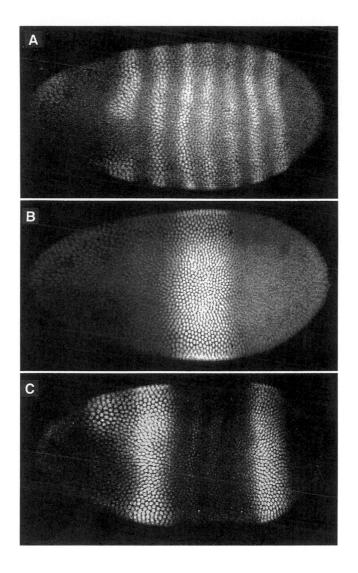

Fig. 1. Simultaneous localization of three segmentation proteins in a *Drosophila* embryo. (**A**) Detection of the seven stripes of hairy protein using α-mouse α-hairy primary antibody and a Cy5-conjugated donkey α-mouse IgG. The image was taken at 647 nm, and pseudocolored in green. (**B**) Detection of the block of Krüppel protein using α-rat α-Krüppel primary antibody, α-goat α-Rat IgG secondary antibody, and a rhodamine-conjugated donkey α-goat IgG. The image was taken at 568 nm and pseudocolored in red. (**C**) Detection of giant protein, using α-rabbit α-giant primary antibody, a biotinylated donkey α-rabbit IgG, and a fluorescein-conugated streptavidin. The image was taken at 498 nm and is pseudocolored in blue. For versions of the three-color merged image of A, B, and C, *see* Color Plate I C,D.

6. Slides should be viewed in a timely fashion after their preparation, as the fluoro-chrome signals tend to degrade over time. Fluorescein is the most labile and should preferably be viewed within a few hours after mounting. Signal lifetime can be preserved by storing the slides in the dark and in a fridge.

7. The choice of objective lens is important for gaining the best resolution. A Zeiss 16× oil lens or a Nikon 10× dry lens is good for viewing whole embryos and a Nikon 40× NA 1.2 or a Nikon 60× NA 1.4 lens is good for viewing individual cells and cell nuclei. It is always better to choose a higher magnification/NA lens than to zoom using the microscope or by cropping the image to gain the best resolution.

8. The following problems can cause the failure to detect one or more of the fluoro-chromes. The most common, and most readily curable, is in waiting too long after mounting before viewing (*see* **Note 6**). Another problem can be old and degraded antibodies. Antisera typically have a life of 1–2 years at 4°C. Old wash reagents, particularly old solutions with bovine serum albumin (BSA) can also degrade signal. Make up reagents fresh at least weekly. Another problem can be crossreactivity of secondary and tertiary antibodies; this will result in the same signal on two or more channels. It is essential to choose host species carefully, and to use reagents specifically formulated for minimal cross-reactivity.

References

1. Williams, J. A., Langeland, J. A., Thalley, B., Skeath, J. B., and Carroll S. B. (1994) Production and purification of polyclonal antibodies against proteins expressed in *E. coli*, in *DNA Cloning: Expression Systems*, IRL Press, Oxford.
2. Ashburner, M. (1989) Drosophila: *A Laboratory Handbook*. Cold Spring Harbor Press, Cold Spring Harbor, NY.
3. Paddock, S. W., Langeland, J. A., Devries, P., and Carroll, S. B. (1993) Three color immunofluorescence imaging of Drosophila embryos by laser scanning con-focal microscopy. *BioTechniques* **14,** 42–47.
4. Paddock, S. W., Hazen, E. J., and DeVries, P. J. (1997) Methods and applications of three color confocal imaging. *BioTechniques* **22,** 120–126.

11

Confocal Microscopy on *Xenopus laevis* Oocytes and Embryos

Denise L. Robb and Chris Wylie

1. Introduction

There are many problems affecting microscopy in *Xenopus*. *Xenopus* oocytes and embryos are fragile and lack connective tissue. They are also full of yolk platelets, which prevents frozen sectioning because the yolk crystallizes and tears the sections. In addition, the yolk autofluoresces, making whole-mount immunocytochemistry possible, but difficult due to background from out-of-focus fluoresecence. All of these problems can be solved through confocal microscopy. Optical sectioning eliminates the need for manual sectioning and makes background fluorescence and autofluorescence negligible. It is still difficult to image the deep structures within the embryo, but thick wax sections can be cut and confocal microscopy again applied.

For years the field of cell biology has relied heavily on microscopy to study cellular organization and function. Most microscopic techniques, however, were optimized for small or flattened cells. More recently, the advent of confocal microscopy *(1)* has granted cell and developmental biologists the luxury of analyzing large cells or embryos, that have been fluorescently labeled. In tissues or embryos fluorescence from outside the focal plane can completely cover the details of interest by reducing image sharpness and contrast. Thus many embryos were previously impervious to immunofluorescent microscopy except through time-consuming and sometimes destructive procedures of embedding and thin-sectioning. The confocal microscope diminishes out-of-focus information and enhances image contrast by illuminating and then detecting an identical small region of the sample. Confocal microscopes are capable of "optically" sectioning through

From: *Methods in Molecular Biology, vol. 122: Confocal Microscopy Methods and Protocols*
Edited by: S. Paddock © Humana Press Inc., Totowa, NJ

large cells, small embryos, or parts of larger embryos, by collecting a series of images from various focal planes of the specimen. Individual captured images can then be compiled to construct a detailed three-dimensional view of the specimen. We have used confocal laser-scanning microscopy quite successfully to analyze the distribution of organelles, proteins, and filaments in the *Xenopus laevis* oocyte and developing embryo. This chapter will detail procedures, some of which have been adapted and modified from the work of David L. Gard and others *(2–4)*, and some of which we have devised, but all with the goal of imaging the cellular and molecular complexities of development.

Preparation of specimens for fluorescent antibody staining and confocal imaging in *Xenopus* always begins with specimen fixation. Most importantly the fixative used needs to be compatible with the specific antibodies or stain being used in the procedure. This must be determined experimentally for individual antibodies. For example, a vimentin antibody we use works only with 2% trichloroacetic acid as a fixative *(5)*, whereas an anti-cytokeratin antibody works with ethanol *(5)*. Another major concern during fixation is the preservation of cellular structures of interest. Much of our work involves imaging microtubules that are best preserved using the crosslinking agents formaldehyde and glutaraldehyde. Because of these factors, choosing a suitable fixation protocol is important. This problem becomes compounded when more than one antibody is being used so different options should be tried until staining is optimized. Lastly, owing to the large size of *Xenopus* oocytes and embryos (~1.2 mm diameter) almost all fixes require at least 2 h for adequate penetration.

Our laboratory has been utilizing two types of sample preparation to take advantage of confocal microscopy. We either process oocytes and embryos as whole-mounts or use the more traditional procedures of embedding and sectioning, but the samples cut as "thick" sections of about 30 µm (rather than the usual 5–10 µm) and are then optically sectioned under the confocal microscope. There are advantages and disadvantages to both of these techniques and the time commitment is about equal.

Preparing embryos for thick sectioning requires that after fixation they are embedded in wax, a process that may decrease antigenicity. Samples are then sectioned rather easily at 30-µm intervals, but must be affixed to specially prepared gelatin-coated slides. Standard antibody incubations of only a couple of hours can be performed, because we are working with sections and not whole embryos. Fluorescence outside the focal plane in these thick sections would be a problem using an epifluorescent microscope, but is eliminated during confocal microscopy. By cutting serial sections the distribution of a particular antigen can be analyzed through an entire embryo.

The major advantage of whole-mount immunocytochemistry on oocytes and embryos is to avoid embedding and sectioning specimens. However, because *Xenopus* oocytes and embryos are so large, long antibody incubations are necessary to achieve adequate penetration. Penetration of antibodies is indeed the limiting factor during whole-mount immunofluorescence, as reports suggest that antibodies penetrate a maximum of 150 µm into fixed *Xenopus* oocytes during an overnight incubation *(4)*. Likewise, washes following antibody incubation need also be very lengthy. We have improved incubation conditions by routinely bisecting oocytes and embryos after fixation and prior to incubations, and including detergent in fixatives and buffers. As with thick sections, fluorescence outside the focal plane is substantially reduced using confocal microscopy. Whole-mount technology and confocal microscopy have allowed us to examine the pattern of germ plasm movement in the embryo (**Fig. 1**), as well as analyze the three-dimensional filamentous microtubule network associated with the germ plasm (**Fig. 2**) *(6)*.

For effective observation under confocal microscopy the two sample preparations are treated differently. Thick sections on slides are mounted in an aqueous medium of phosphate-buffered saline (PBS) and glycerol. Whole-mount embryos are immersed in a clearing solution. Because early *Xenopus* embryos are filled with yolk and are absolutely opaque, a method was devised by Murray and Kirschner using a mixture of benzyl alcohol and benzyl benzoate (1:2 v/v) (Murray's Clear) to make embryos transparent. This technique has been used for fluorescence microscopy of the superficial cytoplasm of early-stage whole-mount embryos *(7)*. Bisected or whole embryos can then be mounted in glass cavity slides or in aluminum slides with bored holes and coverslips on either side to view from the top or bottom.

2. Materials

2.1. Thick Section Immunocytochemistry

2.1.1. Embedding and Sectioning Oocytes and Embryos

1. Fixative (*see* **Note 1**)
2. 4-mL glass vials
3. 100% Ethanol
4. 39°C incubator or warming receptacle
5. Embedding wax [poly(ethylene glycol) distearate + 1% w/w cetyl alcohol]. Once mixed the wax can be stored at 39°C for up to 2 weeks.

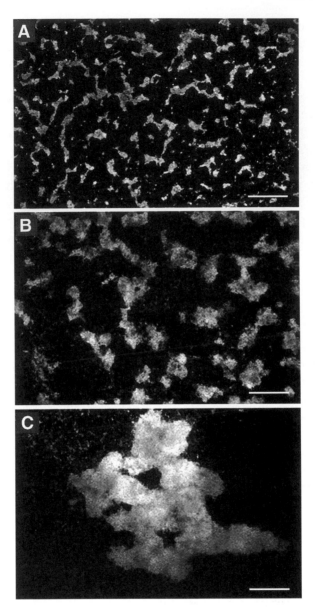

Fig. 1. Progression of germ plasm aggregation in *Xenopus laevis* prick-activated oocytes as viewed with confocal microscopy. The view is of a region of the vegetal hemisphere of the oocyte. Germ plasm is stained with chloromethyl X-rosamine (MitoTracker, Molecular Probes, Inc.), a vital dye for mitochondria found packed in the germ plasm. (A) Germ plasm prior to aggregation; (B) Germ plasm after 2 h of aggregation; (C) Germ plasm that has completely aggregated 4 h after prick activation. Scale bars = 25 μm.

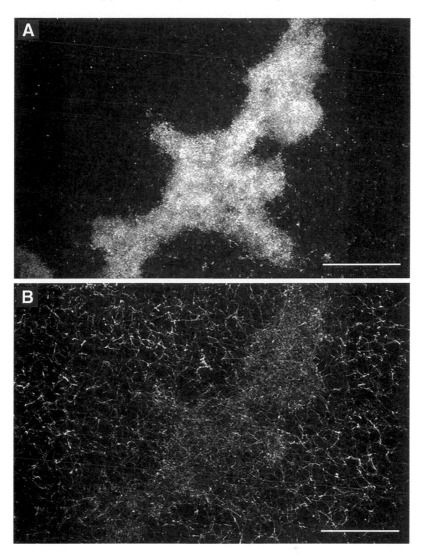

Fig. 2. Microtubules are closely associated with aggregating germ plasm. Microtubules in oocytes were whole mount labeled with an anti-α-tubulin antibody (DM1A, ICN Biomedical) subsequent to germ plasm staining. **(A)** Aggregated germ plasm stained with chloromethyl X-rosamine and viewed with confocal microscopy. **(B)** In viewing only the microtubule network, the location of the aggregated germ plasm can be predicted by a denser network of microtubules. Scale bars = 40 μm.

6. Plastic embedding trays
7. Microtome
8. Distilled water

9. Gelatin-coated slides (*see* **Note 2**)
10. Hot plate
11. Desiccator

2.1.2. Staining Thick Sections

1. Acetone series: 100%, 90%, 70%, 50% (v/v in distilled water)
2. PBS: To make a 10× stock dissolve 80 g of NaCl, 2.0 g of KCl, 14.4 g of Na$_2$HPO$_4$, and 2.4 g of KH$_2$PO$_4$ in 800 mL of distilled water and adjust pH to 7.4 with HCl. Add water to 1 L and autoclave. Dilute 10-fold with distilled water for use.
3. Horse or goat serum. Store frozen at –20°C in 10 mL aliquots.
4. Blocking buffer (10% serum, 4% bovine serum albumin in 1× PBS). Store frozen at –20°C in 1 mL aliquots.
5. Primary antibody
6. Fluorescently conjugated secondary antibody
7. Wet box
8. Eriochrome black (1:100 dilution of 1% stock in water or PBS)
9. Distilled water
10. Aqueous mounting medium made of 70–90% glycerol in 1× PBS with antifade agents (*see* **Note 4**). Store wrapped in foil.
11. Confocal microscope

2.2. Whole-Mount Immunocytochemistry

1. Fixative (*see* **Note 1**)
2. 4-mL glass vials
3. 10% Hydrogen peroxide in ethanol or methanol (optional) (*see* **Note 5**)
4. PBS
5. Ethanol and methanol
6. Fine forceps
7. Pasteur pipets
8. Scalpel
9. Petri dishes
10. 100 m*M* NaBH$_4$ in 1× PBS (optional) (*see* **Note 6**). Make fresh for each use.
11. TBSN: 1× Tris-buffered saline + 0.1% Nonidet P-40 (NP-40). To make a 10× stock of TBS dissolve 80 g of NaCl, 2.0 g of KCl, and 30 g of Tris base in 800 mL of distilled water and adjust pH to 7.4 with HCl. Add distilled water to 1 L and autoclave. Dilute 10-fold with distilled water for use and add 0.1% (v/v) NP-40. Solution can be stored at 4°C for up to a year.
12. Rocker or nutator.
13. 2-mL glass vials (1.5-mL Eppendorf vials can also be used).
14. Primary antibody diluted in TBSN + 2% bovine serum albumin (BSA).
15. Fluorescently conjugated secondary antibody diluted in TBSN + 2% BSA.
16. Cavity slides.
17. Murray's Clear Benzyl benzoate/benzyl alcohol (1:2 v/v).

18. Nail polish
19. Confocal microscope

3. Methods

3.1. Thick Section Immunocytochemistry

3.1.1. Embedding and Sectioning Oocytes and Embryos

1. Fix samples accordingly (*see* **Introduction** and **Note 1**) in 4-mL glass vials.
2. Do two changes of 100% ethanol at room temperature for 30 min each.
3. Do two changes of 100% ethanol at 39°C for 30 min each.
4. Change 100% ethanol to a 50% embedding wax [poly(ethylene glycol) distearate with 1% cetyl alcohol]/50% ethanol mix and let stand at 39°C for 90 min.
5. Change to 100% embedding wax at 39°C and let stand for 1 h.
6. Do two more changes of 100% embedding wax at 39°C for 1 h each.
7. Place samples in embedding trays with fresh wax and allow to harden.
8. Once the wax has hardened trim excess wax away with a razor blade and mount specimens, in the desired orientation for sectioning, onto small wooden blocks (*see* **Note 7**).
9. Cut 20–30 µm sections in ribbons with a microtome.
10. Float ribbons in a small puddle of distilled water on gelatin-coated slides (*see* **Note 2**). Place slides on a 29°C hot plate for 10 min. Ribbons will expand in water. Wick away remaining water.
11. Once slides have air dried place them in a desiccator overnight. This will affix sections tightly to the slides so they are not lost during staining.

3.1.2. Staining Thick Sections

1. Dewax sections using an acetone series. Slides are sequentially submerged in 100%, 90%, 70%, and 50% acetone (v/v with distilled water) for 5 min each.
2. Wash sections in 1× PBS + 1% horse serum for 15 min.
3. Block slides for 30 min to an hour in blocking buffer (10% horse or goat serum, 4% BSA in 1× PBS).
4. Incubate sections in primary antibody (diluted to the appropriate concentration in blocking buffer) in a wet box (*see* **Note 8**). The typical volume of antibody used is a puddle of about 50–100 µL per slide depending on how many sections are on the slide. Antibody solutions cannot be allowed to dry on the sections. Incubations can be at room temperature for 1–2 h or at 4°C overnight.
5. Wash sections in 1× PBS + 1% horse serum 3× 15 min each.
6. Incubate sections in fluorescently conjugated secondary antibody (diluted 1:50 in blocking buffer) for 1 h at room temperature, again in a wet box.
7. Wash sections in 1× PBS + 1% horse serum 3× 15 min each.
8. Yolk autofluorescence (*see* **Note 3**) of sections can be blocked somewhat by immersing slides in a 1:100 dilution of 1% eriochrome black (in PBS or water) for 5–10 min. Eriochrome black can be used with fluorescein isothiocyanate (FITC)-conjugated secondary antibodies but should NOT

be used with tetramethyl rhodomine isothiocyanate (TRITC)-conjugated antibodies.

9. Rinse the sections in distilled water and mount in aqueous mounting medium (70–90% glycerol in 1× PBS) containing antifade agents (*see* **Note 4**).

10. View the specimens on a confocal laser scanning microscope. Take Z-series through 10–30 μm of an oocyte or embryo at 1–5 μm intervals. Project the images for a three-dimensional view. The program NIH Image is useful for manipulating confocal images.

3.2. Whole-mount Immunocytochemistry

(Adapted and modified from Gard and Kropf, **ref. 4**)

1. Fix oocytes or embryos accordingly in 4-mL glass vials (*see* **Introduction** and **Note 1**).

2. Bleach samples in 10% hydrogen peroxide in ethanol or methanol for 6–48 h at room temperature if you want to remove pigmentation from oocytes or embryos (*7*) (*see* **Note 5**).

3. Rehydrate the samples with a series of PBS washes at room temperature. Wash for 15 min in 50% PBS/50% ethanol or methanol and then wash 2× 15 min each in 1× PBS.

4. Bisect the oocytes or embryos with a scalpel blade in a Petri dish of 1× PBS.

5. Incubate samples in 100 mM NaBH$_4$ made fresh (in 1× PBS only) for 4 h at room temperature or overnight at 4°C to reduce aldheydes if fixed in aldehyde fixative (*8*) (optional, *see* **Note 6**).

6. Wash samples in TBSN (1× Tris-buffered saline + 0.1% NP-40) 3× 30 min each at room temperature. Place vials horizontally on rocker or nutator for washes.

7. Carefully transfer samples to 2-mL glass vials using a Pasteur pipet and incubate for 16–24 h at 4°C in 150–250 μL of primary antibody diluted appropriately in TBSN + 2% BSA. Place vials vertically on rocker or nutator for gentle motion.

8. Wash samples in 500 μL of TBSN for 24–36 h at 4°C, changing the wash buffer every 8–12 h. Again use gentle agitation.

9. Incubate samples at 4°C for 16–24 h in 150–250 μL of fluorochrome-conjugated secondary antibody diluted (usually 1:50) in TBSN + 2% BSA. Incubate with gentle motion.

10. Wash samples with gentle agitation in 500 μL TBSN for 24–36 h at 4°C, changing wash buffer every 8–12 h.

11. Dehydrate the samples. Remove wash buffer and add 1 mL of 50% methanol (in 1× TBS or PBS) at room temperature for 15 min. Then proceed with three 30-min rinses in 100% methanol at room temperature. Samples can be stored in methanol either at –20°C or at room temperature in a dark box.

12. Mount samples in Murray's Clear (benzyl benzoate/benzyl alcohol)(1:2 v/v) (*see* **Introduction** and **Note 9**). Carefully remove three to five bisected embryos with a Pasteur pipet and place them in a 0.5-mm cavity slide. Manipulate and

maneuver embryos with a fine forceps or hairloop. Dab methanol away and allow the rest to evaporate. Fill cavity with Murray's Clear and let sit briefly as floating embryos will eventually sink. Carefully add a coverslip and then remove excess Murray's Clear by wicking it away with a paper towel. Seal edges of coverslip by applying nail polish. Slides can be stored in the dark at room temperature or 4°C.

13. View the specimens on a confocal laser scanning microscope. Take Z-series through 10–50 μm of an oocyte or embryo at 1–5 μm intervals. Project images for a three-dimensional view. The program NIH Image is useful for manipulating confocal images.

4. Notes

1. As discussed in the introduction it is important to try a range of different fixatives. A number of fixatives have proven successful for immunostaining in *Xenopus* oocytes and embryos. All fixatives should be made fresh for each use. Following is a list of these fixatives:

   a. 100% MeOH
   b. 2% Trichloroacetic acid (in distilled water)
   c. MEMFA: 0.1 M MOPS (pH 7.4), 2 mM EGTA, 1 mM MgSO$_4$, and 3.7% formaldehyde
   d. FG fix: 80 mM K PIPES (pH 6.8), 1 mM MgCl$_2$, 5 mM EGTA, 0.2% Triton X-100, 3.7% formaldehyde, 0.25% glutaraldehyde *(4)*
   e. 4% paraformaldehyde, 0.1% glutaraldehyde, 100 mM KCl, 3 mM MgCl$_2$, 10 mM HEPES, 150 mM sucrose, and 0.1% Triton X-100 (pH 7.6) *(9)*

2. Coated slides for thick sections need to be prepared in advance. Degrease slides by submerging in a carefully made solution of 10% potassium dichromate/concentrated sulfuric acid (1:10) for at least 24 h. Wash slides well in distilled water. Rinse slides in 95% ethanol and allow to completely air dry. Next place slides in section adhesive for 10 min and then remove, drip dry briefly, and place in a desiccator for overnight drying. Prepare section adhesive as follows. Dissolve 7.5 g of gelatin powder in 150 mL of water with heating. Dissolve 0.5 g of chromic potassium sulfate in 200 mL of water. Add 35 mL acetic acid and 145 mL of 100% ethanol to chromic potassium sulfate. Lastly, while stirring add gelatin mix.

3. *Xenopus* oocytes and cells of embryos contain many yolk platelets that autofluoresce under many commonly used excitation wavelengths. Under epifluorescent microscopy yolk autofluorescence is greater in thick sections than in whole-mounts. Under confocal microscopy, however, yolk autofluorescence becomes less apparent in early-stage embryos.

4. Antifade agents such as DABCO *(10)*, *N*-propyl gallate *(11)*, and *p*-phenylenediamine *(10)* can be added to aqueous mounting mediums. Murray's Clear itself seems to be sufficient at keeping samples from fading too rapidly. Most photobleaching occurs during confocal image collection, so care should be taken to avoid excessive laser exposure.

5. If pigmentation affects immunofluorescence, oocytes and embryos can be bleached in a solution of 10% hydrogen peroxide in ethanol for 6–48 h prior to antibody incubation *(7)*. The longer immersion times appear to make embryos more brittle.

6. Using glutaraldehyde as a fixative can often cause autofluorescence. In such cases embryos can be soaked in 100 mM sodium borohydride ($NaBH_4$) in PBS to reduce the unreacted aldehydes *(8)*. Care should be taken as $NaBH_4$ is very reactive with water and creates much effervescence. This treatment however, has not always been effective in our laboratory and is most often left out.

7. Sectioning wax-embedded specimens can be a difficult task if the of the room becomes warm, as specimens tend to melt. To prevent this, once samples are hardened and mounted onto cutting blocks they can be placed at 4°C for approximately an hour and removed just before cutting. Sectioning can be performed with a bucket of dry ice near by or with small chunks of dry ice placed carefully on the edges of the blade.

8. Wet boxes can be created by putting damp towels in the bottom of Tupperware containers. Slides are then placed on some sort of support system. For processing many slides we have converted commercial lucite desiccator boxes into wet boxes simply by placing wet towels in the bottom.

9. Murray's Clear is a skin irritant, so gloves should be worn and caution should be used when handling it. Inhalation should also be minimized. The solution is also detrimental for microscope lenses, so slides must be sealed properly.

References

1. White, J. G., Amos, W. B., and Fordham, M. (1987) An evaluation of confocal versus conventional imaging of biological structures by fluorescence light microscopy. *J. Cell Biol.* **105,** 41–48.

2. Gard, D. L. (1991) Organization, nucleation, and acetylation of microtubules in *Xenopus laevis* oocytes: a study by confocal immunofluorescence microscopy. *Dev. Biol.* **143,** 346–362.

3. Gard, D. L. (1992) Microtubule organization during maturation of *Xenopus* oocytes: Assembly and rotation of the meiotic spindles. *Dev. Biol.* **151,** 516–530.

4. Gard, D. L. and Kropf, D. L. (1993) Confocal immunofluorescence microscopy of microtubules in oocytes, eggs, and embryos of algae and amphibians, in *Methods in Cell Biology*, Vol. 37, Academic Press, San Diego, pp. 147–169.

5. Torpey, N. P., Heasman, J., and Wylie, C. C. (1992) Distinct distribution of vimentin and ctyokeratin in *Xenopus* oocytes and early embryos. *J. Cell Sci.* **101,** 151–160.

6. Robb, D. L., Heasman, J., Raats, J., and Wylie, C. (1996) A kinesin-like protein is required for germ plasm aggregation in *Xenopus. Cell* **87,** 823–831.

7. Dent, J. and Klymkowsky, M. W. (1989) Whole-mount analysis of cytoskeletal reorganization and function during oogenesis and early embryogenesis in *Xenopus*, in *The Cell Biology of Development* (H. Schatten, and G. Schatten, eds.), Academic Press, San Diego, pp. 63–103.

8. Weber, K., Rathke, P. C., and Osborne, M. (1978) Cytoplasmic microtubular images in glutaraldehyde-fixed tissue cells by electron microscopy and immunofluorescence microscopy. *Proc. Natl. Acad. Sci. USA* **75,** 1820–1824.

9. Larabell, C. A., Torres, M., Rowning, B. A., Yost, C., Miller, J. R., Wu, M., Kimelman, D., and Moon, R. T. (1997) Establishment of the dorso-ventral axis in *Xenopus* embryos is presaged by early asymmetries in β-catenin that are modulated by the Wnt signaling pathway. *J. Cell Biol.* **136,** 1123–1136.

10. Johnson, G. D., Davidson, R. S., McNamee, K. C., Russell, G., Goodwin, D., and Holborow, E. J. (1982) Fading of immunofluorescence during microscopy: a study of the phenomenon and its remedy. *J. Immunol. Methods* **55,** 231–242.

11. Giloh, H. and Sedat, J.W. (1982) Fluorescence microscopy: reduced photobleaching of rhodamin and fluorescein protein conjugates by *n*-propyl galate. *Science* **217,** 1252–1255.

12

Analyzing Morphogenetic Cell Behaviors in Vitally Stained Zebrafish Embryos

Mark S. Cooper, Leonard A. D'Amico, and Clarissa A. Henry

1. Introduction

Owing to its extremely rapid rate of development, as well as its optical transparency, the zebrafish embryo provides an excellent experimental system for analyzing the cellular dynamics that underlie vertebrate body axis formation. Moreover, the zebrafish (*Danio rerio*) is very amenable to genetic manipulation. Recent large-scale saturation mutagenesis screens have isolated several thousand strains of *Danio rerio* that possess recessive mutant alleles for genes involved in pattern formation or morphogenesis (*1,2*). These mutant strains of zebrafish represent a wealth of experimental material from which the patterns of cell division, cell intercalation, cell migration, and coordinate cell shape changes that underlie zebrafish morphogenesis can be analyzed. To help determine the genetic and epigenetic mechanisms that choreograph these events, it is often very useful to follow the morphogenetic behaviors of single cells and cell populations within the living zebrafish embryo.

In this chapter, we discuss how the behavior of cellular populations within intact zebrafish embryos can be analyzed using a combination of vital staining techniques and time-lapse confocal microscopy. In particular, we discuss how a neutral fluorescent dye, Bodipy 505/515, and a related sphingolipid derivative, Bodipy-ceramide, can be used respectively as vital stains for yolk-containing cytoplasm and interstitial space throughout the entire zebrafish embryo. These fluorescent probe molecules allow all of the cells within a living zebrafish embryo to be rapidly stained and then visualized *en masse* using a scanning laser confocal microscope. In addition to Bodipy 505/515 and Bodipy-ceramide, we also discuss a unique fluorescent probe molecule, SYTO-

From: *Methods in Molecular Biology, vol. 122: Confocal Microscopy Methods and Protocols*
Edited by: S. Paddock © Humana Press Inc., Totowa, NJ

11, that can be used to identify the location of the organizer region (i.e., the dorsal marginal zone or DMZ) in late-blastula stage blastoderms *(3)*.

Finally, we outline experimental parameters that are useful for making single-level and multilevel confocal time-lapse recordings of vitally stained zebrafish embryos. Using these time-compression approaches, one can detect and analyze genetically encoded sequences of cell behaviors that underlie the formation of the zebrafish germ layers and organ rudiments.

1.1. Contrast Enhancement

Confocal imaging of cell movement and cell shape changes within living tissue is intimately linked to the selective placement or accumulation of fluorescent probe molecules (i.e., contrast-enhancing agents) at specific locations within a cell or tissue. The contrast produced by the localization of a fluorescent contrast-enhancing agent can be defined mathematically as:

$$C = \Delta I / I_0, \quad \text{(Eq. 1)},$$

where DI represents the change in fluorescence intensity at a given region of the specimen with respect to the mean intensity of its background fluorescence, I_0 (i.e., the rest of the cell or tissue). To first approximation, the fluorescence intensity I of a given volume element of cytoplasm is proportional to the concentration of fluorophores located within that specific volume element. Thus, once sufficient numbers of exogenous fluorophores are inserted into specific compartments of a living embryo, these objects of interest stand out from their background and can be easily detected with a confocal microscope. However, to obtain a successful confocal time-lapse recording, the contrast-enhancement mechanism must also be robust, such that multiple images of vitally stained cells or tissues can be acquired over an extended period of time.

Below, we outline an imaging strategy that is based on inserting large numbers of photostable fluorescent probe molecules into specific compartments of the living zebrafish embryo. Once these fluorescent probe molecules have been inserted, the embryo can be repeatedly scanned using moderate laser illumination intensity without compromising the image quality of individual scans during a time-lapse recording.

1.2. General Aspects of Vital Staining

In general, vital staining involves several procedural steps: (1) solubilization of the vital stain in a physiological labeling medium; (2) permeation, intercalation, or absorption of the vital stain into (or onto) the embryo; (3) localization/accumulation of the probe molecule within specific cellular or subcellular compartments; and (4) washout of unbound stain.

Because the "universal solvent" dimethyl sulfoxide (DMSO) has low toxicity on living tissues, it is an excellent choice for solubilizing and applying vital stains to zebrafish embryos. Embryo Rearing Medium (ERM) containing 1–2% DMSO and a variety of vital stains can be applied to zebrafish embryos for up to 1 h without producing toxic or teratogenic effects (3).

Many vital stains become localized in specific cellular compartments of cells and tissues through diffusion-trap mechanisms. As vital stain molecules from the external medium enter a cellular compartment (e.g., the lipid phase of a cell membrane or the lumen of an organelle) by diffusion, physiochemical characteristics of the molecules cause them to become retained (or trapped) within the compartment. For example, the compartment may be hydrophobic (e.g., $diIC_{18}$ partitions into the lipid phase of cell membranes), electronegative (e.g., Rhodamine 123 accumulates inside active mitochondria owing to the large membrane potential across their inner membrane), or possess multiple binding sites for the probe molecule (e.g., Bodipy-ceramide accumulation in *trans*-Golgi elements of vertebrate tissue cells) *(4,5)*.

Cells and tissues of zebrafish embryos have only weak endogenous fluorescence. Exogenous fluorescent probes molecules must therefore be inserted into zebrafish embryos in preparation for high-resolution confocal imaging. We recommend labeling concentrations on the order of 100 μM when inserting bath-applied fluorescent probes into embryonic zebrafish tissues. The high concentration produces a larger diffusive flux of the vital stain into the embryo, and a faster accumulation of the fluorescent probe into the embryo's constituent tissues. If a fluorescent probe is internalized into cells by endocytosis, submicron endosomes will contain a sufficient concentration of fluorophores to be imaged clearly.

In the next section, we describe the staining characteristics of three complementary vital stains/labels that are quickly and easily applied to zebrafish embryos in preparation for confocal imaging.

1.3. Vital Stains and Vital Labels for Zebrafish Embryos

1.3.1. Bodipy 505/515

Bodipy is an abbreviation that refers to a very versatile set of neutral, boron-containing fluorophores that are derived from a diazaindacene backbone. Bodipy fluorophores have been conjugated to a variety of probe molecules including lipids, hormones, neurotransmitters, dextrans, and the actin-binding probe phallacidin *(5)*. The notable charge neutrality of Bodipy fluorophores minimizes their effect on the properties of biomolecules to which they are conjugated *(5)*.

Bodipy fluorophores are frequently referred to by their excitation and emission maxima. Thus, Bodipy 505/515 indicates that fluorophore is most strongly excited with visible radiation centered at 505 nm (blue light), and emits a spectrum of longer wavelength light that peaks at 515 nm. Bodipy 505/515 excitation maximum lies close to the 488 nm line of an argon/krypton laser. In addition, Bodipy 505/515 possesses a high quantum yield of nearly 0.9, and a relatively low photobleaching rate *(5)*. These fluorescence characteristics make the Bodipy 505/515 fluorophore and its conjugate molecules ideally suited for scanning laser confocal microscopy.

We have found that unconjugated Bodipy 505/515 (4,4-difluoro-1,3,5,7-tetramethyl-4-bora-3a,4a-diaza-s-indacene; mol wt 248) is an excellent vital stain for yolky cytoplasm in zebrafish embryos *(3)*. Bodipy 505/515 has a high oil–water partition coefficient, which allows it to rapidly cross cell membranes and accumulate within lipidic yolk platelets. These lipidic yolk platelets are distributed throughout the cytoplasm of the blastoderm and nearly completely fill the volume of the zebrafish's yolk cell. In contrast, Bodipy 505/515 does not stain nucleoplasm, nor does it remain within interstitial space. These staining characteristics of Bodipy 505/515 allow individual cell boundaries and cell nuclei to be imaged clearly in time-lapse recordings. Thus, karyokinesis, cytokinesis, and cell rearrangement can be followed in great detail throughout gastrulation, neurulation and organ rudiment formation (**Figs. 1**, **2**, and **3**).

1.3.2. Bodipy-Ceramide

Bodipy-ceramide {Bodipy FL C_5-Cer/C_5-DMB-Cer [N-(4,4-difluoro-5,7-dimethyl-4-bora-3a,4a-diaza-s-indacene-3-pentanoyl) sphingosine]; mol wt 631} is a fluorescent sphingolipid that has been used for many years as a vital stain for the Golgi apparatus in cultured vertebrate tissue cells *(4,6,7)*. When applied to intact zebrafish embryo, Bodipy-ceramide stains the plasma membrane, Golgi apparatus and cytoplasmic particles within the superficial envel-

---

Fig. 1. *(facing page)* Cytoplasmic and nuclear dynamics during early cleavage in a zebrafish embryo, vitally stained with Bodipy 505/515. Shown is a confocal time-series through an animal–vegetal plane of an early cleavage (four-cell) stage zebrafish embryo. Bodipy 505/515 preferentially binds to yolk platelets and yolky cytoplasm, leaving nucleoplasm and interstitial space devoid of the fluorophore. The animal pole is toward the top of the image. Scale bar = 100 μm. (**A**) An interphase cell nucleus (**arrow**) of a specific blastomere is visible in this optical section. Large yolk platelets located in the cytoplasm of the cells are brightly labeled (e.g., **arrowhead**). (**B**) Nuclear membranes break down as blastomeres synchronously enter mitosis. An anaphase spindle is visible (**arrow**). At the onset of prophase, a mass of yolk platelets fragment from the surface of the yolk cell and pass via cytoplasmic bridges into the base of uncellularized blastomeres. Yolk platelets located in the cytoplasm of the cells fragment at the same time (e.g., **arrowhead**). (**C**) Nuclear membranes reform (e.g., **arrow**) as the dividing blastomeres enter telophase. The previously fragmented yolk

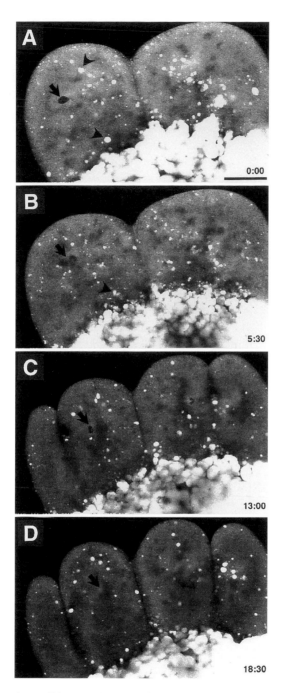

platelets begin to refuse. **(D)** The cleavage furrow between the daughter cells is now complete. Nuclear membranes of the blastomeres have broken down in preparation for another round of cell division.

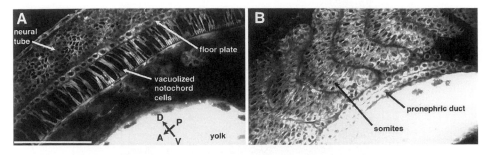

Fig. 2. Cellular detail of organ rudiments in the trunk region of a 24-h zebrafish embryo vitally stained with Bodipy 505/515. (**A**) Sagital sections through the median plane and (**B**) a lateral plane, respectively. Dorsal–ventral (D–V) and anterior–posterior (A–P) axes are shown. The embryo was imaged with a 40x/1.3 NA oil objective. Scale bar = 100 μm.

oping layer (EVL) cells of the embryos. However, once the fluorescent lipid percolates through the enveloping layer epithelium, the fluorescent lipid remains localized within the interstitial fluid of the embryo (hereafter referred to as the interstitum) and freely diffuses between cells (**Fig. 4**).

The insertion of large numbers of fluorophores into the interstitum allow hundreds of cells to be imaged simultaneously, as well as large-scale tissue movements to be observed. Because Bodipy-ceramide is able to diffuse through the interstitum, photobleached molecules are quickly exchanged with unbleached molecules, thus replenishing fluorescence in the scanned field of view. The large reservoir of mobile fluorescent lipid present in the embryo's segmentation cavity (part of the interstitum) allows single-level or multilevel time-lapse recordings to be acquired over extended periods of time (up to 10 h).

Interestingly, there is very little partitioning of Bodipy-ceramide into deep cells. In particular, the Golgi apparatus in deep cells remains essentially devoid of the lipid probe. To explain the lack of permeation of Bodipy-ceramide inside the embryo, we hypothesize that there may be high density of lipid-carrier proteins within the interstitum that can bind endogenous lipoproteins, as well as exogenously inserted fluorescent probe lipids. Binding of Bodipy-ceramide to mobile, impermeant lipid-carrier proteins could thus account for the apparent

---

Fig. 3. *(facing page)* Optical sections of the vegetal pole of a 100% epiboly stage zebrafish embryo showing the position of the forerunner cells in relation to the germ ring and developing body axis. The embryo was double-stained with SYTO-11 and Bodipy 505/515 at 50% epiboly. Scale bar = 100 μm. (**A**) Surface view showing Bodipy-labeled EVL cells. (**B**) A deeper plane of focus (15 μm) into the embryo reveals deep cells and the yolk syncytial layer (YSL). (**C**) 45 μm deeper into the embryo.

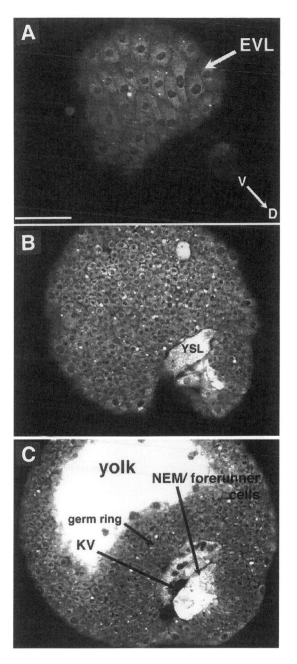

Forerunner cells, brightly labeled with SYTO-11, as well as the germ ring are prominent features. Kupffer's vesicle (KV) is forming between the YSL and the NEM/forerunner cells. The NEM/forerunner cell cluster is located at the posterior limit of the embryo's dorsal midline.

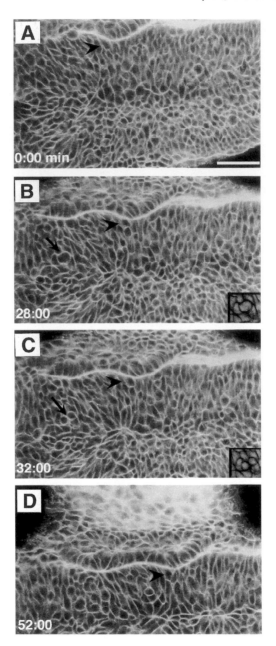

Fig. 4. Patterns of cellular convergence and infolding in the neuroepithelium of the hindbrain region of a five-somite stage (11.6-h) embryo. The embryos has been vitally stained with Bodipy-ceramide. Bodipy-ceramide molecules remain localized to the interstitial space of the embryo, outlining the boundaries of deep cells. Anterior is to the left. Scale bar = 50 μm. (**A**) Lateral folds of the neural plate converge toward the

change in permeation characteristics of Bodipy-ceramide molecules after they enter the embryo's interstitum.

1.3.3. SYTO-11

During the late-blastula stage, a unique group of cells with increased endocytic activity appears at the presumptive dorsal margin (organizer region) of the zebrafish blastoderm *(3)*. The longitudinal position of this cellular domain in late blastula stage embryos (30–40% epiboly) accurately predicts the site of embryonic shield formation at the onset of gastrulation (50% epiboly) (**Fig. 5**). Unlike other blastomeres around the circumference of the blastoderm, these marginal cells do not involute during germ ring formation or blastoderm epiboly. Instead, this group of noninvoluting endocytic marginal (NEM) cells remains at the border where the dorsal EVL and dorsal yolk syncytial layer (YSL) come into contact (**Fig. 4**). During mid- to late epiboly, deep cells within the NEM cell cluster move to the leading edge of the dorsal blastoderm. At this point, these cells are referred to as "forerunner cells" *(3,8)*.

NEM/forerunner cells can be easily visualized in late blastula to late gastrula stage embryos because they exhibit accelerated endocytic activity and can be selectively labeled by applying membrane-impermeant fluorescent probes, such as SYTO-11 (methane, sulfinylbis; ~400 mol wt), to pre-epiboly and early epiboly embryos *(3)*. Because this particular vital labeling method can identify the location of the organizer region in late blastula stage embryo, it is potentially useful for experimental studies of dorsal cell fate specification and embryonic axis formation.

[Because SYTO-11 is internalized into zebrafish embryos by endocytosis *(3)*, as opposed to passive permeation, it is more appropriate to refer to SYTO-11 as a vital label, rather than a vital stain].

2. Materials

1. Bodipy 505/515, Bodipy-C$_5$-ceramide, or SYTO-11 (Molecular Probes, Eugene, OR).
2. Anhydrous DMSO.

---

neural midline. An arrowhead points to a cell at the posterior margin of the fifth rhombomere. The cell remains at this marginal location during convergence and infolding (A–D). Rounded cells (e.g., arrow) within the neural plate are preparing to divide. (**B,C**) A rounded cell in the neural plate undergoes mitosis (**arrow**). The division plane is at 45° with respect to the embryonic axis. (**D**) The right otic vesicle of the embryo is centrally positioned at the lateral margin of the fourth and sixth rhombomeres of the developing hindbrain. The cell at the lateral margin of the neural fold (marked by the **arrowhead**) has converged further toward the neural midline. Note that the neural midline has moved toward the lower portion of the field of view. This is due to a torsional rotation of the embryo.

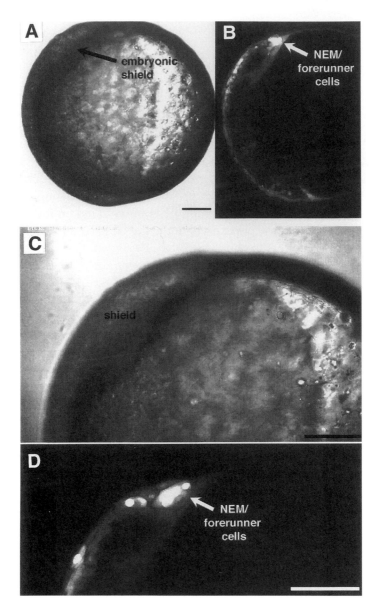

Fig. 5. The position of NEM/forerunner cells (labeled with SYTO-11) in early gastrula stage embryos correlates with the site of embryonic shield formation. Scale bars
= 100 μm. **(A,B)** A Nomarski and confocal image pair of an early gastrula stage
zebrafish embryo showing that the brightly labeled NEM/forerunner cell cluster is
located at the leading edge of the dorsal marginal zone (DMZ). **(C,D)** Higher magnification of the DMZ and the NEM/forerunner cell cluster.

3. 35-mm Plastic culture dishes and/or 1.5-mL microfuge tubes.
4. Drierite.
5. Embryo Rearing Medium (ERM), buffered with HEPES.
6. Fire-polished Pasteur pipets (with bulbs) for transferring embryos.
7. 10- or 20-mL micropipets (VWR Scientific) for molding agarose.
8. Hairloop mounted on a Pasteur pipet.
9. Agarose.
10. Dechorinated zebrafish embryos [see *The Zebrafish Book* (Westerfield, M., ed.)], for manual and enzymatic methods. These methods can also be accessed on the WWW at the Fish Net Website: http://zfish.uoregon.edu/zf_info/zfbook/zfbk.html).
11. Embryonic Rearing Medium for Applying Vital Stains *(9)*: 1.0 mL of Hank's Stock no. 1, 0.1 mL of Hank's Stock no. 2, 1.0 mL of Hank's Stock no. 4, 95.9 mL of double-distilled (dd) H_2O, 1.0 mL of Hank's Stock no. 5, 1.0 mL of fresh Hank's Stock no. 6. Add 0.24 g of HEPES [N-(2-hydroxyethyl)piperazine-N'-(2-ethanesulfonic acid)] to obtain a pH buffer capacity of 10 mM in the final solution (modification of original recipe).
    Use 1M NaOH to pH 7.2.
    Hank's Stock no. 1: 8.0 g of NaCl, 0.4 g of KCl, in 100 mL of dd H_2O.
    Hank's Stock no. 2: 0.358 g of Na_2HPO_4 anhydrous, 0.60 g of $K_2H_2PO_4$, in 100 mL of dd H_2O.
    Stock no. 4: 0.72 g of $CaCl_2$ in 50 mL of dd H_2O.
    Stock no. 5: 1.23 g of $MgSO_4 \cdot 7H_2O$, in 50 mL dd H_2O.
    Stock no. 6: 0.35 g of $NaHCO_3$, 10.0 mL of dd H_2O.

3. Methods

3.1. Vital Staining

As a general procedure, zebrafish embryos are placed in a vital staining solution for 30 min, then washed 3x with HEPES-buffered ERM, in order to remove excess or unbound fluorescent probe molecules. The intensity of staining can be varied by changing the duration of staining. Detailed steps for this procedure are given below for vitally staining embryos with Bodipy 505/515. (This same procedure, with slight modifications, is used for vital staining with Bodipy-ceramide or SYTO-11; *see* **Note 5**).

1. A stock solution of the unconjugated fluorophore Bodipy 505/515 (4,4-difluoro-1,3,5,7-tetramethyl-4-bora-3*a*,4*a*-diaza-s-indacene) was made by dissolving the fluorescent dye in anhydrous DMSO to a stock solution concentration of 5 mM (*see* **Note 1**). Aliquots (20-µL) of the stock solution are placed into microfuge tubes and stored in a light-tight container with Drierite at –20°C. Bodipy 505/515 aliquots are usually stable for up to 6 mo.
2. Preparation of chambers for vital staining. All labeling chambers should be covered with agarose to prevent the dechorionated embryos from sticking to plastic

surfaces. Molded agarose chambers can also be created to minimize the volume of staining solution needed to vitally stain a group of embryos. For very small volumes (50–100 µL) of labeling solution, a labeling chamber can be constructed from the detached cap of a 1.5-mL microfuge tube. Molten agarose can be dropped into the cap using a Pasteur pipet until the agarose completely covers the bottom of the cap.

3. Transfer Pasteur pipets for moving embryos between solutions are made by removing the tip, as well as most of the shank, of the glass pipet (a diamond scribe is useful for this procedure). The remaining 3–5 mm diameter opening is fire-polished using an alcohol flame. When moving embryos between solutions, avoid having the embryos contact the air–water interface at the mouth of the transfer pipette. This is done by tilting the transfer Pasteur pipet to a more horizontal position once embryos have been drawn into the pipet.

4. A hairloop Pasteur pipet for moving and positioning embryos is made by first heating the middle shank of a Pasteur pipet in an alcohol flame. As the glass softens, the pipet is drawn out of the flame. Once the shank is out of the flame, the two ends of the pipet are then quickly pulled in opposite directions, thus stretching and narrowing the middle shank. The middle shank is then broken off. The two ends of a human hair (5–7 cm long) are placed into the tapered pipet and pushed inward until a 0.5 cm diameter loop of hair remains outside. A dab of molten paraffin applied to the pipet tip will secure the hairloop in place.

5. Thaw an aliquot of Bodipy 505/515. The dye is then diluted into HEPES-buffered ERM to a labeling concentration of 100 µ*M*. The final concentration (v/v) of DMSO in the labeling solution is 2%.

6. Embryos are transferred to a labeling chamber and stained with a 100 µ*M* Bodipy 505/515 staining solution for 30 min (*see* **Note 2**). During vital staining, the solution in the labeling chamber should be circulated by gentle swirling every 10 min (*see* **Note 3**). The yolk cell of each embryo will become visibly green if the vital staining is working (*see* **Note 4**). As a precaution during the staining procedure, we recommend that the staining chamber be covered with aluminum foil, to keep vitally stained embryos from being exposed to excess light. However, we have found that exposure to room light for up to 30 min does not harm or overly photobleach vitally stained embryos (*see* **Note 5**).

7. After vital-staining, the embryos are passed through three successive washes of HEPES-buffered ERM, using a fire-polished transfer Pasteur pipet. These washes are performed in separate agarose-coated 35-mm culture dishes. After washing, the embryos are ready to be mounted for observation under the confocal microscope (*see* **Notes 6** and **7**).

3.2. Mounting Embryos for Imaging

3.2.1. Imaging Chambers

Confocal time-lapse recordings of zebrafish embryos require that the vibration of the specimen be minimized. Live embryos are most easily imaged using an inverted microscope mounted on an airtable, because gravity pulls the embryo toward the coverslip and helps maintain its position with respect to the objective. A useful chamber to image teleost embryos is a 0.7 cm thick piece of plexiglass with a 24 mm diameter hole in its center. A 30 mm diameter circular glass coverslip (no. 1 thickness) can be secured with high-vacuum silicone grease (Dow Corning, Midland, MI) to serve as the bottom of the bath well. During time-lapse recordings, a plastic culture dish lid can be placed over the well to prevent air currents from producing displacements of the culture medium and unsecured embryos.

3.2.2. Stabilizing the Spatial Orientation of Embryos

To prevent rolling from occurring during extended time-lapse recordings, we have found it useful to place fluorescently labeled embryos in an agarose holding well made from ERM plus agarose (Type IX) (Sigma, St. Louis, MO) (modified from the procedure in Westerfield, 1995) (*see* **Notes 8** and **9**).

1. Apply a thin coat of molten solution of 1.2% agarose in ERM to the coverslip. Let the agarose cool and harden to form a thin layer of agarose gel on the coverslip.
2. To make holding wells in the agarose layer, start with a 10- or 20-μL glass micropipet in which the end has been pulled out using an alcohol flame. After breaking off the tip of the pipet, heat the remaining end until it melts into a small bead of glass, approximately 0.8 mm in diameter. Use the rest of the pipette as a handle. After heating this glass bead/ball with the alcohol flame, quickly plunge it into the agarose to melt a hole. The holes should be small enough so that the embryo will not be able to roll, but large enough so that epiboly of the blastoderm is not impeded. Because an embryo will be secured inside this hemispherical-shaped hole, the hole should be as round and smooth as possible.
3. Use room temperature ERM or water to wash out any melted agarose so that it does not refill the holes.
4. Repeat this procedure to create multiple holes in the agarose layer. You may have to make a new glass bead, as repeated heating will cause the bead to increase in size as more glass melts.
5. After the agarose wells are made, add ERM buffer to cover the agarose hole and then add the embryos. Gently position the embryos into the agarose holes using a nonsticky implement, such as a hairloop mounted on the end of a Pasteur pipet with a dab of molten paraffin.
6. Carefully position the embryo so that the area of interest is facing toward the objective lens. This operation is most easily accomplished on an inverted microscope with an open holding chamber. However, repositioning the embryos once they are in the wells is difficult and must be done with care. It is useful to secure 5–10 embryos in a given agarose sheet in preparation for a time-lapse recording. This increases the likelihood that a well-stained embryo, in an appropriate orientation, can be located for imaging (*see* **Note 8**).

3.3. Imaging Procedures

3.3.1. Selection of Optics

The choice of objective for imaging zebrafish embryos is determined by several considerations: (1) numerical aperture (NA); (2) working distance; and (3) magnification. The NA must be large enough to adequately collect fluorescent light from the specimen. In addition, the working distance of the objective lens must be great enough to reach the desired plane of focus within the embryo. To examine cell behaviors in zebrafish embryos, we have found a versatile set of objective lenses to be: (1) a dry 20x/0.75 NA; (2) a dry 40x/0.85 NA; (3) an oil 40x/1.3 NA. A 10x/0.5 NA objective is also useful for locating the organizer region (dorsal marginal zone) of zebrafish embryos labeled with SYTO-11.

Appropriate filter sets for the confocal microscope are determined from the excitation and emission spectra of the fluorophores. In **Figs. 1–5**, embryos stained either with Bodipy 505/515, Bodipy-ceramide or SYTO-11 were imaged using an excitation wavelength of 488 nm. Fluorescent light collected from the specimen was filtered by a 515 nm long-pass filter before it was transmitted to the confocal microscope's photomultiplier tube.

3.3.2. Illumination Intensity, Optimization of Gain, and Offset

To generate large numbers of photons for high-quality confocal images with single scans, it is necessary to illuminate the embryo with moderately intense laser light. To obtain optimum contrast for time-lapse recordings, system gain should be increased until saturation begins to occur. At this point, offset is added to bring these pixels below a value of pure white (i.e., 255 on a 0-255 grayscale). A "pleasing image" generally contains a spectrum of gray values that span the entire grayscale. With offset (or black level) adjustments, the background should set at the equivalent of a slightly negative grayscale value. With this offset adjustment, the background will remain pure black throughout the recording. When imaging a zebrafish embryo with Bodipy 505/515, the yolk cell will often be extremely bright (*see* **Note 10**). To achieve optimum contrast of blastomeres, the highly fluorescent yolk platelets in the yolk cell will appear pure white (i.e., fully-saturated on the video monitor).

3.3.3. Parameters for Time-Lapse Recordings

Time-lapse recordings of developing zebrafish embryos are best made from single slow scans of the specimen, as opposed to time-averaged images (e.g., Kalman averages). Because the scanning laser beam of the confocal microscope passes rapidly over the sample, a stroboscopic illumination of the specimen is produced. This greatly reduces the motional blurring of fluorescent objects that are being displaced within cells by active transport or by diffusion

(11). At a slow scan rate, enough photons are collected in a single pass over the specimen to generate adequate signal-to-noise level in the image. By adjusting neutral density filters, it is useful to use as intense laser light as possible without producing substantial photobleaching in the specimen over a desired time-lapse interval. For an embryo labeled with Bodipy 505/515 or Bodipy ceramide, 100–300 frames can be generally recorded with a 300–700-μm field of view over a 1–10-h time period. A minimum of 50 frames is usually needed to determine the cellular dynamics that comprise a morphogenetic event (*see* **Note 11**).

3.3.4. Single-Level Time-Lapse Recordings

Most confocal microscopes now have internal macros for obtaining and transferring single-level and multilevel time-lapse recordings to their host computer's hard disk. Although a computer hard disk is a fast and convenient means for immediate data storage, an excellent high-capacity, random-access storage medium for time-lapse recordings is an optical memory disk recorder (OMDR).

A QuickBasic program and a Bio-Rad macro necessary to activate a Panasonic OMDR in time-lapse form have been published *(12)*. This program, called "OMDR," is compiled into an executable (.exe) file using Microsoft QuickBasic.

```
OMDR
open "com2:2400,n,2,1" for random as #1
com(2) on
print #1, chr$(2) + command$ + chr$(3)
close
```

This function can then be adapted into a macro for the collection of Z-series to an OMDR.

The following macro or "command file" (.cmd) for the Bio-Rad confocal microscope can be used to make a time-lapse recording with a Panasonic OMDR via a serial port connection.

```
TLAPSE.CMD
for i = 1, %2
wait %1
clear
collect 1
$omdr gs
next i
```

Before the above program initiates scanning and recording, the program queries the user for two external inputs, %1 and %2. %1 is the number of seconds between the collection of images, whereas %2 is the number of images to be collected. The "gs" command (after $omdr) signals the Panasonic OMDR to record a single video frame using the OMDR.exe file.

3.3.5. Multilevel Time-Lapse Recordings

If desired, algorithms for multilevel time-lapse recordings can also be implemented using a stepping-motor coupled to the focusing apparatus of the confocal microscope. The following program has been modified slightly from that published previously *(12)*.

```
MLTLAPS.CMD
echo -
$cls
$omdr on
box size 1 hor
clear
print MACRO FOR A MULTILEVEL TIME-LAPSE RECORDING
print
input reps, NUMBER OF MULTILEVEL SAMPLES (Z-SERIES)
input sects, NUMBER OF SECTIONS PER Z-SERIES
input lapse, TIME INTERVAL BETWEEN MULTILEVEL SAMPLES (Z-SERIES)
input vstep, VERTICAL STEP SIZE, (number of motor steps)
input rewind, MOTOR REWIND AMOUNT, (number of motor steps)
print
motor off on
for k = 1, #reps
motor inc #vstep
for i = 1, #sects
clear
print COLLECTING IMAGE #i OF MULTILEVEL SAMPLE #k
collect
$omdr gs
motor step
next i
wait 0
wait #lapse
motor inc #rewind
motor step
next k
motor off
echo +
```

3.3.6. Analysis of Time-Lapse Recordings

Many optical memory disk recorders (OMDRs) allow a 100-fold range of speeds over which the time-lapse recording can be sped up or slowed down.

Because the visual perception of human observers is limited in bandwidth, this flexibility in OMDR playback rates greatly facilitates data-to-brain coupling. To trace the trajectory of individual cells, or to digitize individual OMDR images (analog video), the video output of the OMDR can be routed to the input of a video frame buffer. A variety of software programs can then be used for either digitizing analog OMDR images or performing morphometric/kinematic analyses of cells in the time-lapse recording.

Saving confocal time-lapse images as digital files is the best procedure to preserve their spatial resolution. However, one must load these digital image files into a suitable program for viewing on a computer. NIH Image is a useful program, as it will import Bio-Rad .PIC files using a specific macro (NIH Image and Bio-Rad macros can be downloaded from the following URL: http://rsb.info.nih.gov/nih-image/more-docs/docs.html). One can load a Z-series and scan through the data set using arrow keys. If desired the scan-through can be animated and saved.

Finally, acquired confocal time-lapse recordings can be compressed into digital movie files in either a QuickTime or a MPEG format for distribution over the Internet. Examples of such compressed time-lapse recordings from our laboratory, showing zebrafish embryos stained with Bodipy 505/515 and Bodipy-ceramide, are located on the WWW at the following URL: http://weber.u.washington.edu/~fishscop/.

4. Notes

1. Bodipy 505/515 will appear yellow-green when solvated in DMSO. Once the DMSO solution is dispersed into aqueous ERM, the color of Bodipy 505/515 often changes to orange-red. This solution is appropriate for vital staining, unless large particles of Bodipy 505/515 start to precipitate out of solution. If this occurs, one should remake a new Bodipy 505/515 stock solution using anhydrous DMSO. Because DMSO is hydroscopic, it is possible that Bodipy 505/515 aliquots may accumulate H_2O in the $-20°C$ freezer over time, causing the DMSO in the Bodipy 505/515 aliquot to lose its solvating activity.

2. The yolk cell of zebrafish embryos becomes visibly green within several minutes after the embryo is placed in an aqueous labeling solution containing 100 μM Bodipy 505/515 and 2% DMSO. The yolk cell continues to accumulate dye as the embryo remains in the labeling solution, and becomes intensely green after 30 min of staining. To the naked eye, the blastoderm of the Bodipy-labeled embryo appears colorless. However, when these embryos are examined under the scanning laser confocal microscope, the embryo's blastomeres are well labeled with Bodipy 505/515 (**Fig. 1**). In young embryos (< 24 h), Bodipy 505/515 is retained within the yolk cell and the yolk-containing cytoplasm of cells.

3. In primordia-stage embryos, as in younger embryos, Bodipy 505/515 does not stain the nucleoplasm of cells. The dye is also absent from fluid-filled organelles,

such as vacuoles within the extending notochord. Initially, Bodipy 505/515 is absent from both cerebral-spinal fluid and the bloodstream. However, in later stage embryos (> 24 h), Bodipy 505/515 becomes highly concentrated in the blood and cerebrospinal fluid (data not shown). We speculate that this change in staining pattern results from Bodipy 505/515 remaining bound to lipoproteins as they are exported from the yolk cell into the bodily fluids of the embryo.

4. Another small, neutral fluorophore, Bodipy 564/591, represents an alternative fluorescent vital stain for yolky cytoplasm in zebrafish embryos *(3)*. Because Bodipy 564/591 can be excited with the yellow 568 nm line of the argon/krypton laser, this fluorophore can be used for dual-labeling purposes with other vital stains, such as SYTO-11 (*see* **Fig. 3** and **Note 4.6**). Bodipy 564/591 can be obtained from Molecular Probes (Eugene, OR) on a custom synthesis basis. Embryos can be labeled with Bodipy 564/591 using the same procedure as Bodipy 505/515.

5. Embryos can be vitally stained with either Bodipy-ceramide or SYTO-11 using the same procedure used for staining with Bodipy 505/515. Recommended labeling concentrations (and labeling times) for Bodipy-ceramide and SYTO-11 are 100 μM (30 min), and 75 μM (15 min), respectively.

6. Costaining with Bodipy fluorophores. Using a sequential application of the standard vital staining procedure (*see* **Fig. 3**), embryos can be first vitally stained with Bodipy 505/515, and then vitally-labeled with SYTO-11. Dual-staining with these two fluorescent probes allows NEM/forerunner cells to be observed along with all other blastomeres. Cell nuclei and endosomes labeled with SYTO-11, or cellular cytoplasm labeled with Bodipy 564/591, can be viewed independently using 488 nm and 568 nm excitation wavelengths, respectively. When desired, both fluorophores can be viewed simultaneously using dual-wavelength excitation (488 nm and 567 nm) and 585 nm long-pass emission.

7. Bodipy 505/515, Bodipy 564/591, Bodipy-ceramide, and SYTO-11 do not appear to have any teratogenic effects on developing zebrafish embryos. Normal somitogenesis and neurulation proceed on schedule. Hatched fry do not exhibit any noticeable morphological malformations or behavioral abnormalities.

8. pH stabilization: One of the most important experimental variables to control in making time-lapse recordings is the pH of the embryonic or tissue culture medium. Bicarbonate-containing media, in particular, are subject to extreme changes in pH with temperature, as carbon dioxide exchange with the atmosphere alters the carbonic acid/bicarbonate equilibrium of the medium. CO_2, pH, and HCO_3^- are interrelated through the following equilibrium:

$$CO_2 + H_2O = H_2CO_3 = H^+ + HCO_3^- \tag{Eq. 2}$$

Atmospheric CO_2 tension will alter the concentration of dissolved CO_2 in a temperature-dependent fashion *(10)*. Increased HCO_3^- concentration pushes the reaction to the left. Equilibrium will be reached only in conditions where atmospheric CO_2 is stable at a given temperature. In the absence of CO_2 and temperature control, it is necessary to add a buffering agent at twice the concentration of HCO_3^- to achieve pH stabilization of the experimental solution. Therefore, 10–

20 m*M* HEPES should be added to all experimental salines or media to stabilize their pH during extended time-lapse recordings.

9. Tricane (MS-222) Anesthesia: To prevent muscle twitching during time-lapse recordings, embryos at 20-somite stage or later can be anesthetized with 0.1 mg/mL of Tricaine (also known as MS-222; Sigma, St. Louis, MO) dissolved in ERM *(9)*.

10. Light absorption in vitally stained embryos: Owing to absorption by superficial layers of fluorescently labeled cells, the excitation light is attenuated as it passes deeper into the embryo. This results in a progressive loss of fluorescent light emanating from deeper optical sections of embryos labeled with Bodipy 505/515. Darkness occurs because of light absorption by labeled cells and/or organ rudiments in the path of the scanning laser beam. However, because many of the cellular movements that produce body axis formation take place on the surface of the yolk cell, embryos can be usually rotated into an appropriate orientation for viewing an area of interest.

11. Compensations for photobleaching. The rate of photobleaching that occurs during a time-lapse recording is inherently linked to the choice of fluorophore, as well as laser illumination intensity. Even when using a fairly photostable fluorophore, such as Bodipy 505/515, it is inevitable that some photobleaching will take place over the duration of the recording. One can add extra gain at the beginning of a recording, anticipating that a certain rate of photobleaching will occur. Alternatively, one can gradually increase the imaging system's gain by hand during the recording. This helps keep the average black level of the image constant during the recording. The imaging tradeoff produced by this mode of compensation is that a gradual increase in pixel noise will occur during the recording, as the gain of the imaging system is steadily increased. It is sometimes useful to employ this mode of compensation when recording cellular dynamics with Bodipy 505/515. This mode of compensation for photobleaching is not usually necessary when using Bodipy-ceramide, as Bodipy-ceramide is able to diffuse through interstitial fluid and replenish fluorescent molecules in scanned regions where they have been bleached.

Acknowledgments

M. S. C. is indebted to R. E. Keller, M. Schliwa, J. P. Miller, S. E. Fraser, M. V. Danilchik, and S. J. Smith for sharing their theoretical and practical insights, over the years, on how to image the dynamics of living cells and embryos. This work was supported by a NSF Presidential Young Investigator Award IBN-9157132, and a University of Washington Royalty Research Fund Grant 65-9926. M. S. C. gratefully acknowledges equipment and software donations from the Bio-Rad Corporation, Meridian Instruments, and the Universal Imaging Corporation through the NSF PYI program. L. A. D. and C. A. H. were supported by a NIH Molecular and Cellular Biology Training Grant PHS NRSA P32 6M07270 from NIGMS. L. A. D. was also supported through a NIH Developmental Biology Training Grant 5T32HD07183-18.

References

1. Driever, W., Solnica-Krezel, L., Schier, A. F., Neuhauss, S. C. F., Malicki, J., Stemple, D. L., Stainer, D. Y. R., Zwartkruis, F., Abdelihah, S., Rangini, Z., Belak, J., and Boggs, C. (1996) A genetic screen for mutations affecting embryogenesis in zebrafish. *Development* **123,** 37–46.
2. Hafter, P., Granato, M., Brand, M., Mullins, M. C., Hammerschmidt, M., Kane, D. A., Odenthal, J., van Eeden, F. J. M., Jiang, Y.-J., Heisenberg, C.-P., Kelsh, R. N., Furutani-Seiki, M., Vogelsang, E., Beuchle, D., Schach, U., Fabian, C., and Nüsslein-Volhard, C. (1996) The identification of genes with unique and essential functions in the development of the zebrafish, *Danio rerio. Development* **123,** 1–36.
3. Cooper, M. S. and D'Amico, L. A. (1996) A cluster of noninvoluting endocytic cells at the margin of the zebrafish blastoderm marks the site of embryonic shield formation. *Dev. Biol.* **180,** 184–198.
4. Pagano, R. E., Sepanski, M. A., and Martin, O. C. (1989) Molecular trapping of a fluorescent ceramide analogue at the Golgi apparatus of fixed cells: interactions with endogenous lipids provides a *trans*-Golgi marker for both light and electron microscopy. *J. Cell Biol.* **109,** 2067–2080.
5. Haugland R. (1996) *Handbook of Fluorescent Probes and Research Chemicals*, Molecular Probes, Eugene OR.
6. Lipsky, N. G. and Pagano, R. E. (1985a) A vital stain for the Golgi apparatus. *Science* **228,** 745–747.
7. Lipsky, N. G. and Pagano, R. E. (1985b) Intracellular translocation of fluorescent sphingolipids in cultured fibroblasts: endogeneously synthesized sphingomyelin and glucocerebroside analogues pass through the Golgi *en route* to the plasma membrane. *J. Cell Biol.* **100,** 27–34.
8. Melby, A. E., Warga, R. M., and Kimmel, C. B. (1996) Specification of cell fates at the dorsal margin of the zebrafish gastrula. *Development* **122,** 2225–2237.
9. Westerfield M. (1995) *The Zebra Fish Book*, 3rd ed., University of Oregon Press, Eugene, OR
10. Freshney, R. I. (1987) *Culture of Animal Cells*, 2nd ed., Wiley-Liss, New York.
11. Cooper, M. S., Cornell-Bell, A. H., Chernjavsky, A., Dani, J. W., and Smith, S. J. (1990) Tubulovesicular processes emerge from *trans*-Golgi cisternae, extend along microtubules, and interlink adjacent *trans*-Golgi elements into a reticulum. *Cell* **61,** 135–145.
12. Terasaki, M. and Jaffe, L. A. (1993) Imaging endoplasmic reticulum in living sea urchin eggs, in *Cell Biological Applications of Confocal Microscopy*, Vol. 38 (Matsumoto, B., ed.), Academic Press, San Diego, pp. 211–220.

13

Confocal Imaging of Living Cells in Intact Embryos

Paul M. Kulesa and Scott E. Fraser

1. Introduction

How individual cells and tissues organize and coordinate to form the structure of a developing embryo is a major question in developmental biology. Equally intriguing is how cells in adult tissue respond during wound healing and tissue regeneration. In these morphogenetic processes there can be a wide variety of cell behaviors: individual cells may change shape or divide, migrate to different areas, or differentiate to form a pattern. Entire tissue layers can spread across a surface, fold up or cavitate to form a three-dimensional structure. For example, during formation of the neural tube in chick embryos, the neural epithelium folds up from a flat sheet of cells to form a tube with a hollow lumen. As the neural tube becomes shaped into various regions of the nervous system, the neural crest cells emerge from the dorsal midline of the neural tube and migrate to pattern the peripheral nervous system. A main theme in these examples is that the cell and tissue events are dynamic, on a time scale from minutes to days, covering spatial regions ranging in size from microns to millimeters.

It remains largely unknown how individual cells "know" where to go, when to move, and when to differentiate, or how tissue shape changes are coordinated. Some hints of the underlying processes have been obtained with light microscopy; however, this has been primarily limited to imaging thinly sectioned tissue or dissociated cells in culture. In these systems, it is difficult to know whether cells removed from their natural environment behave the same as they would in the intact embryo. The more relevant scenario is to image the living intact embryo, allowing the full interplay of cell movements, tissue interactions, forces, and signaling events. In those cases where the whole embryo cannot be studied, thick slice cultures offer a means to follow cells in a more natural setting. Both the intact embryo and thick slice cultures offer the chal-

From: *Methods in Molecular Biology, vol. 122: Confocal Microscopy Methods and Protocols*
Edited by: S. Paddock Humana Press Inc., Totowa, NJ

lenge of imaging through thick, highly scattering tissue while maintaining viability.

Because of its ability to image labeled structures deep within a thick specimen, the confocal microscope permits the imaging of cells and tissue within their own environment. Many different workers have used the optical sectioning ability of the confocal microscope to explore a wide variety of biological phenomena in living tissue ranging from the calcium dynamics in a living marine invertebrate embryo to the growth of axons and dendrites in cultured tissue slices of mammalian nervous tissue *(1,2)*. Because confocal microscopy does not require fixing and clearing of the specimen it is ideal for recording cell and tissue dynamics in time-lapse. Just as a better understanding of construction would result from repeated observation of a rising building, rather than from merely examining the finished product, time-lapse data establish a spatial and temporal reference frame needed to link invaluable gene expression data with cell and tissue dynamics.

Performing confocal microscopy of living cells does not come without its share of difficulties, and special challenges arise in imaging explanted tissues or entire embryos. Issues such as maintaining the health of the tissue need to be balanced with short and clean optical paths. Many embryos, depending on their size, can be maintained intact and imaged in their own natural environment, such as zebrafish or sea urchin embryos. It is formidable to view and sustain other embryos that are less visually accessible, such as those vertebrates growing in an opaque egg or inside a host animal, such as a chick, mouse, or rat. In these cases, a special chamber may have to be constructed, and staining techniques may have to be developed to deliver a fluorescent dye to a specified area, perhaps deep within an embryo. Finally, although the confocal microscope can optically section through thick tissue, there is a trade-off between how deep into the tissue a particular objective can be used, its numerical aperture (NA), and the clarity of the image, making the choice of the objective lens and orientation of the embryo important. Here, we elaborate on the important aspects that should be considered for the imaging of live cells in intact embryos. We illustrate this with some example techniques we have used in our own research on imaging cell and tissue dynamics in early avian embryos.

2. Materials

2.1. Embryo Preparation

1. Howard Ringer's solution
2. Albumen waste container
3. Scotch tape (3M no. 810)
4. 18- and 25-gage needles; 1- and 3-mL syringes
5. Set of dissecting tools

6. India ink (Pelikan Fount no. 211143)
7. Filter paper (Whatman no. 1001-150; 150 mm diameter)
8. Pipetmen (P20, P200, P1000)
9. Sterile Pasteur pipet
10. Sterile plastic transfer pipet
11. 70% and 100% Ethanol
12. Sucrose (Fisher no. S5-500)
13. Fast Green FCF (Fisher no. 42053)
14. Culture medium
    a. Neurobasal medium (Gibco no. 21103-015)
    b. B-27 Supplement (Gibco no. 17504)
    c. L-glutamine (Sigma no. 3126)

2.2. Dyes

1. DiIC18(3) (Molecular Probes no. D-282)
2. DiI-CM (Molecular Probes no. C-7000; fixable)

2.3. Culture Chamber Components

1. Six-well plate (Falcon no. 3046)
2. Culture Insert (Millipore; Millicell-CM, PICM 030 50)
3. Circular glass coverslips (Fisher no. 48380 080; 25-mm diameter)
4. Fibronectin (Sigma no. F-2006)
5. Silicone grease (Dow Corning no. 79810-99)
6. Soldering tool

2.4. Microscopes and Thermal Insulation Components

1. Bio-Rad MRC-500 upright confocal microscope
2. Bio-Rad MRC-600 inverted confocal microscope
3. Optical Magnetic Disk Recorder (Panasonic no. TQ-3038F)
4. Rewritable Optical Disk (3M no. 15175; 590MB)
5. Incubator heaters (Lyon no. 115-20)
6. Cardboard (~4 mm thick)
7. Thermal insulation (Reflectix; 5/16 inch thick)
8. Velcro (McMaster-Carr no. 9489K65; 3/4 inch stick-on)

3. Methods

3.1. Whole Embryo Chick Explant Technique

We have developed a method of explanting the entire chick embryo including some of the surrounding blastoderm, which appears successful with embryos as early as five somites (**Fig. 1**). By keeping the embryo intact and spreading out the surrounding blastoderm onto a membrane surface, whole embryo explants develop at a growth rate similar to that of intact embryos incubated near the microscope stage, and maintain tissue integrity for approxi-

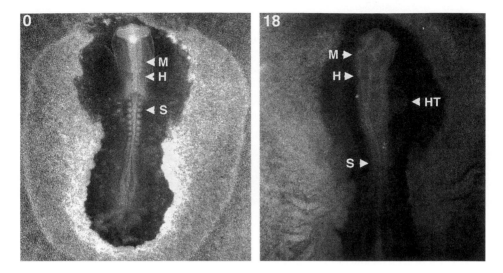

Fig. 1. Whole embryo chick explant culture. A typical whole embryo chick explant was positioned dorsal side down onto the culture insert membrane: (0 h) the embryo shown is a nine somite stage embryo; (18 h) after a time-lapse imaging session, the culture insert was removed from the six-well plate and flipped over to view the dorsal side of the embryo explant. The embryo explant has developed to 21 somites. Some of the anatomical features are labeled, including the midbrain (M), hindbrain (H), heart (HT), and somites (S). Scale bar = 1 mm.

mately 1–2 days in culture. The embryo explant can be placed either dorsal side up or down on the membrane. We describe this culture technique in detail.

1. About 1 h prior to removing the embryo from the egg, prepare the culture insert membrane. Coat the membrane surface with 200 µL of a fibronectin solution (20 µg/mL of fibronectin in Ringer's solution). The fibronectin solution helps to keep the embryo in place and makes the membrane surface transparent. The membrane is 200 µm thick and has a pore size of 0.4 µm. Place the culture insert in a Petri dish and cover to keep the fibronectin solution from evaporating. Rest the Petri dish slightly tilted so that the excess fibronectin solution runs to one side.
2. To remove the embryo from the egg, cut out an oval paper ring from filter paper. Make the inside diameter of the ring large enough to cover the entire embryo and the outside diameter to cover most of the surrounding blastoderm. Place the oval paper ring over the embryo and cut around the outside of the ring. With a pair of forceps, remove the oval paper ring from the egg, with the embryo attached, and place into a Ringer's solution. Separate the oval paper ring from the embryo with the forceps. Clear excess yolk platelets and India ink from the embryo by gently blowing Ringer's solution across the embryo with a P200 pipetman (set between 100 and 150).

3. Before placing the embryo on the insert membrane, pipet any excess fibronectin solution from the membrane with a pipetman. Use a plastic transfer pipet to transfer the whole embryo onto the culture membrane.
4. Gently spread out the surrounding blastoderm and the embryo onto the membrane with forceps. Remove excess solution with a pipetman placed at the rostral or caudal ends of the explanted embryo so that as the solution drains, the embryo maintains a straight line posture along the rostrocaudal axis. This naturally spreads out the explanted tissue without flattening the embryo and mimics the tension of the blastoderm that is normally stretched over the yolk sac of the intact embryo.
5. The culture medium that best supports the overall integrity and health of the chick explants is a defined medium consisting of Neurobasal Medium (Gibco), 98 mL: B27 supplement (Gibco), 2 mL; and 0.5 mM L-glutamine *(3)*.

3.2. Culture Chambers for Chick Embryo Explants

One method we have used for chick embryo explants is based on a closed culture chamber design originally developed for light microscope imaging of trunk neural tube explants in chick *(3)*. We have modified this chamber design for imaging chick morphogenesis with both inverted and upright confocal microscopes (*see* **Notes 1–3**). The closed chamber is made from a plastic six-well plate (Falcon) that together with the culture insert (Millipore) and embryo make up the culture system (**Fig. 2A**). We discuss our modifications in relation to the design criteria outlined previously.

1. To increase the resolution of the explant tissue during confocal imaging, the plastic bottom of one of the wells of the six-well plate is replaced with a glass coverslip (*see* **Note 4**). To make a neatly cut hole, press a soldering tool against a 5-cent coin placed on the outside wall of the plastic well. After the hole is made, use a fine sandpaper to smooth the area and wipe any debris away with a paper towel dipped in 70% ethanol. On the inside of the well with the hole, spread silicone grease around the circumference of the hole and attach a circular glass coverslip over the hole.
2. To prepare the six-well plate to support the embryo, the well with the glass coverslip is filled with 1.5 mL of tissue culture medium and the other wells are filled to two thirds with sterile water. Three support legs underneath the culture insert raise the membrane 1 mm off the culture dish bottom, allowing culture medium to be soaked up through 0.4 µm pores in the membrane. Sterile water in the other wells provides a hydrated atmosphere when the dish is sealed.
3. After the culture insert is placed in the well plate, a parafilm strip (5 cm × 40 cm) is wrapped around the sides of the dish. When sealed, the well plate provides a well oxygenated and hydrated environment for the embryo, which can be sustained for 1–2 days (*see* **Note 5**).
4. To maintain the temperature of the microscope stage at 37°C, the stage area is enclosed in a thermal insulation box (*see* **Note 6**). The walls of the box are made of cardboard pieces, cut to fit the individual microscope. Each piece of cardboard

Fig. 2A

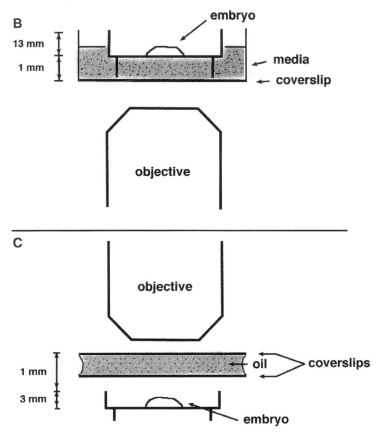

Fig. 2B,C

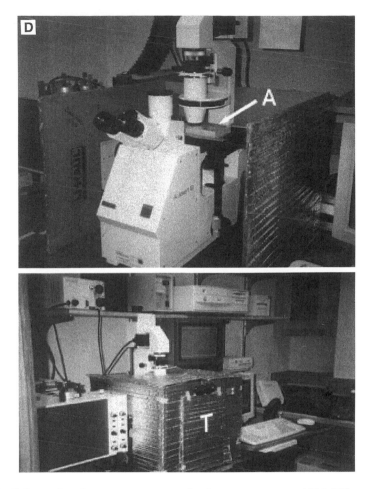

Fig. 2. Culture chamber components and microscopy setups. **(A)** Millipore Culture Insert and Falcon 6-well plate setup. **(B)** Schematic of microscopy setup for inverted microscope. **(C)** Schematic of microscopy setup for upright microscope. **(D)** Confocal inverted microscope with the six-well plate (A) on the microscope stage and thermal insulation box (T).

is covered with thermal wrapping (Reflectix Insulation) and Velcro strips are placed in strategic positions for easy assembly/disassembly. The insulation box is built large enough to enclose one or two incubator heaters (Lyon). The heaters are turned on approx 1–2 hours prior to imaging to allow a gradual warming of the stage and objectives. The insulation box has holes cut in the front, covered by flaps, to provide viewing through the eyepieces and adjusting the focus or position of the embryo explant.

5. With an inverted microscope, the light path to the embryo passes first through the glass coverslip, then through the tissue culture medium and membrane, and on to the embryo (**Fig. 2B**). Thus, the working distance between the embryo explant and an objective is slightly more than 1 mm (*see* **Note 7**). This means that we are limited to using low-magnification objectives (less than or equal to 10X) or long working distance high magnification objectives. Because we are interested in observing long duration events and preserving the morphology of the embryo, we sacrifice some working distance in exchange for growing the tissue on the membrane of a culture insert (*see* **Note 8**).

3.3. Adaptation of Culture Chamber for Upright Confocal Microscopy

To image from above we utilize the same closed chamber design but raise the culture insert so that the embryo is close to the roof of the well plate, minimizing the working distance. We describe the modifications to the basic chamber design below.

1. Complete **step 1** of **Subheading 3.2** in the same manner. The first change to the culture chamber design is to melt a hole in the top cover over one of the wells of the six-well plate and attach a glass coverslip over the hole on the inside. This will increase the image resolution when using the upright microscope.
2. Within this well, fit a plastic Petri dish, with a diameter slightly less than the inner diameter of the well, snuggly into the well. Make sure that the rim of the Petri dish is flush with the top of the well.
3. Shave off the top two thirds of the culture insert, leaving only the membrane legs and about 2 mm of plastic wall above the membrane. Use a fine sandpaper to smooth off the top of the wall and gently blow off any debris with canned air. Fill the Petri dish with 1 mL of culture medium and the other wells with sterile water. We use only 1 mL of culture medium as the walls of the modified culture insert are only 2 mm high.
4. The membrane can now be prepared and the embryo positioned on the membrane following **Subheading 3.1., steps 1–5**. The culture insert can then be placed into the well plate.
5. One of the basic problems with using an upright microscope, when viewing a live specimen requiring a warmed environment, is the formation of condensation on a surface over the embryo in the imaging pathway. In our design, condensation is likely to form on the glass coverslip attached to the cover of the six-well plate. To minimize condensation on the glass coverslip, we use a glass coverslip sandwich with a layer of heated oil in between. Secure the corners of a square 24 mm coverslip on top of the culture insert with silicone grease. Place a drop of warmed microscope immersion oil on top of this coverslip and then place the lid of the well plate gently on top. This should spread the drop of oil on the glass coverslip but not over the sides of the culture insert. This keeps the glass warm and prevents condensation from forming on the surface.
6. The working distance of approximately 2 mm is nearly the same for the inverted microscope chamber; however, the advantage is that in imaging from above, the

imaging pathway is only through glass and a culture membrane (**Fig. 2C**).

We have used the techniques described in **Subheadings 3.1–3.3** to image cell and tissue dynamics in time-lapse confocal microscopy during early chick nervous system development. We describe some of our imaging results (*see* **Notes 9–11**) and then present a summary, including some of the lessons we have learned (*see* **Note 12**).

4. Notes

1. *Tissue culture and chamber design considerations.* The more important aspects of designing a culture chamber are to maintain the health and integrity of an embryo or tissue, while allowing the best possible resolution from the confocal microscope. Details such as providing a surface for holding an embryo in place while supplying oxygen, water, and tissue culture medium in a temperature-controlled environment can be particularly challenging to incorporate simultaneously. Different sizes and shapes of embryos and the requirements of maintaining living cells within the framework of the confocal microscope have led to the design of many different culture techniques for live imaging. Chambers ranging from simple, sealed Petri dishes to sophisticated, manufactured setups that allow the simultaneous measurement of physiological parameters have been constructed (for a summary of different chamber designs, *see* **ref. 2**). Often a chamber may have to sustain tissue over long periods of time to capture a complete biological event. Thus, it is worth the time it takes to design a chamber that is flexible and simple enough to set up and use to perform many experiments, without extensive labor or cost. We describe some aspects of culture chamber design below.

2. *Surface for growing embryo or tissue.* There are three important criteria for a culture chamber. First, it must provide mechanical support of the embryo or tissue so that it can grow and remain in an orientation for imaging. Second, it must permit a supply of nutrients to the tissue. Third, the chamber must offer a clear light path to the specimen. A culture membrane, a collagen gel, or an agarose bed (supporting the embryo within a well or groove cut into the agarose) meet all three criteria. Make sure the culture membrane has pores large enough to allow nutrients to pass and is transparent when moistened and thin enough to image through. Agarose can be mixed up with a nutrient medium.

3. *Holding the embryo or tissue in place.* Tissue should be kept from drifting in the X–Y-plane or moving out of focus in the Z-direction, yet permitted the freedom to undergo the natural movements and experience the forces to which it might normally be subjected. For example, an embryo or tissue may drift due to thermal expansions, become unsettled by the flow of tissue culture media, or move because of natural growth movements. To minimize drift, a membrane surface or agarose bed may be precoated with a substance to which cells can attach. This coating may be a fibronectin solution or a plasma clot, or as simple as double-stick tape on a glass surface. An embryo or tissue can also be held in place with extra materials, such as filter paper strips or a blanket of overlying tissue.

Embedding tissue in agarose can hold the tissue in place, and may be used if the goal is to inhibit cell movement or tissue interaction.

4. *Keep a clean light path to the embryo.* The light path between the embryo and the objective lens should be as optically friendly as possible, with minimal light path scattering or distortion. Any medium in which you embed an embryo or use for nutrient supply should be optically clear. Indicator dyes can increase background fluorescence. Bubbles, yolk platelets, or any unnatural debris near the embryo should be removed. It is best if any plastic in the optical path is replaced with glass.

5. *Keep the tissue oxygenated and hydrated.* A reservoir of water and nutrients can be provided for in a closed chamber or can be perfused through the chamber during the imaging session. Choices concerning recycling or perfusing might be based on aspects such as the length of time necessary to maintain the tissue and whether the tissue requires a steady, new supply of nutrients for viability. These needs must be integrated with maintaining the position of the embryo and avoiding excessive condensation on imaging surfaces, which can be a major problem when the chamber is warmed.

6. *Temperature control.* Many embryos need to be maintained at a constant temperature to develop normally. Designs vary from a custom built, thermally insulated box surrounding both the microscope and a stand-alone heater to commercially available stage heaters with microsensors hooked up to an automated temperature control system. It is important that the design minimize thermal fluctuations and vibrations that could cause drift or decrease image resolution, especially at higher magnification.

7. *Objective lens magnification and working distance.* It is important to build the chamber so that it can accommodate a variety of microscope objectives. The maximum distance between the objective front surface and the specimen is called the working distance. A chamber design should try to maximize tissue viability and minimize working distance. Many open chamber designs allow high-magnification water objective lenses to be lowered into the culture medium directly above the specimen. Closed chambers, however, usually have a restriction such as a plastic lid that can limit working distance. If visualizing small details of cell behavior is desired, it is important to consider that high magnification objectives have much smaller working distances. A 10X (NA 0.25–0.30) objective lens can have a working distance of 2–3 mm, whereas a 40X Neofluar NA 0.75 can have a working distance of less than 1 mm. Long working distance (LWD) objective lenses can be used in sacrifice of image quality. The Nikon 40X LWD objective has a working distance of about 1 cm, allowing more flexibility with the chamber design, but the 0.5 NA reduces image quality. Zooming in with a low-power objective will magnify detail, and offers a different compromise, but there is a limit to the image quality.

8. *Other types of culture chambers.* In the case of aquatic embryos, such as sea urchin or zebrafish, methods for holding the embryo in place underwater have been developed. Sea urchin eggs have been attached to a glass coverslip by coating the coverslip with a thin layer of protamine sulfate, which rests above a hole cut in a plastic Petri dish (*4*). Zebrafish embryos have been placed in a triangular

well cut into an agarose bed, with a thin agarose blanket gently draped over the egg to hold it in place *(5)*.

9. *Confocal imaging of chick embryo explants.* In the chick, there are two distinct segmentation processes which pattern the embryo during days 2–3 of development. In the rostral part of the embryo, the neural tube is shaped into seven repeated segments, called rhombomeres. The rhombomeres are thought to provide a spatial groundplan by which cells sense their position and differentiate into specific neurons. The rhombomeres are also the location for the emergence of a sub-population of migratory cells, known as the cranial neural crest cells, which emigrate from the neural tube to pattern the peripheral nervous system. It is largely unknown what cellular mechanisms shape the rhombomeres and how neural crest cells "know" when to leave the neural tube and where to go. Although molecular gene expression data are rapidly becoming known, there is a tremendous need to provide a spatial and temporal reference frame that could link the molecular and cellular data. We describe below examples from typical confocal imaging sessions that use the whole embryo chick explant technique and the culture chambers we described in **Subheadings 3.2** and **3.3** for both inverted and upright confocal microscopes.

10. *Inverted microscope.* Using the culture chamber design described in **Subheading 3.2**, we have imaged the shaping of the neural tube into the rhombomere segments. The inverted confocal microscope is outfitted with a thermal insulation box which surrounds an incubator heater and microscope stage area (**Fig. 2D**). By shining a light through the embryo, which is oriented dorsal-side down on the membrane, the transmitted light image reveals the outline of the neural tube walls. **Figure 3** shows the neural tube of a typical embryo explant before and after 10 h of time-lapse imaging. Notice how the neural tube walls have appeared to expand and constrict in places to shape the tube into segments.

The shaping of the rhombomere segments takes just over 15 h, after which the neural tube begins to return to a less obviously segmented structure. Bright field images start to become hazy after 10–15 h in culture for reasons we do not yet understand. With this time constraint, we can watch the entire rhombomere segmentation in two separate time-lapse sequences. We typically record digital images every 5 min to optical disk (Pinnacle); NIH Image 1.60 *(6)* is used to play back the images in movie form. With a 10X Neofluar NA 0.30 and the aperture fully open, the Z-resolution is approx 30–40 μm.

We have also imaged the movement of dye-labeled neural crest cells, using fluorescence confocal microscopy. An example of neural crest cell migration in a typical chick embryo explant is shown in **Fig. 4**. By injecting a fluorescent dye into the lumen of the neural tube, premigratory neural crest cells are labeled. Notice in **Fig. 4** that we can follow the trajectory of individual neural crest cells as they emerge from the neural tube and migrate into surrounding unlabeled tissue. We use a 10X Neofluar NA 0.30 objective lens and zoom in by a factor between 1 and 2 to observe individual or small groups of cells. The haziness we encountered after 10–15 h of time-lapsing does not seem to affect the confocal

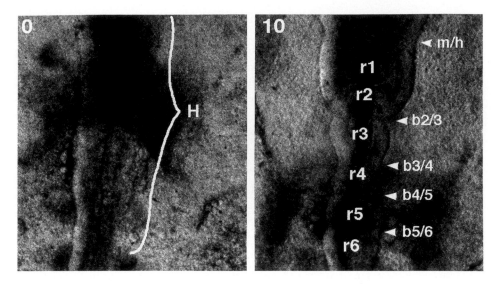

Fig. 3. Bright field images of hindbrain segmentation: (0 hrs) the hindbrain (H) of the neural tube appears as fairly unsegmented; (10 h) the neural tube shape changes give rise to the segmented structure of the hindbrain. The rhombomeres, r1–r6, are labeled as well as the boundary regions between them. For example, the boundary between r2 and r3 is labeled as b2/3, and the boundary between the midbrain and hindbrain is m/h. Figures 3 through 5 are examples from time-lapse confocal imaging sessions of typical whole embryo chick explants. Each image represents one confocal section taken with an inverted microscope.

fluorescence imaging, where time-lapse imaging sessions have lasted up to 18 h. The second segmentation process in the early developing chick embryo produces repeated clumps of cells or somites that form in pairs adjacent to the neural tube. The somite pattern is laid down in a regular sequence moving caudally down the anteroposterior axis of the embryo. Somites give rise to muscle, vertebrae, and skin. With the same imaging methods as described previously, we are able to watch the outline of a somite take shape from the onset of its budding into an individual clump and formation of the characteristic circular cross-sectional shape. **Figure 5** shows an example of somite formation in a typical embryo explant. Notice in the figure that three new somite pairs form in 4.5 h, which is the same rate as an ovo. In a 10-h time-lapse sequence we can observe approximately seven new somites and record digital images every 5 min to optical disk. Using a 10X Neofluar NA 0.30 provides a field of view large enough to capture the most newly formed somites and a portion of the unsegmented region.

11. *Upright microscope.* We have found that culturing a chick embryo dorsal side up on the membrane allows the embryo to better mimic its normal rotation. In the cranial region, this rotation turns the embryo so that the dorsal neural tube lays

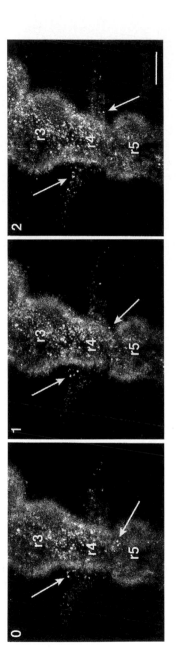

Fig. 4. Fluoresecent DiI injected into the neural tube labels cells within the neural tube and neural crest cells which are shown emigrating into the surrounding unlabeled tissue. The arrows follow individual neural crest cells leaving the neural tube from the region of rhombomeres r3, r4, and r5: (0 h) the **top arrow** identifies a neural crest cell after it has left the neural tube to join the population adjacent to r4, while the **bottom arrow** identifies a cell near the boundary between r4 and r5; (1 h) the **bottom arrow** shows that the neural crest cell has moved into r4, while the **top arrow** shows that the other neural crest cell has moved further away from the neural tube; (2 h) the **bottom arrow** shows that the neural crest cell is emigrating away from the neural tube having joined the stream exiting adjacent to r4. Scale bar = 100 μm.

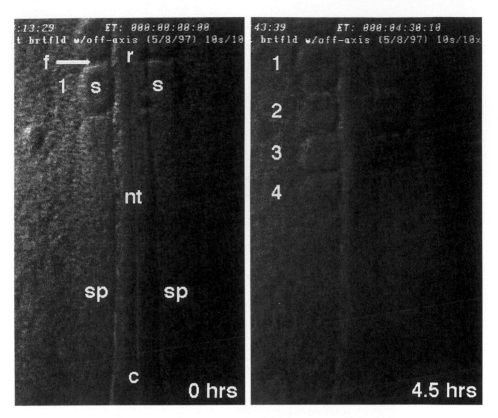

Fig. 5. Bright field images of somite formation: (0 h) the most newly formed somite pair is labeled by 1; (4.5 h) the fourth somite pair has formed. The rostral (r) portion of the embyro is at the top of the image, showing the intersomitic furrow (f), the unsegmented region called the segmental plate (sp). The bottom of the image is the caudal (c) portion of the embryo explant.

on its side. Although the neural crest cells move fairly planar as they exit the dorsal neural tube, their full migration pathway is from the dorsal side of the embryo all the way around to the ventral side. The rotation naturally exposes one side of the embryo, which allows us to follow more of the migration pathway of cranial neural crest cells as they reach the branchial arches. By using the optical sectioning capability of the confocal microscope, we have taken as many as 10 Z-sections of 10 μm each at each time point to capture a majority of neural crest cells and time-lapse sessions have lasted as long as 24 h. In **Fig. 6**, an example of a typical embryo explant shows the patterning of the cranial neural crest cells as the embryo explant begins rotating to the left. By 4 h into the time-lapse, a large number of neural crest cells have left the neural tube from the midbrain and rhombomere 1, r1, region. At 8 h, the natural rostral progression of the growing

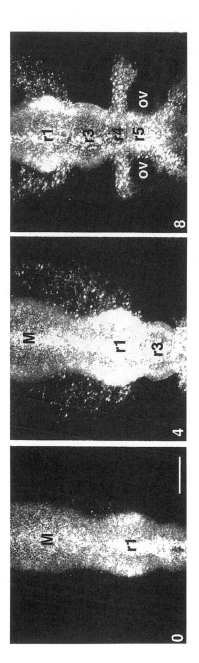

Fig. 6. DiI-labeled neural crest cells and cells within the neural tube of a typical embryo explant dorsal side up on a culture membrane. Each image represents 10 images, at 10 μm Z-increments projected onto one image, taken from a time-lapse confocal imaging session on an upright microscope: (0 h) the fluorescently labeled neural tube in the hindbrain region before many of the neural crest cells have left the neural tube; (4 h) neural crest cells are migrating into the surrounding unlabeled tissue from the midbrain and rhombomere 1 regions; (8 h) neural crest cell migration continues from the midbrain and rhombomere 1 regions. The natural growth of the embryo explant exposes the caudal rhombomeres and the other migrating neural crest streams adjacent to r4 and from r5 and r6. The midbrain (M), rhombomeres r1–r6, and the otic vesicle (ov) are labeled. Scale bar = 100 μm.

embryo explant exposes the pattern of neural crest cells adjacent to r4 and caudal to the otic vesicle.

We recorded fluorescence images to an optical magnetic disk recorder (OMDR; Panasonic) at 2-min intervals for time-lapse sessions that lasted as long as 24 h. Using an OMDR allows the rapid playback of frames and has the advantage of determining whether any adjustments to either the microscope or specimen need to be made during the time-lapse. We also recorded digital images every 2 min to optical disk (Pinnacle); NIH Image 1.60 *(6)* is used to play back the images in movie form.

12. *Summary.* We have outlined the important considerations for imaging living cells in intact embryos. This included addressing aspects of how to maintain the health and morphology of the tissue in combination with confocal microscopy.

As an example of the coordination between live embryos and imaging we discussed several examples from our work on imaging cell and tissue dynamics in the developing chick nervous system. Specifically, we discussed how chick embryos growing on a culture membrane maintain normal development until the period of natural rotation of the embryo, lasting from 1 to 2 days. Precoating the culture membrane with a fibronectin solution aids cell and tissue attachment to the surface and the pores of the membrane. Glass coverslips were used in place of plastic wherever appropriate and the culture membrane was transparent when moistened so that the imaging pathway was as clear as possible. The additional wells of the culture chamber held excess sterile water while tissue culture media was placed underneath the culture membrane to provide a moist and nutrient-rich environment. Although the closed culture chamber design limited our choices of objective lenses to low magnification or high magnification with long working distances, we were able to visualize the global and local cell movements we were interested in. In addition, the ease and simplicity of the culture chamber preparation decreased our experiment time. The six wells of the plate often provided a spare well to culture an additional embryo explant, which allowed us to choose between two embryo explants to image, once we visualized the explants on the microscope. Data from the nonimaged embryo explant would then be used for growth comparison to intact embryos, allowing two experiments to be performed during one time-lapse imaging session.

We utilized off-axis illumination to view the chick neural tube and somites in brightfield. This significantly improved the quality of an image by providing a 3D aspect to the outline of the tissue. For fluorescence imaging, we tried as often as possible to use the low-power setting on the laser which can help to prevent photobleaching of the specimen and prolong imaging sessions. We typically started each confocal imaging session using the highest neutral density filter and adjusted the gain and contrast so that the image displayed the widest possible spectrum on the 0–255 black–white scale. If the image appeared faint we would increase the amount of laser light to the specimen by decreasing the attenuation of the laser beam. We frequently used a 10X Neofluar NA 0.30 to visualize cell movements and found better resolution by increasing the zoom rather than switching to a 20X

LWD objective. We set the aperture to be fully open; although this translated into less confocal effect, it increased the sensitivity of the microscope and allowed for a maximum optical section thickness so that we could capture a majority of cell trajectories for longer periods of time.

There are several fluorescent labeling techniques ranging from iontophoretic labeling of small numbers of cells to bulk soaking of tissue, as well as different mixtures for effective delivery to various parts of the cell. We fluorescent labeled groups of neural crest cells by pressure injecting a large amount of dye into the neural tube and allowing the dye to be absorbed by cells. We found that buffering our DiI in warm sucrose [10 µL of DiI solution (0.5 mg of DiIC18(3) in 1 mL of 100% ethanol) in 90 µL of 0.3 M sucrose] worked best for labeling the neural crest cells and cells in the neural tube. We added a small amount (< 1 mg in 100 µL of dye solution) of Fast Green to the dye solution to make it easier to see the dye we were injecting into the neural tube.

During an imaging session, there are several ways to determine whether the tissue culture and chamber design are working effectively. First, one way to check the health of the embryo is to compare the embryo or cell movements at various spatial and temporal points to intact embryos. We used the formation of somites as a guide to the growth of the embryo explants. From time-lapse imaging data, embryo explants formed one new somite pair every 1.5 h, which is the same as in the intact embryo. In the formation of rhombomeres, we noted that the rhombomere boundaries formed in the same spatial order in the embryo explants as in intact embryos. The second way to check the health of the embryo is to watch for unexpected signs. The sudden appearance of uncharacteristic cell behavior or movements may be due to the onset of cell death, loss of health of the tissue, or tissue drift. Expansion or compression of cells or tissue may be due to thermal fluctuations. Tissue may start to go out of focus or the images become opaque as a result of tissue movement, condensation on a surface in the imaging pathway, or some other factor. We encountered an opacity of the image after approx 15 h during chick embryo explant imaging, which may be due to yolk platelets dissolving into the culture media.

In summary, advanced planning to address the challenges of imaging living tissue can save hours of laborious trial and error. The simpler the system, the easier it may be to troubleshoot. In the end, finding a robust culture chamber and confocal microscope setup leads to the flexibility to do many experiments and allows perturbations to the animal system with the benefit of watching cell and tissue dynamics in a living embryo.

References

1. Matsumoto, B. (1993) *Methods in Cell Biology,* Vol. 38: *Cell Biological Applications of Confocal Microscopy.* Academic Press, San Diego.
2. Terasaki, M. and Dailey, M. E. (1995) Confocal microscopy of living cells, in *Handbook of Biological Confocal Microscopy,* (Pawley, J. B., ed.), Plenum Press, New York.
3. Krull, C. E. and Kulesa, P. M. (1998) Embryonic explant and slice preparations for studies of cell migration and axon guidance, in *Current Topics in Developmental Biology,* Vol. 36, Academic Press.

4. Summers, R. G., Stricker, S. A., and Cameron, R. A. (1993) Applications of confocal microscopy to studies of sea urchin embryogenesis, in *Methods in Cell Biology,* Vol. 38: *Cell Biological Applications of Confocal Microscopy.* Academic Press, San Diego.
5. Shih, J. and Fraser, S. E. (1995) Distribution of tissue progenitors within the shield region of the zebrafish gastrula. *Development* **121(9),** 2755–2765.
6. Rasband, W. S. and Bright, D. S. (1995) NIH Image: a public domain image processing program for the Macintosh. *Microbeam Analysis Soc. J.* **4,** 137–149.

14

Live Confocal Analysis
with Fluorescently Labeled Proteins

Helen Francis-Lang, Jonathan Minden, William Sullivan, and Karen Oegema

1. Introduction
1.1. General

The development of laser scanning confocal microscopy provides a powerful means to observe structures and components within the cell. Equally important advances have also occurred in the development of reagents and techniques for generating functional fluorescently tagged proteins. When these labeled proteins are used in conjunction with confocal and other advanced fluorescence microscopes, the dynamics of a given protein within the living cell can be readily analyzed in real time. Live analysis provides an appreciation of the cellular and developmental dynamics that is virtually impossible to obtain through observation of fixed samples. In addition, tracking down the primary defect and determining causal relationships in mutant and drug-mediated phenotypes often can be accomplished only through live fluorescence analysis. Finally, live analysis has been used to confirm the existence of structures in the embryo that were previously contested as possible fixation artifacts (1).

Although live fluorescence analysis was rarely performed only a few years ago, it is becoming much more routine, in large part due to the widespread use of the green fluorescent protein (GFP; and see Chapter 15). Live fluorescence analysis has also benefitted from the commercial availability of excellent fluorescence reagents and increased access to advanced fluorescence microscopes, such as the confocal. The direct tagging of purified proteins with fluorescent moieties in vitro provides an alternative strategy to fusion with GFP for analyzing protein dynamics in living embryos. Direct fluorescent labeling of pro-

From: *Methods in Molecular Biology, vol. 122: Confocal Microscopy Methods and Protocols*
Edited by: S. Paddock © Humana Press Inc., Totowa, NJ

teins may be the best approach in cases in which it is difficult to express the GFP fusion protein (e.g., during early stages of embryogenesis) or when fusion with GFP interferes with protein structure or function. In addition, direct fluorescent labeling allows the simultaneous visualization of multiple fluorescent probes.

In our experience, generating a functional fluorescently labeled protein and injecting it into the cell without damage are the most technically demanding aspects of this approach. Consequently, this chapter describes a specific protocol for directly attaching fluorescent probes to antibodies for live analysis and a section on applying this procedure more generally to purified and bacterially overexpressed proteins. We then describe general principles of microinjecting fluorescently labeled proteins, using *Drosophila* embryos as a specific example. Finally, we describe the successful application of these techniques in *Drosophila* and other model organisms.

1.2. Strategies for Creating Fluorescent Conjugates

Proteins can be covalently modified using several types of chemistry *(2,3)*. Most commonly used are fluorophore derivatives that modify amino or sulfhydryl groups. Succinimidyl esters and isothiocyanates are two types of fluorophore derivatives that have been widely used to label primary amines of proteins (lysine residues and the N-terminal amino group). Succinimidyl esters are now more frequently used than isothiocyanates because they react more specifically, more quickly, at lower pH, and result in more stable products that do not deteriorate during storage. Sulfonyl chlorides, such as lissamine rhodamine B sulfonyl chloride and Texas red sulfonyl chloride are highly reactive and unstable in water, especially at the higher pH required to react with aliphatic amines, making reproducible conjugation difficult. Succinimidyl ester derivitives of these, which are more suited to the conditions required for reproducible conjugation reactions, have now been developed. The sulfhydryl groups of cysteine residues can be labeled using iodoacetic acid derivatives. Because cysteines are relatively rare in proteins, cysteine labeling can be more uniform than labeling of protein primary amines. In addition, cysteines can be engineered into proteins to provide a convenient target for labeling *(4)*.

However, because cysteines are rare and often critical for protein function, random lysine labeling using succinimidyl esters of fluorophores has proven more generally applicable. Although preparing the optimal conjugates suitable for the most critical assays may require extensive experimentation, it is often relatively easy to conjugate fluorophores to proteins in a manner sufficient for visualization in vivo. Described below are two labeling protocols that we have successfuly employed to label antibodies and expressed fusion proteins with the succinimidyl esters of commonly used fluorophores.

2. Materials

1. $0.1M$ Na_2CO_3 pH 9.3
2. $2M$ potassium glutamate (pH 8.0). (This can be made by adjusting the pH of glutamic acid with potassium hydroxide.)
3. Small spin columns containing about 850–1000 µL of desalting gel (such as Bio-Gel P-6 from Bio-Rad) and *see* **Note 1**.
4. 25 mM stock of fluorophore dissolved in dimethyl sulfoxide (DMSO). For long-term storage of unreacted fluorophore at –80°C, the DMSO must be of high quality and dry. The fluorophore concentration should be determined by absorbance using the appropriate extinction coefficient because the values obtained by dissolving a weighed amount of dye are often inaccurate.
5. SM-2 polystyrene beads (Bio-Rad; cat no. 152-3920). Beads are prewashed in methanol, before equilibriating in the protein buffer.
6. Random IgG for pilot reactions (*see* **Subheading 3.1.**) can be purchased from many sources such as Jackson ImmunoResearch or Sigma.
7. Injection buffer: 50 mM K-glutamate (~pH 7.0), 0.5 mM $MgCl_2$, 1 mM glutathione.
8. Here are some useful sites on the Internet.
   a. The home page for Molecular Probes, provider of fluorescent reagents:
      http://www.probes.com/
   b. A very comprehensive listing of confocal microscopy resources, many movie and image links:
      http://www.cs.ubc.ca/spider/ladic/confocal.html
   c. Confocal images and movies of living cells from the Terasaki lab home page:
      http://www2.uchc.edu/htterasaki
   d. Integrated Microscopy Resource at University of Wisconsin, useful microscopy links, several images, and movies:
      http://www.bocklabs.wisc.edu/imr/home.htm
   e. Collection of images from the Biosciences Imaging group, University of Southampton:
      http://www.neuro.soton.ac.uk/research.info.html
   f. Images and movies from the Advanced Microscopy Imaging Laboratory, SUNY:
      http://corn.eng.buffalo.edu/www/confocalImages/readme.html
   g. Sedat lab home page including movies of nuclei and chromosomes:
      http://util.ucsf.edu/sedat/
   h. Bowerman lab home page with *Caenorhabditis elegans* confocal movies:
      http://eatworms.swmed.edu/Worm_labs/Bowerman/
   i. Home page of the Strome lab, with fluorescent *C. elegans* confocal movies:
      http://sunflower.bio.indiana.edu:80/~sstrome
   j. Home page of the Sullivan lab, with fluorescent *Drosophila* confocal movies:
      http://www-biology.ucsc.edu/people/billllab/main.html
   k. Home page of the Mitchison lab with a tour of cell motility:
      http://skye.med.harvard.edu/

l. The kinesin home page containing movies of cell motility:
http://www.blocks.fhcrc.org/~kinesin/

3. Methods

3.1. Generating Fluorescently Labeled Antibodies

The protocol described in this section utilizes commercially available succinimidyl esters of available fluorophores that provide a simple and efficient means of covalently attaching the fluorophores to proteins. Several fluorophores are available with excitation/emission maxima closely matching the krypton/argon laser emission lines (488 nm, 568 nm, 647 nm) and optical filters of commonly used confocal microscopes. The properties of several of these fluorophores are described in **Table 1**.

The required buffers and materials are listed in **Subheading 2**. Because the reactivity of the fluorophores varies from batch to batch (and may decline during storage), a pilot reaction with random IgG should first be performed to determine the appropriate labeling conditions. The protocol described below is for a small-scale labeling reaction (100 µL of antibody at concentrations between 0.5 and 20 mg/mL) and is performed at room temperature. This reaction can be scaled up as required (for **steps 1** and **5**, larger prepacked desalting columns are available (*see* **Subheading 2**).

1. Equilibrate a centrifuge desalting column (*see* **Subheading 2**) into coupling buffer (0.1M Na$_2$CO$_3$, pH 9.3). Apply 100 µL of the antibody solution (between 0.5 and 20 mg/mL) to the column and spin to elute the antibody into a microfuge tube. This procedure results in minimal dilution and excellent recovery of the antibody in coupling buffer.
2. Add 1 µL of fluorophore solution (in DMSO; *see* **Subheading 2**) to the antibody solution. Mix by gently flicking the tube (avoid bubbles) and *see* **Note 2**.
3. Incubate at room temperature for 30 min.
4. Add 10 µL of 2M potassium glutamate (pH 8.0). The free amino group on the glutamate quenches the reaction.
5. Apply the antibody to a second centrifuge column preequilibrated in injection buffer if the antibody will be used for live analysis or phosphate-buffered saline (PBS) if the antibody is to be used for immunofluorescence of fixed samples. This separates the labeled antibody from the uncoupled fluorophore and exchanges the antibody into a buffer suitable for injection or storage and *see* **Note 3**.
6. Remove a 10-µL aliquot from the labeled-antibody solution and store in the dark at 4°C. As described in the next step, this will be used to determine the stoichiometry of labeling. For live analysis, divide the remainder of the labeled-antibody into 2-µL aliquots freeze in liquid nitrogen, and store at –80°C. If using for immunofluorescence of fixed samples, divide into 10 ul aliquots and snap freeze.
7. Dilute the 10 µL of labeled antibody to a final volume of 100–150 µL with injection buffer and take an absorption spectrum of this sample. **Figure 1** depicts

Table 1
Fluorophores Compatible with the Krypton/Argon Laser
(Laser Lines at 488, 568, and 647 nm)

Fluorophore	Excitation	Emission	Extinction coefficient	$\varepsilon_{D280}/\varepsilon_{DM}$
Fluorescein[a]	495 nm	519 nm	74,000 M^{-1}cm^{-1}	0.19
Oregon green 488	495	521	76,000	0.19
Cy3	550	570	150,000	0.08
Tetramethylrhodamine[b]	546	576	95,000	0.21
Rhodamine red-X	560	580	129,000	0.17
X-Rhodamine[b]	574	602	78,000	0.20
Texas red-X	583	603	116,000	0.15
Cy5	649	670	250,000	0.05

All data was obtained from Molecular Probes (2) and from Amersham.
5- (and 6-) Carboxyfluorescein, succinimidyl ester (Molecular Probes, C-1311)
Oregon green 488 carboxylic acid, succinimidyl ester 5-isomer (Molecular Probes, O- 6147)
5- (and 6-) Carboxytetramethylrhodamine, succinimidyl ester (Molecular Probes, C-1171)
5- (and 6-) Carboxy-X-rhodamine, succinimidyl ester (Molecular Probes, C-1309)
Rhodamine red-X, succinimidyl ester, mixed isomers (Molecular Probes, R-6160)
Texas red-X, succinimidyl ester, mixed isomers (Molecular Probes, T-6134)
Cy3-OSu Monofunctional reactive fluorophore (Amersham PA13100).
Cy5 -OSu Monofunctional reactive fluorophore (Amersham PA13600)
[a]The absorption and fluorescence of fluorescein are pH dependent, for details see pp. 552,553 of the Molecular Probes Catalog (2). The values given here are at pH 9.0. Both absorption and emission are significantly reduced below pH 7.0.
[b]The values for these dyes given here are based on spectra taken in MeOH. The absorption and emission spectra in pH 8.0 buffer are red shifted ~8 nm and the extinction coefficients are ~ 10% lower (2).
Also *see* **Notes 10–13**.

absorption spectra for a protein labeled at two different stoichiometries. The stoichiometry of labeling is calculated by determining the ratio of the molar concentration of the fluorophore and the molar concentration of the protein (described in detail in **Fig. 2**).

Although the optimal stoichiometry will depend on the precise application and the fluorophore, we have found that for live analysis of *Drosophila* embryos or immunofluorescence of fixed samples, a fluorophore to antibody stoichiometry between 1 and 4 produces satisfactory results. If the stoichiometry is too high or low, vary the concentration of fluorophore added in **step 2** accordingly.

3.2. Generating Fluorescently Labeled Proteins

Fluorescent labeling of purified proteins can be accomplished using conditions similar to those described for antibody labeling. Because individual proteins behave differently, labeling protocols vary from protein to protein. A key issue in covalent labeling of proteins is to ensure that the labeled protein retains

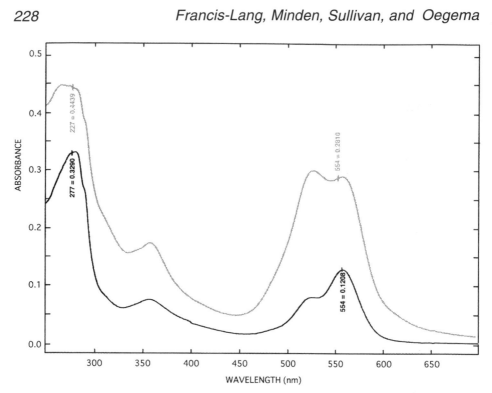

Fig.1. The absorption spectra of the same protein labeled at a fluorophore to protein molar ratio of 1 (**solid line**) and 5 (**stippled line**), respectively.

functionality (*see* **Note 5** on quality control). As many proteins are less stable than antibodies, the conditions described in the protocol below are milder than those used for antibody labeling.

1. We begin with protein in 50 mM Na phosphate pH 8.0, 250 mM NaCl, 1 mM 2-mercaptoethanol. The last step of our fusion protein purification was Superose-6 gel filtration chromatography (Pharmacia). This step served to buffer exchange the protein, to remove aggregated protein and proteolysis products, and was essential to obtain good localization of our fusion proteins. Other buffers and salt concentrations, optimized for your protein, can be used in the labeling reaction, but avoid buffers that contain primary amines (such as Tris) or strong reducing agents [such as dithiothreitol (DTT)]. In buffers below pH 8.0, the labeling reaction will proceed more slowly and it is also possible to get side reactions between the fluorophore and cysteine thiols.

2. To 75 μL of fusion protein (1–5 mg/mL) in 50 mM phosphate pH 8.0, 250 mM NaCl, 1 mM 2-mercaptoethanol, add 0.75 μL of 25 mM fluorophore dissolved in

Fig. 2. (*facing page*) Formulae for calculating fluorophore: protein labeling stoichiometry and extinction coefficients of novel proteins.

Figure 2

Calculation of Labeling Stoichiometry

Labeling stoichiometry is calculated by determination of the fluorophore to protein molar ratio. The value reflects the average number of fluorophore molecules conjugated to each protein molecule.

The fluorophore concentration in the sample is determined by dividing the absorbance of your sample at the absorbance maximum for the fuorophore by the extinction coefficient of the fuorophore at that wavelenght.

$$[\text{Fluorophore}] = A_{DM}/\varepsilon_{DM}$$

where A_{DM} is the absorbance of your sample at the absorption maximum for the labeling fluorophore (e.g., 495 nm for fluorescein) and ε_{DM} is the extinction coefficient of the fluorophore at its absorbance maximum (e.g., 74,000 $M^{-1}cm^{-1}$ for fluorescein; see Table 1).

The protein concentration in your sample is determined by taking the absorbance of your somaple at 280 nm, subtracting the absorbance of the fluorophore at 280 nm, and then dividing this by the extinction coefficient of your protein at 280 nm (details on calculating the extinction coefficient of your protein from its sequence will be given below).

$$[\text{Protein}] = [A_{280} - ADM \times (\varepsilon_{D280}/\varepsilon_{DM})]/\varepsilon_P$$

where A_{280} is the absorbance of your sample at 280 nm, ADM is the absorbance of your sample at the absorption maximum for the labeling fluorophore, ε_{D280} is the extinction coefficient of the fluorophore at 280 nm, ε_{DM} is the extinction coefficient of the fluorophore at its absorbance maximum, and ε_P is the extinction coefficient of your protein at 280 nm.

Figure 1 depicts the spectra for the same protein labeled with rhodamine at stochiometries of approx 5 (stipple) and 1 (solid). Note the appearance of the characteristic absorption doublet of fluorophore in the overlabeled spectrum (stipple). The absorbance at 522 nm is caused by the fluorescence quenching due to noncovalent dimers of rhodamine bound to the same protein molecule. When labeling with rhodamine, this effect can occur even at relatively low labeling stoichiometries (less than 1) because this fluorophore is particularly prone to dimer formation, even in solution. The absorbance peak shoulder at about 355 nm is another rhodamine absorbance peak.

Determining the Extinction Coefficient of Your Protein at 280 nm

Gil and Von Hippel (24) have shown that it is possible to estimate the extinction coefficient of native proteins from their amino acid composition with a standard error of about 5% in most cases. For native proteins:

ε_{280} nm = [no. of tryptophan molecules) $\times$ 5690 + (no. of tyrosine molecules) $\times$ 1280
+ (no. of cysteine molecules) $\times$ 120] $\times$ 1.05 $M^{-1}cm^{-1}$

DMSO. As in the antibody labeling protocol, the amount of fluorophore added is varied to attain the desired labeling stoichiometry.

3. Incubate on ice for 5 min.
4. Add 7.5 µL of potassium glutamate (pH 8.0) and 0.75 µL of 0.5M DTT (optional) to stop the reaction.
5. Apply the labeling reaction to a centrifuge column equilibrated in 50 mM HEPES, 100 mM NaCl (pH 7.6), to remove free fluorophore and to transfer the protein to the buffer used for injection and *see* **Note 4**.
6. The stoichiometry of labeling is assayed by spectroscopy (**Figs. 1** and **2**). If the protein is over- or underlabeled, repeat the procedure varying the amount of fluorophore added accordingly. Proteins were labeled to a stoichiometry of approximately 0.5 fluorophore molecules per protein. We generally label proteins to lower stochiometries to our antibodies to prevent interference with function.
7. Freeze the labeled protein in 2-µL aliquots in liquid nitrogen and store at –80°C.

3.3. Preparing and Loading Injection Needles

1. Injection needles are constructed from 75-mm glass tubing with an outer and inner diameter of 1.21 mm and 0.90 mm, respectively (Drummond Scientific, Broomall, PA cat. no. N-51A).
2. A mechanical needle puller is used to draw the tubing to a fine closed point. Typically three or four needles are drawn and examined on a compound microscope. Those that form a smooth unbroken tip are choosen for injection. These are stored on a clay mount in a Petri dish maintained at 4°C.
3. The needle is backloaded using a 10-µL Hamilton syringe precooled to 4°C. As described in the previous section, the labeled proteins are stored in 2-µL aliquots at –80°C. Immediately prior to injection an aliquot is rapidly thawed and microfuged for 3 min at 10,000 rpm. This pellets denatured or aggregated protein and reduces the chance of the needle clogging. A Hamilton syringe is used to draw the labeled protein from the microfuge tube and backload it into the syringe. From this point, as much as possible, the loaded needle is maintained in the dark at 4°C.
4. The most technically demanding aspect of the injection procedure is generating a small hole, approximately 3–5 µm in diameter, at the tip of the needle. For reference syncytial embryonic nuclei are 5 µm in diameter. To achieve this the needle is mounted in the microinjection apparatus and pressurized. As shown in **Fig. 3A,B**, the needle is gently brought in contact with a glass slide covered with halocarbon oil (Series 77, CAS no. 9002-83-9, Halocarbon Products Corp.). A small break is readily detected because fluid immediately flows from the needle and is clearly visible in the oil.
5. Once this is achieved, the needle is removed from the injection apparatus and stored in the dark at 4°C. Kept in this manner, the loaded needle will suffice for a full day of microinjection.

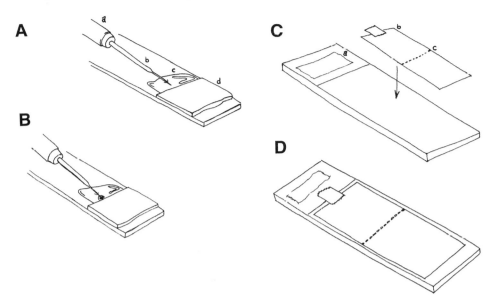

Fig. 3. Setup for breaking the tip of microinjection needles (**A,B**). a: needle holder; b, microinjection needle; c, halocarbon oil. Preparation of samples for microinjection (**C,D**). a, double-stick tape; b, coverslip; c, glue.

3.4. Preparing and Injecting Samples

Our experience has been primarily with *Drosophila* embryos studied with an inverted microscope. As many aspects of this procedure will more generally apply to other systems we describe it here (*see* **Note 5**).

1. So that the samples remain stably attached to the coverslip during injection and recording, we prepare a solution of clarified glue. From 5–10 mL of heptane and 12 or more inches of double-stick tape (3M) are added to a 50-mL conical tube and allowed to rotate overnight. The solution is aliquoted into 1.5-mL centrifuge tubes and microfuged to remove particulate matter. The consistency of the solution should be such that it can be easily micropipeted. Over time the heptane evaporates and the solution becomes viscous. Additional heptane is required to thin the solution.

2. To prepare embryos for injection, a slide is constructed with a small piece of double-stick tape at one end and a temporarily taped coverslip at the other end. A pipetman is used to trace 2 μL of glue solution along a thin line on the coverslip (**Fig. 3C,D**).

3. Twenty to thirty, 1–2-h-old *Drosophila* embryos are removed from a collection plate and placed on the piece of double-stick tape. These embryos are hand dechorionated and 10–15 embryos are placed in a line along the film of glue.

4. In order to prevent bleeding when the embryos are injected, they are placed in a drying chamber for 8–14 min (a sealed plastic dish containing Drierite serves as

an effective drying chamber). The drying time varies from 8 to 15 min owing to day-to-day variation in temperature and humidity. One must be careful to not to overdry the embryos because this will create recording artifacts. One sign of overdrying is stretch marks and indentations along the embryo.

5. Immediately after the embryos are dried, they are covered with oil and microinjected. The most critical aspect of manual microinjection is needle pressure. To obtain the correct needle pressure, the tip of the needle is placed in the oil near the embryos and pressure is applied until a small bubble of injection fluid is observed in the oil. Pressure is adjusted until one can observe a steady gradual increase in bubble volume. The needle is then pulled away from the bubble. If the pressure is correct no fluid flows from the needle. When the needle is reintroduced into the bubble fluid again flows and the bubble increases in volume. With this pressure, the injected embryos absorb enough fluid to replace that lost by dehydration. Using this procedure, we typically inject approximately 1–5% egg volume with fluorescently conjugated protein.

6. Once injected, the coverslip is removed from the slide and is ready for microscopy (*see* **Notes 6–9**).

4. Notes

1. Columns are sold prepacked (Bio-Rad; cat no. 732-6002) and used according to the manufacturer's instructions. Alternatively, you can buy the small Bio-Spin disposable plastic columns (Bio-Rad; cat no. 732-6008) and resin (Bio-Rad; cat no. 150-0738) separately and pack them yourself. A stock of desalting gel can be prepared by autoclaving 16 g of dry Bio-Gel P-6 resin in 200 mls of distilled water. This slurry may be stored at room temperature for extended periods of time. To pour a small centrifuge column load 1600–2000 μL of the autoclaved slurry (about 850–1100 μL of packed resin) into the column. During this process the columns can be supported by placing them in 13 × 100-mm glass test tubes. Equilibrate the column by pipetting 2 × 1.5 mL of the desired buffer onto the resin and allowing it to flow through by gravity. Put the column into a 17 × 100 mm plastic snap cap tube (with the cap removed) that contains a 1.5-mL microfuge tube. Remove excess buffer by spinning at top speed in a clinical centrifuge for 4 min. Load your sample onto the resin and centrifuge as before, collecting the flowthrough (your protein) in a fresh 1.5-mL microfuge tube.

   If the reactions in **steps 1–5** are to be scaled up for larger volumes, Econo-Pac 10G desalting columns (Bio-Rad, cat. no. 732-2010) may be used. These are convenient for sample volumes between 1 and 3 mL.

2. To determine the appropriate fluorophore concentration for the labeling reaction, we first set up three test reactions using a concentration of random IgG similar to that of the antibody to be labeled. To these reactions we add 1 μL of either 25 mM, 8 mM, or 3 mM fluorophore solution. We select a concentration that results in a 1:4 molar ratio of fluorophore to protein (*see* **Subheading 3.1.7**). If none of the initial fluorophore concentrations result in the desired stoichiometry, further test reactions are performed.

3. (Optional) If free flurophore has not been adequately removed for your application, the antibody can be incubated at 4°C for 1 h with an equal volume of SM-2 polystyrene beads that have been first washed in MeOH and then equilbrated in injection buffer. The labeled antibody can be tested for the presence of free fluorophore by addition of sodium dodecyl sulfate (SDS) to 0.5% followed by precipitation of the protein with five volumes of cold ethanol and analyzing the supernatant for the presence of flurophore (*see* determination of absorbance spectra below).

4. (Optional) If free fluorophore has not been adequately removed, the protein can be incubated at 4°C for 1 h with an equal volume of SM-2 polystyrene beads, as described in the previous section for separating antibody from uncoupled fluorophore.

5. Labeling damage to the protein and phototoxicity due to fluorescent excitation are two main concerns when using fluorescently labeled proteins in vivo. With well-characterized proteins, functional assays can often be used to determine if the activity or biochemical characteristics of a protein are perturbed by labeling. Often this is not possible with uncharacterized proteins, so we ensure that the labeling stoichiometry is less than one fluorophore per polypeptide. We routinely centrifuge the protein prior to microinjection (**step 3**, preparing and loading needles), to remove precipitated, denatured protein. A number of methods have been applied to ensure that only active protein is injected into the embryo and where applicable these assays have been employed to determine protein activity after labeling *(5)*. In the case of cytoskeletal proteins, such as actin and tubulin, this has been accomplished by selecting active protein by multiple cycles of assembly and disassembly after labeling *(6,7)*. The DNA binding proteins, histones and topoisomerase II, were labeled while bound to DNA cellulose to protect functionally important sites from modification *(5,8)*.

The concentration of injected protein is another concern; we have found that injecting proteins with concentrations greater than 10 mg/mL may perturb normal development. This concentration artefact and photo-induced damage can be detected by mitotic failures and aberrations in other cellular processes, most often centered around the injection area where protein concentration is the highest. To reduce the extent of photo-induced damage, one can reduce the frequency of image acquisition and attenuate the beam intensity by increasing the neutral density filter. Finally, we verify the results of our live recordings through fixed analysis, as these techniques complement one another *(9)*. Although it is difficult to analyze dynamic processes using fixed material, large sample numbers are readily obtained. These images should correspond closely to the live studies. Applying fixed analysis in conjunction with live studies is especially important when studying mutant and drug-induced abnormal phenotypes.

Live fluorescence analysis of the early embryonic divisions in the *Drosophila* embryo has proven a useful tool understand to unravel the primary defects that occur due to mutation mutants or drug treatments *(9–12)*. Below we give three examples taken from our research that illustrate the application of this technique.

6. We injected rhodamine-labeled tubulin *(7)* into *Drosophila* embryos and cap-
    tured confocal images every 10–30 s to study microtubule dynamics during the
    synchronous cortical embryonic divisions (**Fig. 4**). During prophase of nuclear
    cycle 11 bright asters of centrosome nucleated microtubules are observed either
    side of the embryonic nuclei, which appear as dark circles (**A**) due to the exclu-
    sion of tubulin monomers by the intact nuclear envelope. As the nuclear enve-
    lope breaks down during prophase, an influx of tubulin into the nuclear space
    occurs (**B**). As metaphase proceeds, increased labeling of the mitotic spindle is
    observed, owing to the increase in density and length of microtubules between
    the pairs of sister centrosomes (**C–E**). As chromosomes separate during anaphase
    the labeled spindles increase in length (**F**). As the embryo enters telophase and
    the nuclear envelopes surrounding the sister nuclei reform, tubulin is again
    excluded from the nuclear space (**G**). Strong tubulin labeling of prominent
    midbodies between sister nuclei and of duplicating centrosomes at the poles is
    observed. As each nucleus enters interphase the astral microtubule staining moves
    out the plane of focus as a consequence either of centrosomal migration or nuclear
    rotation (**H**).
7. We have also used these techniques to examine the behaviour of chromosomes
    during the *Drosophila* cortical nuclear cycles, using rhodamine-labeled histones
    (8,13,14) as fluorescent probes (**Fig. 5**). During late prophase the DNA condenses
    (**A**) and the chromosomes begin to align on the metaphase plate (**B**). During
    metaphase the chromosomes are maximally compacted (**C**), before beginning to
    separate synchronously during anaphase (**D**). During late anaphase sister chro-
    mosomes are completely separated and individual chromosome arms may be dis-
    cerned (**E**). The DNA decondenses during telophase as illustrated by the more
    diffuse histone labeling (**F**). The DNA remains decondensed during the short
    intervening interphase (**G**), before the chromosomes begin condensation once
    more, as the nuclei synchronously enter the next prophase (**H**).
8. CP190 is a 1096 amino acid *Drosophila* centrosomal protein that oscillates in a
    cell-cycle-specific manner between the nucleus during interphase, and the cen-
    trosome during mitosis *(15–17)*. To characterize the regions of CP190 respon-
    sible for its nuclear and centrosomal localizations, we made a series of 6 X His
    fusion proteins spanning CP190. The purified fusion proteins were fluorescently
    labeled with rhodamine using the protocol described in this chapter, injected into
    syncytial *Drosophila* embryos, and followed using time-lapse fluorescence con-
    focal microscopy *(18)*. Injection of labeled fusion proteins resulted in one of three
    localization patterns, examples of which are shown in **Fig. 6**. In each case, micro-
    graphs were selected from a time-lapse series taken of a portion of the embryo's
    surface. Fusion protein CP190 a (containing CP190 amino acids 167–321) local-
    izes to nuclei during interphase and to the cytoplasm during mitosis (**Fig. 6**,
    leftmost panels). The fusion protein CP190 b (CP190 amino acids 266–608)
    localizes to centrosomes throughout the cell cycle (**Fig. 6**, middle panels) and a
    fusion protein spanning both of these regions (CP190 amino acids 167–608)
    localizes to nuclei during interphase and to centrosomes during mitosis (**Fig. 6**,

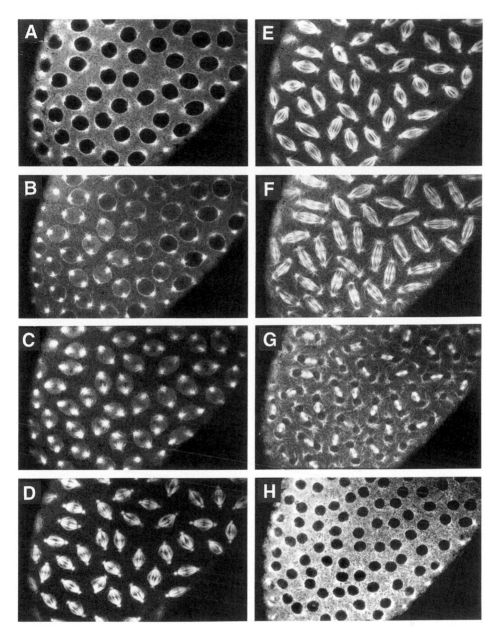

Fig. 4. Analysis of tubulin dynamics in living *Drosophila* embryos (Reprinted from *The Journal of Cell Biology*, by copyright permission of the The Rockerfeller University Press).

CP190 a+b, rightmost panels). Using this type of analysis, we were able to identify the 19 amino acid nuclear localization signal responsible for the nuclear

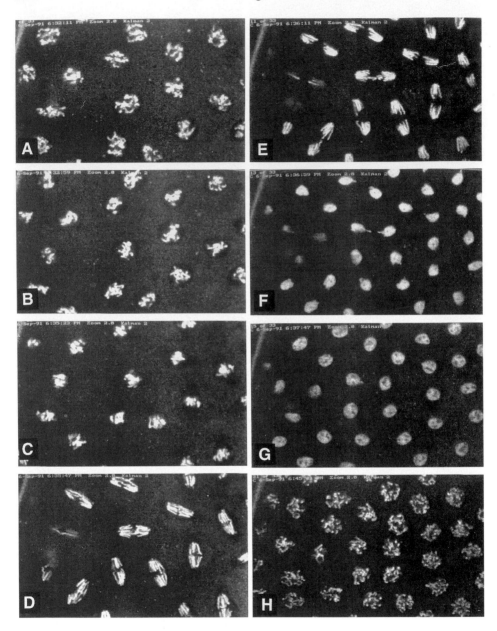

Fig. 5. Analysis of chromosome behaviour using labeled histones in living *Droso-phila* embryos. (Reproduced from *Molecular Biology of the Cell* [1993], Vol.4, pp. 885–8960, with permission of the American Society for Cell Biology.)

localization of our fusion proteins and a region of 124 amino acids near the center of the protein that is sufficient to confer robust centrosomal localization.

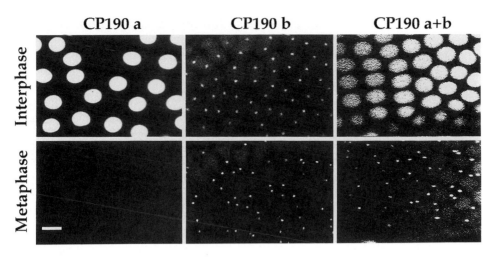

Fig. 6. Domain analysis of the centrosomal protein, CP190 in living *Drosophila* embryos.

9. The technique of microinjecting fluorescently labeled proteins into living samples and following their dynamics by confocal microscopy has also been performed in other model systems. The reorganization of the cytoskeleton during meiosis and premature cytoplasmic streaming has been studied by microinjection of *Drosophila* oocytes *(19,20)*. The behavior of cortical actin *(1)* and the segregation of germ granules *(21)* have been examined in living *C. elegans* embryos by microinjection of oocytes with labeled actin and an anti-P granule antibodies, respectively. In the *Xenopus* embryo, live confocal analysis of eggs injected with rhodamine tubulin was used to examine the relationship between microtubules and cortical rotation *(22)*. Labeled tubulin has also been injected into the zebrafish zygote and used to monitor cytokinesis of early blastomeres using live confocal microscopy (S. Jesuthasan, personal communication), and into the giant squid axon to monitor slow axonal transport *(23)*.

10. Oregon green 488 is a new fluorophore that has the same absorption and emission spectra as fluorescein, but is more photostable, has a greater resistance to photobleaching and is less pH dependent.

11. Other fluorophores such as the Molecular Probes series of BODIPY™ fluorophores, have been developed but may be significantly quenched upon conjugation to proteins. Derivitives of these fluorophores containing spacer arms may prove valuable for preparing conjugates with proteins in the future.

12. For single labels, we prefer Oregon green, Cy3, and Cy5. Tetramethylrhodamine is a less expensive alternative to Cy3 that is also commonly used. For double labels, we prefer Oregon green with Texas red, X-rhodamine, or Cy5. The emission spectra of Cy5 and fluorescein are better separated, but Cy5 is barely visible to the naked eye so for many applications Texas red or X-rhodamine is preferable. Tetramethylrhodamine or Cy3 can also be used for double labeling with

fluorescein or Oregon green, but the emission spectra of these dyes overlap more significantly, making signal separation difficult.

13. Although very useful because of their spectral properties, we have found Texas red-X and X-rhodamine the most difficult to use for antibody labeling because of their hydrophobicity. Texas red-X appears to work better than X-rhodamine and is fine if used at labeling stoichiometries less than 1.2 (dye/ antibody).

Acknowledgments

We would like to thank Doug Kellogg for providing the spectra in **Fig. 1** and NickDompé for the illustrations in **Fig. 3**.

REFERENCES

1. Hird, S. (1996) Cortical actin movements during the first cell cycle of the *Caenorhabditis elegans* embryo. *J. Cell Sci.* **109,** 525–533.
2. Haugland, R.P. *Handbook of Fluorescent Probes and Research Chemicals.* Molecular Probes. Eugene, OR.
3. Means, G. E. and Feeney R. E. (1971) *Chemical modification of proteins.* Holden-Day, San Francisco.
4. Post P. L., Trybus K. M., and Taylor D. L. (1994) A genetically engineered, protein-based optical biosensor of myosin II regulatory light chain phosphorylation. *J. Biol. Chem.* **269,** 12,880–12,887.
5. Swedlow, J. R., Sedat, J. W., and Agard, D. A. (1993) Multiple chromosomal populations of topoisomerase II detected in vivo by time-lapse, three-dimensional wide-field microscopy. *Cell* **73,** 97–108.
6. Hyman, A., Drechsel, D., Kellogg, D., Salser, S., Sawin, K., Steffen, P., Wordeman, L., and Mitchison, T. (1991). Preparation of modified tubulins. *Methods Enzymol.* **196,** 478–485.
7. Kellogg, D.R., Mitchison, T. J., and Alberts, B. M. (1988). Behavior of microtubules and actin filaments in living *Drosophila* embryos. *Development* **103,** 675–686.
8. Minden, J. S., Agard, D. A., Sedat, J. W., and Alberts, B. M. (1989) Direct cell lineage analysis in *Drosophila melanogaster* by time-lapse three-dimensional optical microscopy of living embryos. *J. Cell Biol.* **109,** 505–516.
9. Debec, A., Kalpin, R. F., Daily, D. R., McCallum, P. D., Rothwell, W. F., and Sullivan, W. (1996) Live analysis of free centrosome behavior in normal and aphidicolin-treated *Drosophila* embryos. *J. Cell Biol.* **134,** 103–115.
10. Fogarty, P., Kalpin, R. F., and Sullivan, W. (1994) The *Drosophila* maternal-effect mutation grapes causes a metaphase arrest at nuclear cycle 13. *Development* **120,** 2131–2142.
11. Sullivan, W., Minden, J., and Alberts, B. M. (1990) daughterless-abo-like, a *Drosophila* maternal-effect mutation that exhibits abnormal centrosome separation during the late blastoderm divisions. *Development* **110,** 311–323.
12. Sullivan, W., Daily, D. R., Fogarty, P., Yook, K., and Pimpinelli, S. (1993) Delays in anaphase initiation occur in individual nuclei of the syncytial *Drosophila* embryo. *Mol. Biol. Cell* **4,** 885–896.

13. Valdes-Perez, R. E. and Minden, J. S. (1995) *Drosophila melanogaster* syncytial nuclear divisions are patterned: time-lapse images, hypothesis and computational evidence. *J. Theor. Biol.* **175,** 525–532.
14. Kalpin, R. F., Daily, D. R., and Sullivan, W. (1994) Use of dextran beads for live analysis of the nuclear division and nuclear envelope breakdown/reformation cycles in the *Drosophila* embryo. *Biotechniques* **17,** 730–733.
15. Callaini, G. and Riparbelli, M. G. (1990) Centriole and centrosome cycle in the early *Drosophila* embryo. *J. Cell Sci.* **97,** 539–543.
16. Whitfield, W. G. F., Chaplin, M. A., Oegema, K., Parry, H., and Glover, D. M. (1995) The 190kDa centrosome-associated protein of *Drosophila melanogaster* contains four zinc finger motifs and binds to specific site on polytene chromosomes. *J. Cell Sci.* **108,** 3377–3387.
17. Whitfield, W. G. F., Millar, S. E., Saumweber, H., Frasch, M., and Glover, D. M. (1988) Cloning of a gene encoding an antigen associated with the centrosome in *Drosophila. J. Cell Sci.* **8,** 467-480.
18. Oegema, K., Whitfield, W. G. F., and Alberts, B. (1995) The cell cycle dependent localization of the CP190 centrosomal protein is determined by the coordinate action of two separable domains. *J. Cell Biol.* **131,** 1261–1273.
19. Matthies, H.J., McDonald, H. B., Goldstein, L. S. B., and Theurkauf, W. E. (1996) Anastral meiotic spindle morphogenesis: role of the non-claret disjunctional kinesin-like protein. *J. Cell Biol.* **134,** 455–464.
20. Theurkauf, W. E. (1994) Premature microtubule-dependent streaming in *cappuccino* and *spire* mutant oocytes. *Science* **265,** 2093–2096.
21. Hird, S. N., Paulsen, J. E., and Strome, S. (1996) Segregation of germ granules in living *Caenorhabditis elegans* embryos: cell-type-specific mechanisms for cytoplasmic localisation. *Development* **122,** 1303–1312.
22. Larabell, M. A., Rowning, B. A., Wells, J., Wu, M., and Gerhart, J. C. (1996) Confocal microscopy analysis of living *Xenopus* eggs and the mechanism of cortical rotation. *Development* **122,** 1281–1289.
23. Teraski, M., Schmidek, A., Glabraith, J. A., Gallant, P. E., and Reese, T. S. (1995) Transport of cytoskeletal elements in the giant squid axon. *Proc. Natl. Acad. Sci. USA* **92,** 11,500–11,503.
24. Gil, S. C. and Von Hippel, P. H. (1989) Calculation of protein extinction coefficients from amino acid sequence data. *Analyt. Biochem.* **182,** 319–326.

15

Live Imaging with Green Fluorescent Protein

Jim Haseloff, Emma-Louise Dormand, and Andrea H. Brand

1. Introduction

If developmental biologists were given the chance to design the perfect cell- or tissue-specific marker, they would ensure that it had several properties. First, it would function in living animals, eliminating the need for fixation and dehydration and their associated artefacts. Second, it would permit each stage of development to be studied in a single, intact embryo. Third, it would function in all cell types and would reveal their morphology, making it simple to identify different cells without compromising their viability.

In 1994, all cell and developmental biologists won the lottery when it was demonstrated that green fluorescent protein (GFP), a kind gift from the jellyfish *Aequorea victoria*, fulfilled all of these requirements *(1)*. GFP is a naturally fluorescent, nontoxic, protein that functions in a wide variety of transgenic animals *(1–5)*. The GFP protein is small (27 kDa) and can freely enter the nucleus, fill the cytoplasm of the cell, and diffuse into small cytoplasmic extensions. Most importantly, GFP requires no substrate, but fluoresces simply in response to ultraviolet (UV) or blue light. As the excitation and emission spectra of GFP are similar to those of fluorescein isothiocyatenate (FITC), the protein can be visualized with most conventional epifluorescence and confocal filter sets. And, as if we deserved a bonus, GFP is resistant to bleaching and is therefore ideal for long time-course analyses.

GFP is revolutionizing the study of developmental and cell biology in a wide variety of organisms from plants to mammals. Previously, cell labeling required the use of invasive techniques such as dye microinjection or immunohistochemistry. Using GFP as a cell marker, individual cells can be labeled in vivo and followed throughout development without harmful manipulation *(1)*.

From: *Methods in Molecular Biology, vol. 122: Confocal Microscopy Methods and Protocols*
Edited by: S. Paddock © Humana Press Inc., Totowa, NJ

1.1. Properties of GFP

In *A. victoria*, GFP is concentrated in bioluminescent organs, called lumisomes, at the margin of the jellyfish umbrella. Light emanates from yellow tissue masses that consist of about 6000–7000 photogenic cells. The cytoplasm of these cells is densely packed with fine granules that contain the components necessary for bioluminescence *(6,7)*. These include a Ca^{2+} activated photoprotein, aequorin, that emits blue-green light, and GFP, an accessory protein that accepts energy from aequorin and reemits it as green light *(8)*. GFP is an extremely stable protein of 238 amino acids *(3)*: its fluorescent properties are unaffected by prolonged treatment with 6 *M* guanidine-HCl, 8 *M* urea or 1% sodium dodecyl sulfate (SDS), and 2-d treatment with various proteases such as trypsin, chymotrypsin, papain, subtilisin, thermolysin, and pancreatin at concentrations up to 1 mg/mL *(9)* GFP is stable in neutral buffers up to 65°C, and displays a broad range of pH stability from 5.5 to 12.0.

Wild-type GFP has two excitation peaks, with maxima at 395nm (long wavelength UV) and 475 nm (blue light) *(8,10)*, and emits green light with a peak wavelength of 508 nm. The intrinsic fluorescence of the protein is due to a unique chromophore that forms posttranslationally and autocatalytically, by cyclization of amino acids 65–67, Ser-Tyr-Gly *(3,11,12)*. The tyrosine residue is then oxidized *(3,11,12)*. The crystal structure of GFP *(13,14)* shows an 11-strand β-barrel enclosing the central α helix that contains the chromophore. The barrel acts as a solvent cage to protect the chromophore from quenching and photochemical damage.

Several genomic and cDNA clones of GFP were isolated from a population of *A. victoria (3)*, and the coding sequence from one cDNA, pGFP10.1, has been used for protein expression, first in *Escherichia coli*, *Caenorhabditis elegans (1,12,15)*, and *Drosophila melanogaster* [*(16)* and now in many other organisms *(2,4,5)*.

1.2. Imaging GFP

The development of GFP as a marker in many different biological systems has emphasised the need to image GFP at high resolution. Confocal microscopy has played a pivotal role in extending the potential of GFP as a tool in biological research. In this chapter, we describe protocols for imaging expression of GFP in *Arabidopsis* and *Drosophila* .

Although GFP can be detected by conventional epifluorescence microscopy (using most FITC filter sets or a Chroma GFP filter set), autofluorescence can be a problem when examining living whole-mount preparations. In *Drosophila* embryos, e.g., yolk autofluorescence can obscure the signal from GFP, while in larvae reflection from the cuticle can cause difficulties. Confocal microscopy generates optical sections of fluorescently labeled sample and is ideal for imaging cells labeled with GFP. When imaging the ventral side of a *Drosophila* embryo, the background fluorescence emanating from the more dorsally placed yolk can be completely eliminated by optical sectioning.

The resolution of confocal images is much greater than can be obtained by conventional epifluorescence techniques. Furthermore, after collecting a stack of optical sections (a Z-series) the sample can be reconstructed in three dimensions, then rotated or tilted to view cells or structures that would otherwise be obscured. Most confocal software enables time-lapse imaging in one focal plane, and 4D imaging, where Z series are collected over time. These series of images can then be played back as movies.

High resolution optical techniques can be used noninvasively to monitor the dynamic activities of living cells. Using coverslip-based culture vessels, specialized microscope objective lenses and the optical sectioning properties of the confocal microscope *(17)*, it is possible to monitor simply and precisely both the three-dimensional arrangement of living cells, and their behavior through long time-lapse observations. For time-lapse studies, it is very important that GFP fluorescence be bright, to minimize high levels of illumination that can cause phototoxicity and photobleaching during observation (*see* **Subheading 7.2.**). Spectral variants of GFP are now available that allow double-labeling) *(18,19)* (and **Subheading 1.4.**). The precision with which cellular structures can be labeled with GFP, and the ease with which subcellular traffic can be monitored indicates that this approach will be invaluable for cell biological and physiological observations.

1.3. Mutant Forms of GFP

The expression of wild-type GFP was initially reported to be poor or variable in a number of heterologous systems. For example, the GFP mRNA sequence is mis-spliced in transgenic *Arabidopsis thaliana* plants, resulting in the removal of 84 nucleotides from within the coding sequence, between residues 380 and 463 *(20)*. Alteration of the GFP coding sequence and mutation of the cryptic intron are required for proper expression of the gene in *Arabidopsis* and other plants *(20,21)*. A number of groups have been attempting to improve expression of GFP in animal systems, and have produced GFP variants with altered codon usage. For example, GFP variants with "humanized" or other optimised codon usage lead to better translation efficiency. These also show improved levels of expression in plants *(22–24)*, and it is likely that these altered forms confer some degree of immunity from aberrant RNA processing.

Wild-type GFP is temperature sensitive, and fluoresces poorly at temperatures above 25°C *(25)*. Polymerase chain reaction (PCR)-based mutagenesis of the GFP coding sequence has generated a thermotolerant mutant with improved fluorescence *(25)*. The mutant contains two altered amino acids (V163A, S175G) that greatly improve folding of the apoprotein at elevated temperatures. The V163A mutation has been isolated independently by several groups *(19,25–28)* and may play a pivotal role in folding. Further mutations originated in a screen by Cormack et al. *(29)*, who introduced large numbers of random amino acid substitutions into the 20 residues flanking the chromophore of GFP. They used fluorescence-activated cell sorting to select variants that fluoresced 20- to 35-fold more intensely than wild-type and showed that the folding of these mutant proteins in bacteria was more efficient. One of the variants [GFPmut1, *(29)*] changes two amino acids within the central α-helix of the protein, adjacent to the chromophore, F64L and S65T. Recombination of these mutations with V163A and S175G, which are found on the outer surface of the protein *(13,14,25)*, leads to markedly improved fluorescence *(30–32)*. The beneficial effect of both sets of mutations on protein folding, and their apparent additive effect, suggests that they may play separate roles in the folding or maturation process.

1.4. GFP Spectral Variants for Multichannel Confocal Microscopy

Although wild-type GFP can be excited by light of 400 nm and 475 nm wavelengths, the 400 nm excitation peak predominates. This is a useful property for simple detection of the protein using a long-wavelength UV source. However, efficient blue light excitation (approx 470–490 nm) is essential when using microscopes equipped with fluorescein filter sets, or confocal microscopes or cell sorters that use argon lasers.

The relative amplitudes of the excitation peaks of GFP can be altered by mutagenesis *(12,19,33,34)*. A GFP variant containing the I167T mutation *(12)* has dual excitation peaks of almost equal amplitude *(25)* and is highly fluorescent in vivo. This allows the efficient use of techniques that require either UV or blue light excitation of the protein, e.g., when screening GFP-expressing tissues with a UV lamp, or when using blue laser light excited confocal microscopy, respectively. GFP variants that contain the S65T mutation *(29,35)* are now widely used and provide optimized properties for blue light excitation, but are not useful for detection by long-wavelength UV light.

Different colored GFPs are particularly useful for distinguishing cells, organelles, or proteins in the same living organism. It is possible to alter the fluorescence spectra of GFP more extensively by introducing additional substitutions in and around the chromophore. For example, the Y66H substitution dramatically shifts both the excitation and emission spectra of GFP to give a "blue fluorescent protein" (BFP) *(12)* (excitation maximum = 382 nm, emission maximum = 448 nm). There is much interest in using these "color variants" for multichannel imaging with GFP. However, there is considerable spectral overlap between some of the mutants, and the choice of GFP variant and excitation source in an experiment must be carefully considered.

BFP is the variant most easily distinguished from GFP, thanks to its blue-shifted spectrum. However, the first generation of BFP variants *(12,19)* were weakly fluorescent and bleached rapidly compared to GFP. BFP variants with improved fluorescence, owing to the "folding"mutations V163A *(25)* and F64L *(36,37)*, are now available. However, BFP must be excited using a UV light source, which is damaging to living cells, and to detect the protein by confocal microscopy requires the use of a separate UV laser.

An alternative approach is to use the cyan fluorescent variant (CFP) and the yellow fluorescent variant (YFP). CFP bears a tryptophan substitution within the chromophore (Y66W), which gives a broad excitation peak at 440–455 nm and an emission maximum at 483 nm. YFP (excitation maximum 514nm, emission maximum 527 nm) was generated by modifying residues that lie close to the chromophore (S65G, S72A, T203Y) *(13,29)*. Whereas the first reported variants were sometimes poorly expressed, the introduction of altered codon usage and mutations to improve protein folding has improved fluorescence.

Multiline argon ion lasers are relatively inexpensive and emit light mainly at discrete wavelengths of 458 nm, 477 nm, 488 nm, and 514 nm (lasers with about 80 mW or better output are required for bright 458 nm emission). Each of the laser lines falls near to an excitation peak for one of the GFP variants [458 nm = CFP, 477 nm = wild-type GFP and mGFP5 *(25)*, 488 nm = S65T variants of GFP, 514 nm = YFP]. Proteins can be tagged with the various fluorescent colors, and visualized in living cells using a confocal microscope equipped with an argon ion laser and a motorized excitation filter wheel containing laser line excitation filters. The different fluorescent tags can be excited in turn, using the appropriate laser lines, and the signals collected through specialized emission filter blocks.

More recently, it has been found that, on photoactivation by blue light, GFP is switched to a new green-absorbing and red-emitting state *(38,39)*. This property of GFP has already been exploited to measure protein diffusion in the cytoplasm of living bacteria and is likely to have many further applications.

1.5. GFP-Protein Fusions

Wang and Hazelrigg demonstrated that proteins can be fused to either the C- or N- terminus of GFP without inhibiting fluorescence *(16)*. This property has been exploited to fuse GFP to proteins of interest to examine their subcellular localization, or to target GFP to subcellular structures such as the cytoskeleton or organelles.

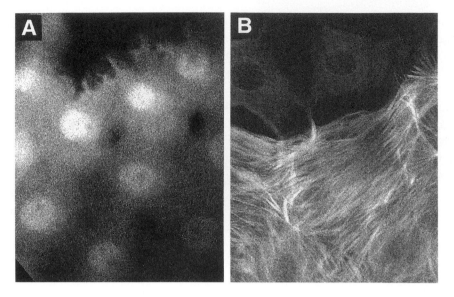

Fig. 1. Comparison of expression of unfused GFP (**A**) with Tau-GFP (**B**) in the epidermis of living *Drosophila* larvae. (A) GFP is detected at high levels in the cell nuclei and at lower levels in the cell cytoplasm. (B) Tau-GFP is excluded from the nucleus and instead highlights the cell cytoskeleton by binding to microtubules (*43*; *see* http://www.welc.cam.ac.uk/brand lab/).

1.5.1. GFP Targeted to the Cytoskeleton

The use of GFP to examine cytoskeletal dynamics, by fusion to cytoskeleton-associated proteins, has recently been reviewed (*40*). Cytoskeleton-associated protein fusions to GFP have also been useful for examining cell morphologies. For example, the microtubule-associated protein, τ (*41,42*) has been fused to GFP to target GFP to the microtubules (*43–45*) (**Fig. 1**). Because axons are particularly dense in microtubules, τ-GFP is an excellent marker for axonal morphology.

1.5.2. GFP Targeted to Organelles

Many organelle-targeted GFP fusions have been engineered by fusing the GFP protein to organelle-specific proteins (reviewed in **ref. 5**). GFP has been targeted to the nucleus (*22,46–49*), cell membrane (*50*), endoplasmic reticulum (*20,51*), mitochondria (*18,48*), and plastids (*22*). One of the cell membrane-targeted GFPs is a fusion to the NMDA receptor, a ligand-gated ion channel found in certain neurons (*52*). The GFP-NMDAR1 chimera is functional and fluorescent, thus enabling studies of the expression, localization, and processing of the ion channel.

1.5.3. GFP-Protein Dynamics

Apart from being used as a marker of live cells or organelles, GFP can be fused to proteins to study their subcellular localization and redistribution during the cell cycle or in response to external stimuli (reviewed in **ref. 5**). For example, transport of a secreted protein, chromogranin B, has been visualised in HeLa cells by fusing GFP to the C-terminus of the protein *(53)*. It is possible to visualize secretion of GFP, allowing analysis of transport through the secretory pathway in living cells. Endoplasmic reticulum-to-Golgi apparatus transport of a viral glycoprotein tagged with GFP has recently been visualized in living COS cells *(54)*.

Although these studies used cultured cells, GFP fusions have also been used to examine protein expression in the whole organism. For example, in *C. elegans*, the expression and subcellular localization of Daf-7, a protein thought to be necessary for the functioning of a subset of chemosensory neurons, was investigated by examining the expression of a Daf-7-GFP fusion *(55)*. Daf-7 is thought to regulate the formation of the dauer larva in response to pheromonal signals. Daf-7-GFP is found to be expressed in the cytoplasm of the appropriate chemosensory neurons to mediate this pheromone response.

The following protocols are for imaging expression of GFP in *Arabidopsis* and *Drosophila*.

2. Materials: Arabidopsis
2.1. Preparation of Arabidopsis Seedlings

1. Ethanol
2. Sterile water
3. Surface sterilizing solution (1% (w/v) sodium hypochlorite and 0.1% (v/v)
4. Nonidet P-40 (NP40) detergent)
5. GM medium:
   1× Murashige and Skoog basal medium with Gamborg's B_5 vitamins (Sigma)
   1% sucrose
   0.5 g/L 2-(*N*-morpholino)ethanesulfonic acid (MES)
   0.8% Agar (adjusted to pH 5.7 with 1*M* KOH)

2.2. Fluorescent Counterstaining of Arabidopsis Seedlings

1. 10 µg/mL aqueous solution of propidium iodide (Sigma)
2. FM 1-143 (Molecular Probes Inc.)

2.3. Double Labeling Arabidopsis Seedlings with GFP Variants (see Subheading 2.1.)

3. Methods for Visualizing GFP in Arabidopsis Seedlings
3.1. Mounting and Observing GFP-Expressing Arabidopsis Seedlings

1. Perform procedures in a laminar flow hood. Place 20–100 transgenic *Arabidopsis* seeds in a 1.5-mL microfuge tube and wash for about 1 min with 1 mL of ethanol.
2. Incubate the seeds with 1 mL of surface sterilizing solution for 15 min at room temperature.
3. Wash the seeds three times with 1 mL of sterile water, and transfer by pipet to agar plates containing GM medium (56). Add 25 mg/L kanamycin if antibiotic selection of transgenic seedlings is necessary.
4. Incubate sealed plates or vessels for 1–3 d in the dark at 4°C, and then transfer to an artificially lit growth room at 23°C for germination.
5. *Arabidopsis* seedlings germinate after 3 d, and can be used for microscopy for several weeks. Remove GFP-expressing *Arabidopsis* seedlings from agar media and mount in water under glass coverslips for microscopy.

We examine specimens using a Bio-Rad MRC-600 laser-scanning confocal microscope equipped with a 25 mW krypton/argon or argon ion laser and filter sets suitable for the detection of fluorescein and Texas red dyes (Bio-Rad filter blocks K1/K2 with krypton/argon ion laser, and A1/A2 with argon ion laser).

We routinely use a Nikon 60x PlanApo numerical aperture (NA) 1.2 water immersion objective lens to minimize loss of signal through spherical aberration at long working distances. For the collection of time-lapse images, the laser light source was attenuated by 99% using a neutral density filter, the confocal aperture was stopped down, and single scans were collected at 2-s intervals.

We transfer the large data files to an Apple® Macintosh computer, and the programs PicMerge, authored by Eric Sheldon, and 4DTurnaround, authored by Charles Thomas, are used with Adobe Photoshop® and Premiere® to produce QuickTime movies for display and analysis.

3.2. Fluorescent Counter-Staining of Arabidopsis Seedlings

Autofluorescent chloroplasts, normally present in the upper parts of the plant, and certain red fluorescent dyes can provide useful counterfluors for GFP.

3.2.1. Labeling Root Meristem Cell Walls with Propidium Iodide

Propidium iodide is a cationic dye that does not readily cross intact membranes, and yet it penetrates throughout the meristem and binds to cell walls, forming an outline of the living cells. The dye is excluded by the Casparian strip present in older parts of the root and does not penetrate shoot tissue well, and thus is best suited for use in the root meristem (**Fig. 2**).

1. Grow *Arabidopsis* seedlings in sterile culture.
2. Remove seedlings from agar media and place in a well of a microtiter dish with 1 mL of staining solution for 10–20 min at room temperature.

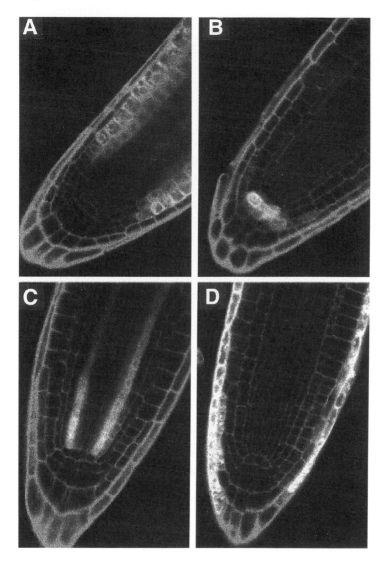

Fig. 2. GFP expression in *Arabidopsis* root tips double labeled with propidium iodide. Propidium iodide labels the root meristem cell walls. **(A–D)** Different patterns of GFP expression can be generated by enhancer detection. (J. P. H. and S. Hodge, unpublished; *see* http://brindabella.mrc-lmb.cam.ac.uk).

3. Mount seedlings in water under a coverslip for direct microscopic observation.
4. Propidium iodide is red fluorescent and can be detected using a filter set suitable for Texas red fluorescence, with little spillover between GFP (fluorescein) and propidium iodide channels.

3.2.2. Labeling Plasma Membranes with FM 1-43

The cationic styrylpyridinium dye FM 1-43 (Molecular Probes) provides a useful stain for the plasma membrane in root and shoot tissue of *Arabidopsis*. It is particularly useful for specifically labeling the plasma membrane of shoot epidermal cells, and we have been using this to characterize GFP expression patterns in *Arabidopsis* cotyledons and leaves.

1. Remove seedlings from sterile culture and place in 1 mL of 1μg/mL of FM 1-43 in water for 10 min at room temperature.
2. Mount seedlings in water under a coverslip for direct microscopic observation.
3. FM 1-43 emits a broad orange fluorescence, and signal can be detected in both red and green emission channels. We generally use 488 nm wavelength laser light to excite both GFP and FM 1-43, and to collect the emissions of both fluors in the same channel. This is possible because of the very localized distribution of FM 1-43 in shoot epidermal cells.

3.3. Double Labeling Arabidopsis Seedlings with GFP Variants

1. Grow plants expressing mCFP, mGFP5, and/or mYFP proteins in sterile culture, and mount in water for microscopy.
2. We use a Bio-Rad MRC-600 microscope equipped with an 80 mW argon ion laser and a motorized excitation filter wheel containing narrow band-pass filters to select laser lines at 458 nm for excitation of mCFP, 477 nm for excitation of mGFP5 (mGFP5), and 514 nm for excitation of mYFP. A multiline argon ion laser of higher power (>50 mW) is generally needed to provide 45 8nm illumination of useful intensity.
3. We use mCFP and mYFP, and mGFP5 and mYFP together for double-labeling experiments. The proteins are sequentially excited using the appropriate laser lines, and the signals collected through specialized emission filter blocks: mCFP/mYFP = 495 nm longpass dichroic mirror, 485±15 nm, 540±15 nm bandpass filters, or mGFP5/mYFP = 527 nm longpass dichroic mirror, 500 nm longpass, and 540 nm±15 nm bandpass filters (Omega Optical). The use of selective monochromatic excitation allows useful discrimination between mGFP5 and mYFP, which have overlapping fluorescent spectra. The greater spectral differences between mCFP and mYFP result in clean discrimination of the fluorescent signals.
4. Sequentially collected images may be merged and pseudocoloured using Adobe Photoshop.

4. Notes

4.1. Imaging Roots of Arabidopsis Seedlings

For extended time-lapse imaging of roots, sterile seeds can be sown in coverslip-based vessels (Nunc) which comprise four wells, each containing about 400 μL of low gelling temperature agarose with GM medium. The roots of these plants grow down through the media and then along the surface of the coverslip. The roots are then ideally positioned for high resolution microscopic imaging through the base of the vessel.

5. Materials: *Drosophila*

5.1. *Live* Drosophila *Samples*

Home-made sieves: Cut the top off a 15 mL polypropylene Falcon tube, about 4 cm from the screw-on cap. Cut a wide hole in the cap. Place fine gauze over the end of the Falcon tube and hold it in place by screwing on the cap.
18 × 18 mm Coverslips (Menzel-Glaser)
22 × 40 mm Coverslips (Menzel-Glaser)
50% Clorox
Glue (double-sided Scotch tape glue dissolved in heptane)
Voltalef oil: 10S
Halocarbon oil: 50% Halocarbon 27 and 50% Halocarbon 700 (Sigma)
Air-permeable Teflon membrane mounted on Perspex frame
Parafilm

5.2. *Fixed* Drosophila *Samples*

PBT [phosphate-buffered saline (PBS), 0.1% Triton X-100 (Sigma)]
50% Clorox
4% Formaldehyde (BDH) in PBT (make up fresh)
Heptane (Sigma-Aldrich)
Methanol (Fisher Scientific International)
Vectashield (Vector Labs)

6. Methods for Visualizing GFP in *Drosophila*

The following protocols are for examining GFP fluorescence in either live (**Fig. 3**) or fixed whole-mount embryos (**Fig. 4**). We have also stained embryos with rabbit anti-GFP antibodies (Clontech) with excellent results, although we have had less success using monoclonal anti-GFP antibodies (Clontech).

6.1. *Live* Drosophila *Samples*

1. Collect embryos in a vial, or on an apple juice agar plate. To ensure that most embryos are at the appropriate stage, it is helpful to do short collections (2–4 h) and then age them until the appropriate stage of development.
2. Wash eggs into a sieve placed on a plastic tray, with water and a paintbrush.
3. Dechorionate in 50% Chlorox for 3 min, then wash thoroughly with water.
4a. For an upright microscope, transfer the embryos with a paintbrush into a drop of Voltalef or Halocarbon oil placed in the middle of an air-permeable Teflon membrane stretched over a Perspex frame (designed by E. Wieschaus).

Place an 18 × 18 mm coverslip on either side of the embryos, 2–3 cm apart, to prevent them being squashed, and cover with a 22 × 40 mm coverslip. The embryos will develop normally through embryogenesis in most cases.

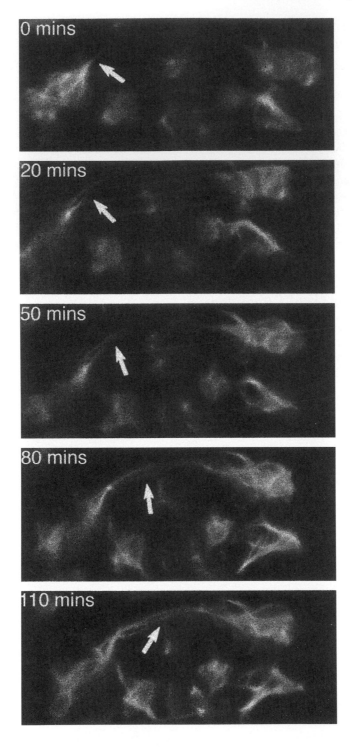

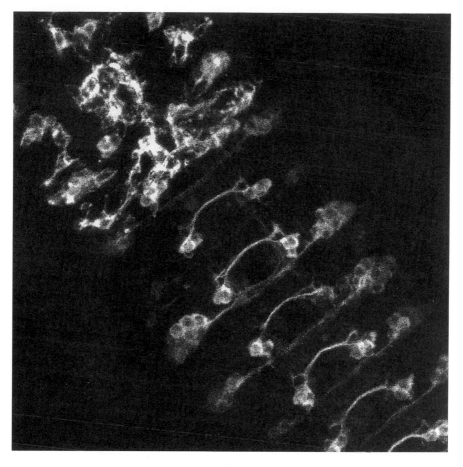

Fig. 4. GFP in neurons and glia of a fixed *Drosophila* embryo. Tau-GFP continues to fluoresce after formaldehyde fixation. In this transgenic line, Tau-GFP labels both neurons and glia in the thorax (**top**), but only neurons in the abdomen (**bottom**) *(45)*.

We do not seal the coverslips as it is often useful to roll the embryos to obtain a dorsal, ventral, or lateral view. In addition, the solvents in many nail varnishes have been reported to inhibit GFP fluorescence *(1)*.

4b. When using an inverted microscope, the embryos can be mounted simply by gluing them to a coverslip. Cut a square hole in the middle of a small piece of parafilm, and place this "frame" around the embryos to prevent oil running off.

Fig. 3. *(facing page)* Time-course of axon outgrowth in a *Drosophila* embryo. Neurons expressing Tau-GFP can be followed in living embryos. Tau-GFP labels cell bodies and is transported into axons *(43)*. The axon projections grow toward and then across the ventral midline in the embryonic central nervous system.

6.2. Fixed Drosophila Samples

GFP remains fluorescent after fixation and, although real-time analysis is sacrificed, GFP expression can be correlated with endogenous gene expression. GFP fluorescence is preserved, if somewhat diminished, after formaldehyde fixation.

1. Transfer dechorionated embryos to an Eppendorf tube, and half fill with heptane.
2. Remove residual PBT and top up with 4% formaldehyde in PBT. The embryos should float at the interface between the formaldehyde and heptane. Fix for 30 min with gentle rolling.
3. Replace the formaldehyde with methanol and remove the vitelline membranes by shaking vigorously for 30 s.
4. Wash in methanol for 10 s, then replace with PBT to rehydrate the embryos. Do not leave the embryos in methanol longer than necessary as it has been reported to lead to a rapid and irreversible loss of GFP fluorescence (10).
5. Proceed with antibody staining (e.g., see ref. 57).
6. Mount embryos in Vectashield. Use two 18 × 18 mm coverslips flanking the embryos as supports, then cover with a 22 × 40 mm coverslip. We do not seal the coverslips (see Subheading 6.1., step 4a).

In our hands, GFP fluorescence does not survive in situ hybridization protocols, which include an overnight incubation in formamide.

6.3. Confocal Microscopy

We collect our images using a Bio-Rad MRC1024 confocal scan head and krypton/argon mixed gas laser, on a Nikon E800 upright microscope. GFP is excited by the 488nm laser line. We use a laser power between 1% and 30%, and the standard setup for imaging FITC, which uses a 522/32 emission filter (i.e., wavelengths between 506–538 nm are transmitted). The Chroma HQ500LP emission filter, which transmits light of wavelengths greater than 500 nm, gives a brighter GFP signal. We have used GFP labeling in conjunction with other vital dyes, such as DiI for cell lineage tracing and acridine orange to detect dying cells.

Embryos can be viewed with a 60× planapochromat, 1.4 NA oil immersion objective lens. To reduce spherical aberration when focusing deep into an aqueous sample, embryos are mounted in water, or another aqueous mounting medium, and viewed with a long-working distance, cover slip corrected, water immersion objective (e.g., a Nikon 60× planapochromat, 1.2 NA, working distance 220 μm) (17). Unfortunately, these objectives are extremely expensive ($7000–8000, £5000–£6000).When imaging more than 60–80 μm deep in the tissue, even under the conditions described previously, image quality degrades rapidly. To reduce noise and sharpen the image, we Kalman average between 2 and 15 frames whenever possible.

7. Notes

7.1. Movement

When movement is rapid, it is not possible to Kalman average several scans, as this blurs rather than sharpens the image. Similarly, it may not be possible to project a Z-series, as each optical section will not be aligned with its neighbors. For example, when imaging the synchronous nuclear divisions in precellular embryos, we scan a single focal plane continuously on a slow setting. When observing macrophage movements, we collect an image every 15–20 s on a slow setting, and Kalman average 2–3 frames.

As mentioned in above, we do not seal our coverslips. For this reason, we collect Z-series from the top of the sample to the bottom, thereby forcing the coverslip against the embryo. If the focus motor is set in the opposite direction, the objective can lift the coverslip, and the embryo with it. It is possible to glue the embryos to the coverslip in an appropriate orientation, using a nontoxic glue such as Scotch tape dissolved in heptane, although this may impair image quality.

7.2. Bleaching

GFP is an ideal marker for time-lapse studies because of its resistance to photo-bleaching. When examining intact embryos in their vitelline membranes, we can collect images in a single focal plane, every 15 s, over a 2-h period without a major loss of signal. When collecting a 4D series, it has been possible to collect a Z stack of optical sections every 5 min over 5 h. To limit bleaching and photo damage to the specimen we try to keep the number of scans for Kalman averaging and the number of sections per Z-series to a minimum. In living samples, cells often continue to express GFP during the period of the time course, thereby replenishing the fluorescent protein.

We do almost all of our imaging on whole-mount embryos within their vitelline membranes. In our experience, animals that have been dissected prior to viewing (e.g., as "flat preps") tend to bleach more readily.

7.3. Making Movies

The Bio-Rad Laser Sharp confocal software allows confocal images to be converted to PICT or TIFF format. When converting large Z- or time-series, it is faster to use Confocal Assistant to generate TIFF files. These can then be exported to a Macintosh computer and assembled using Adobe Photoshop or Illustrator. To make movies that can be transferred to videotape, we assemble Z- or time-series in Adobe Premiere.

References

1. Chalfie, M., Tu, Y., Euskirchen, G., Ward, W. W., and Prasher, D. C. (1994) Green fluorescent protein as a marker for gene expression. *Science* **263,** 802–805.

2. Cubitt, A. B., Heim, R., Adams, S. R., Boyd, A. E., Gross, L. A., and Tsien, R. Y. (1995) Understanding, improving and using green fluorescent proteins. *Trends Biochem. Sci.* **20,** 448–455.
3. Prasher, D. C., Eckenrode, V. K., Ward, W. W., Prendergast, F. G., and Cormier, M. J. (1992) Primary structure of the aequorea victoria green fluorescent protein. *Gene* **111,** 229–233.
4. Prasher, D. C. (1995) Using GFP to see the light. *Trends Genet.* **11,** 320–323.
5. Gerdes, H. and Kaether, C. (1996) Green fluorescent protein: applications in cell biology. *FEBS Lett.* **389,** 44–47.
6. Davenport, D. and Nichol, J. A. C. (1955) Luminsecence in *Hydromedusae. Proc. Royal Society* **Series B, 144,** 399–411.
7. Morin, J. G. and Hastings, J. W. (1971) Energy transfer in a bioluminescent system. *J. Cellular Physiol.* **77,** 313–318.
8. Morise, H., Shimomura, O., Johnson, F. H., and Winant, J. (1974) Intermolecular energy transfer in the bioluminescent system of *Aequorea. Biochemistry* **13,** 2656–2662.
9. Bokman, S. H. and Ward, W. W. (1981) Renaturation of *Aequorea* green fluorescent protein. *Biochem. Biophys. Res. Comm.* **101,** 1372–1380.
10. Ward, W. W., Cody, C. W., Hart, R. C., and Cormier, M. J. (1980) Spectrophotomeric identity of the energy transfer chromophores in *Renilla* and *Aequorea* green fluorescent proteins. *Photochem. Photobiol.* **31,** 611–615.
11. Cody, C. W., Prasher, D. C., Westler, W. M., Prendergast, F. H., and Ward, W. W. (1993) Chemical structure of the hexapeptide chromophore of the *Aequorea* green fluorescent protein. *Biochemistry* **32,** 1212–1218.
12. Heim, R., Prasher, D. C., and Tsien, R. Y. (1994) Wavelength mutations and posttranslational autoxidation of green fluorescent protein. *Proc. Natl. Acad. Sci. USA* **9,** 12,501–12,504.
13. Ormo, M., Cubitt, A. B., Kallio, K., Gross, L. A., Tsien, R. Y., and Remington, S. J. (1996) Crystal structure of the *Aequorea victoria* green fluorescent protein. *Science* **273,** 1392–1395.
14. Yang, F., Moss, L. G., and Phillips, G. N. (1996) The molecular structure of green fluorescent protein. *Nature Biotechnol.* **14,** 1246–1251.
15. Inouye, S. and Tsuji, F. I. (1994) *Aequorea* green fluorescent protein - expression of the gene and fluorescence characteristics of the recombinant protein. *FEBS Lett.* **341,** 277–280.
16. Wang, S. and Hazelrigg, T. (1994) Implications for *bcd* messenger RNA localization from spatial distribution of exu protein in *Drosophila* oogenesis. *Nature* **369,** 400–403.
17. Haseloff, J. and Amos, B. (1995) GFP in plants. *Trends Genet.* **11,** 328,329.
18. Rizzuto, R., Brini, M., De Giorgi, F., Rossi, R., Heim, R., Tsien, R. Y., and Pozzan, T. (1996) Double labelling of subcellular structures with organelle-targeted GFP mutants *in vivo. Curr. Biol.* **6,** 183–188.
19. Heim, R. and Tsien, R. Y. (1996) Engineering green fluorescent protein for improved brightness, longer wavelengths and fluorescence resonance energy transfer. *Curr. Biol.* **6,** 178–182.

20. Haseloff, J., Siemering, D. R., Prasher, D. C., and Hodge, S. (1997) Removal of a cryptic intron and subcellular localisation of green fluorescent protein are required to mark transgenic *Arabidopsis* plants brightly. *Proc. Natl. Acad. Sci. USA* **94,** 2122–2127.

21. Reichel, C., Mathur, J., Eckes, P., Langenkemper, K., Reiss, B., Koncz, C., Schell, J., and Maas, C. (1996) Enhanced green fluorescence by the expression of an *Aequorea victoria* green fluorescent protein mutant in mono- and dicotyledonous plant cells. *Proc. Natl. Acad. Sci. USA* **93,** 5888–5893.

22. Chiu, W., Niwa, Y., Zeng, W., Hirano, T., Kobayashi, H., and Sheen, J. (1996) Engineered GFP as a vital reporter in plants. *Curr. Biol.* **6,** 325–330.

23. Haas, J., Park, E. C., and Seed, B. (1996) Codon usage limitation in the expression of HIV-1 envelope glycoprotein. *Curr. Biol.* **6,** 315–324.

24. Pang, S. Z., DeBoer, D. L., Wan, Y., Ye, G., Layton, J. G., Neher, M. K., Armstrong, C. L., Fry, J. E., Hinchee, M. A. W., and Fromm, M. E. (1996) An improved green fluorescent protein gene as a vital marker in plants. *Plant Physiol.* **112,** 893–900.

25. Siemering, K. R., Golbik, R., Sever, R., and Haseloff, J. (1996) Mutations that suppress the thermosensitivity of green fluorescent protein. *Curr. Biol.* **6,** 1653–1663.

26. Crameri, A., Whitehorn, E. A., Tate, E., and Stemmer, W. P. C. (1996) Improved green fluorescent protein by molecular evolution using DNA shuffling. *Nature Biotechnol.* **14,** 315–319.

27. Davis, S. J. and Viestra, R. D. (1998) Soluble, highly fluorescent variants of green fluorescent protein (GFP) for use in higher plants. *Plant Mol. Biol.* **36,** 521–528.

28. Kohler, R. H., Zipfel, W. R., Webb, W. W., and Hanson, M. R. (1997) The green fluorescent protein as a marker to visualize plant mitochondria *in vivo. Plant J.* **11,** 613–621.

29. Cormack, B. P., Valdivia, R. H., and Falkow, S. (1996) FACS-optimized mutants of the green fluorescent protein (GFP). *Gene* **173,** 33–38.

30. Zernicka-Goetz, M., Pines, J., Ryan, K., Siemering, K. R., Haseloff, J., Evans, M. J., and Gurdon, J. B. (1996) An indelible lineage marker for *Xenopus* using a mutated green fluorescent protein. *Development* **122,** 3719–3724.

31. Zernicka-Goetz, M., Pines, J., Siemering, K. R., Haseloff, J., and Evans, M. J. (1997) Following cell fate in the living mouse embryo. *Development* **124,** 1133–1137.

32. Schuldt, A., Adams, J. H. J., Davidson, C. M., Micklem, D. R., Haseloff, J., St Johnston, D., and Brand, A. H. (1998) Miranda mediates asymmetric protein and RNA localisation in the developing nervous system. *Genes Develop.* **12,** 1847–1857.

33. Delagrave, S., Hawtin, R. E., Silva, C. M., Yang, M. M., and Youvan, D. C. (1995) Red-shifted excitation mutants of the green fluorescent protein. *Bio-Technology* **13,** 151–154.

34. Ehrig, T., O'Kane, D. J., and Predergast, F. G. (1995) Green fluorescent protein mutants with altered fluorescence excitation spectra. *FEBS Lett.* **367,** 163–166.

35. Heim, R., Cubitt, A. B. and Tsien, R. Y. (1995) Improved green fluorescence. *Nature* **373,** 663,664.

36. Stauber, R. H., Horie, K., Carney, P., Hudson, E. A., Tarasova, N. I., Gaitanaris, G. A., and Pavlakis, G. N. (1998) Development and applications of enhanced green fluorescent protein mutants. *Biotechniques* **24,** 462–471.

37. Yang, T. T., Sinai, P., Green, G., Kitts, P. A., Chen, Y. T., Lybarger, L., Chervenak, R., Patterson, G. H., Piston, D. W., and Kain, S. R. (1998) Improved fluorescence and dual colour detection with enhanced blue and green variants of the green fluorescent protein. *J. Biol. Chem.* **273,** 8212–8216.

38. Elowitz, M. B., Surette, M. G., Wolf, P. E., Stock, J., and Leibler, S. (1997) Photoactivation turns green fluorescent protein red. *Curr. Biol.* **7,** 809–812.

39. Sawin, K. E. and Nurse, P. (1997) Photoactivation of green fluorescent protein. *Curr. Biol.* **7,** 606,607.

40. Ludin, B. and Matus, A. (1998) GFP illuminates the cytoskeleton. *Trends Cell Biol.* **8,** 72–77.

41. Butner, K. A. and Kirschner, M. W. (1991) Tau protein binds to microtubules through a flexible array of distributed weak sites. *J. Cell Biol.* **115,** 717–730.

42. Callahan, C. A. and Thomas, J. B. (1994) Tau-b-galactosidase, an axon-targeted fusion protein. *Proc. Natl. Acad. Sci. USA* **91,** 5972–5976.

43. Brand, A. (1995) GFP in *Drosophila. Trends Genet.* **11**, 324,325.

44. Micklem, D. R., Dasgupta, R., Elliott, H., Gergely, F., Davidson, C., Brand, A., Gonzalez-Reyes, A., and St Johnston, D. (1997) *mago nashi* is required for the polarisation of the oocyte and the formation of perpendicular axes in *Drosophila. Curr. Biol.* **7,** 468–478.

45. Dormand, E. L. and Brand, A. H. (1998) Runt determines cell fates in the *Drosophila* embryonic CNS. *Development* **125,** 1659–1667.

46. Davis, I., Girdham, C. H., and O'Farrell, P. H. (1995) A nuclear GFP that marks nuclei in living *Drosophila* embryos; maternal supply overcomes a delay in the appearance of zygotic fluorescence. *Develop. Biol.* **170,** 726–729.

47. Ogawa, H., Inouye, S., Tsuji, F. I., Yasuda, K., and Umesono, K. (1995) Localization, trafficking and temperature-dependence of the *Aequorea* green fluorescent protein in cultured vertebrate cells. *Proc. Natl. Acad. Sci. USA* **92,** 11,899–11,903.

48. Rizzuto, R., Brini, M., Pizzo, P., Murgia, M., and Pozzan, T. (1995) Chimeric green fluorescent protein as a tool for visualizing subcellular organelles in living cells. *Curr. Biol.* **5,** 635–642.

49. Shiga, Y., Tanakamatakatsu, M., and Hayashi, S. (1996) A nuclear GFP beta-galactosidase fusion protein as a marker for morphogenesis in living *Drosophila. Dev. Growth Differ.* **38,** 99–106.

50. Moriyoshi, K., Richards, L. J., Akazawa, C., O'Leary, D. D. M., and Naanishi, S. (1996) Labeling neural cells using adenoviral gene transfer of membrane-targeted GFP. *Neuron* **116,** 255–260.

51. Terasaki, M., Jaffe, L. A., Hunnicutt, G. R., and Hammer, J. A. (1996) Structural change of the endoplasmic reticulum during fertilization: evidence for loss of membrane continuity using the green fluorescent protein. *Develop. Biol.* **179,** 320–328.

52. Marshall, J., Molloy, R., Moss, G., Howe, J., and Hughes, T. (1996) The jellyfish green fluorescent protein: a new tool for studying ion channel expression and function. *Neuron* **14,** 211–215.

53. Kaether, C. and Gerdes, H. (1995) Visualization of protein transport along the secretory pathway using green fluorescent protein. *FEBS Lett.* **396,** 267–271.
54. Presley, J. F., Cole, N. B., Schroer, T. A., Hirschberg, K., Zaal, K. J. M., and Lippincott-Schwartz, J. (1997) ER-to-Golgi transport visualized in living cells. *Nature* **389,** 81–85.
55. Schackwitz, W. S., Inoue, T., and Thomas, J. H. (1996) Chemosensory neurons function in parallel to mediate a pheromone response in *C. elegans. Neuron* **17,** 719–728.
56. Valvekens, D., Van Montagu, M., and Van Lijsebettens, M. (1988) *Agrobacterium tumefaciens*-mediated transformation of *Arabidopsis thaliana* root explants by using kanamycin selection. *Proc. Natl. Acad. Sci. USA* **85,** 5536–5540.
57. Patel, N. H. (1994) Imaging neuronal subsets and other cell types in whole-mount *Drosophila* embryos and larvae using antibody probes, in Drosophila melanogaster: *Practical Uses in Cell and Molecular Biology* (Goldstein, L. S. B. and Fyrberg, E. A., eds.), Academic Press, San Diego, pp. 446–485.

16

Fluorescent Calcium Indicators:
Subcellular Behavior and Use in Confocal Imaging

Donald M. O'Malley, Barry J. Burbach, and Paul R. Adams

1. Introduction

Most animal cells maintain a large gradient of free calcium across the plasma membrane (10,000-fold or more), with intracellular free calcium held at a level of about 100 nM. To accomplish this, cells have an array of molecular machinery including ATP-driven calcium pumps, calcium exchangers, intracellular calcium storage compartments, and an assortment of calcium binding proteins *(1–6)*. This machinery is needed because calcium is used ubiquitously as a signaling molecule, involving numerous receptors and a complex maze of intracellular signaling cascades *(7–10)*. One approach to the study of this biological arena is the tracking of calcium levels in different subcellular regions. Our goal is to consider how fluorescent calcium indicators, in conjunction with confocal microscopy, can best be used to image subcellular calcium dynamics.

Calcium subserves critical functions in nerve and muscle cells. In skeletal muscle, e.g., calcium plays a pivotal role in muscular contraction, whereas in neurons, calcium triggers the release of neurotransmitters from nerve terminals *(11–13)*. In the nervous system, calcium further regulates both gene expression and the excitability of neural networks and can also participate in the long-term storage of information. Fluorescent calcium indicators, which were developed largely through the efforts of Roger Tsien and his colleagues *(14,15)*, provide a remarkable tool for looking at these processes. These indicators are especially useful when combined with a powerful imaging tool such as confocal microscopy. Our lab group has used confocal calcium imaging to visualize the distribution of functioning calcium channels, to study events occurring at the plasma and nuclear membranes, and as an in vivo probe of

From: *Methods in Molecular Biology, vol. 122: Confocal Microscopy Methods and Protocols*
Edited by: S. Paddock © Humana Press Inc., Totowa, NJ

neuronal activity *(16–21)*. Although fluorescent calcium indicators are quite good at revealing the dynamics of intracellular processes, they are, paradoxically, not so good at telling us the absolute levels of free calcium at different subcellular locations.

The difficulty in determining absolute free calcium levels is well illustrated by an exchange of letters concerning findings from Stephen Baylor's group that were published in *The Biophysical Journal (22,23)*. These letters debated the level at which free calcium rests in skeletal muscle and could come to no more agreement than to say that free calcium rests somewhere between 30 nM and 300 nM. A prime cause of this uncertainty is that calcium indicators behave very differently in vivo (inside cells) than they do in vitro (in cuvettes). At issue in these letters was a potentially dramatic shift in the indicator's intracellular dissociation constant (K_D). A further complication is that a calcium indicator's behavior can vary from one subcellular region to another—a situation that has led to the publication of some rather improbable "findings." The thesis of this chapter is that investigators must consider such complications if they hope to reach accurate conclusions about subcellular calcium dynamics. In reality, this is an extremely challenging problem without a robust, practical solution. However, achieving at least a partial understanding of an indicator's intracellular behavior will help, even if only to make clear the limitations of the technique. This chapter provides methods for: (1) assessing an indicator's behavior in "confocal cuvettes," (2) assessing the indicator's intracellular behavior, and (3) interpreting subcellular fluorescence gradients. To illustrate key issues that arise in subcellular calibration, this methodology is applied to some highly controversial nuclear/cytoplasmic fluorescence gradients.

2. Materials
2.1. Intracellular Solutions
2.1.1. Calibration Solutions

1. Zero Calcium:

ATP, disodium salt	2 mM
HEPES sodium	5 mM
$MgCl_2$	2 mM
TEA-chloride (tetraethylammonium)	20 mM
Gluconic acid	100 mM
Cesium hydroxide	100 mM
BAPTA	10 mM

   (Note: adjust final pH to 7.3 with NaOH)
2. High Calcium: (add to zero calcium solution, then check pH):

Calcium chloride	10 mM

   (and reduce gluconic acid and cesium hydroxide to 90 mM)

Table 1
Formulation of Calcium Series

Free Ca^{2+} (nM)	Total Ca^{2+} (mM)	Calibration Solutions 0 Ca^{2+}	Calibration Solutions 10 mM Ca^{2+}	Bound Fluo 3	Predicted Fluor.
0	0.000	1000	0	0.00	100
1	0.062	994	6	0.25	114
7	0.421	958	42	1.74	199
35	1.803	820	180	8.08	561
62	2.807	719	281	13.48	868
177	5.284	472	528	30.78	1854
350	6.911	309	691	46.78	2766
700	8.204	180	820	63.75	3734
3,580	9.666	34	966	89.99	5229
7,330	9.888	12	988	94.85	5506
14,200	10.000	0	1000	97.27	5659
45,600	10.110	0	1000^a	99.10	5749

This formulation is based on the calibration solutions specified in **Subheading 2.1.1**. 10 mM BAPTA and 100 mM fluo-3 are entered into the calculation program and the amount of Total Calcium (mM) is calculated by the program for each Free Calcium level (in nM) specified. The number of microliters of 0 calcium and high calcium solution to make 1 mL of each solution in the calcium series is also specified. At the same time that Total Calcium is determined from the program, the amount of calcium-bound fluo-3 (%) is also recorded. This is used to calculate the predicted fluorescence using $F = (100 - B) + 58B$.

^{a}Because this calcium level exceeds the 10 mM calcium in the stock, add a small amount of extra calcium, 1.2 µL, from a 100 mM CaCl$_2$ stock solution.

2.1.2. Physiological Solution

Same as zero calcium solution, but remove BAPTA and increase gluconic acid and cesium hydroxide to 110 mM.

2.2. Calcium/BAPTA Mixtures

The makeup of a set of solutions with different levels of free calcium, i.e., a *calcium series* for evaluating fluorescent calcium indicators, is described in **Table 1.**

2.3. Imaging System (Main Components)

1. Bio-Rad MRC 600 (or other line-scanning confocal imaging system)
2. Zeiss IM 35 inverted microscope (or other inverted microscope)
3. Leitz 50X 1.0 NA water immersion objective (or other high throughput objective)
4. Newport vibration isolation table
5. UniBlitz external shutter (required for patch-clamping experiments)

2.4. Electrophysiology

2.4.1. Basic Patch Clamping (Main Components)

1. Axoclamp 2A patch-clamp amplifier
2. Tektronix storage oscilloscope
3. Leader monitor oscilloscope
4. WPI interval generator
5. Gould 2 channel chart recorder
6. Narashige aqua micromanipulator
7. Flaming-Brown micropipet puller, Sutter Instruments

2.4.2. Intracellular Perfusion (Additional Items)

1. Modified electrode holder A058-C, E. W. Wright
2. Internal perfusion tube, quartz, 200 μm outer diameter
3. Microcap micropipets, Drummond Scientific Co

2.5. Definitions

Two terms need to be distinguished:

1. *Cytoplasm*—the contents of the cell outside of the cell's nucleus, including *both* organelles and cytosol.
2. *Cytosol*—the "aqueous space" of the cytoplasm, i.e., the solution that surrounds the cytoplasmic organelles.

3. Methods

Fluorescent calcium indicators are typically loaded into cells via one of two main techniques: (1) direct "injection" of the indicator or (2) using the membrane crossing or acetoxymethylester or "AM"-form of the indicator. Within the injection category, we use an *electrophysiological* technique, patch-clamping, to load calcium indicators into nerve cells—either cultured cells or cells in brain slices. Although electrophysiology is not a trivial technique to acquire, it has the advantage of loading the fluorescent indicator directly into the aqueous space of the cell, i.e., the cytosol. The alternative, bathing cells in the membrane permeant "AM" form of an indicator, allows the indicator to cross the plasma membrane, after which it is metabolically trapped inside the cell by deesterification. *AM* indicators cross not only plasma membranes, but also organellar membranes which adds another level of complexity to subcellular calibrations (*24–29*; *see* **Note 1**). This chapter focuses on the simpler case of the non-AM indicators, i.e., the water-soluble forms: free acid, salt, and dextran-linked forms. This chapter also focuses on "single-wavelength" indicators, which are most widely used in confocal imaging, rather than the "ratiometric" indicators, where images are acquired at two wavelengths and used to generate a ratio-image (**Note 2**). The issues discussed pertain, however, to both AM and ratiometric indicators.

3.1. Formulation of Calibration Solutions

The *intracellular* solution (**Subheading 2.1.**) used in our patch-clamp pipettes is designed to mimic, to some extent, the intracellular milieux (**Note 3**). The baseline behavior of a fluorescent calcium indicator is first measured in a cuvette using the calibration version of our intracellular solution (**Subheading 2.1.1,**). The chief components of the intracellular solution are a physiological salt concentration, a pH buffer, ATP, magnesium, a calcium buffer, and the fluorescent calcium indicator. BAPTA is the calcium buffer of choice for our experiments because it is faster and less pH-sensitive than the other main calcium buffer, EGTA *(30,31)*.

For calibration experiments, a high concentration of BAPTA, 10 m*M*, is used to "clamp" the concentration of free calcium either in vitro or inside cells. Free calcium is set at different levels by adding varying amounts of calcium to a 0 calcium/10 m*M* BAPTA solution. This calcium series (**Table 1, Subheading 2.2.**) spans the calcium-sensitive range of the indicator. The fluorescence of the indicator is then measured at these different calcium levels. A fixed concentration of calcium indicator (100 μ*M* in our experiments) is used with the calcium series to determine the K_D and dynamic range (maximal fluorescence response) of the indicator. In practice, the desired set of calcium levels is created by mixing together varying amounts of 0 calcium calibration solution and high calcium solution (10 m*M* total calcium). The required total calcium is first calculated and the relative mix of the 0 and high calcium solutions determined (**Table 1**).

The calculation of the total amount of calcium required to set free calcium at a specific level is not trivial, since multiple equilibria (calcium/BAPTA and calcium/indicator) must be solved. Bers et al. *(31)* describe a program that does this calculation based on an iterative procedure reported by Fabiato and Fabiato *(32; see* **Note 4**). Using this type of program, we enter the total amount of indicator (100 μ*M*) and a guess of its K_D (400 n*M* for fluo 3) plus the total amount of buffer (10 m*M*) and its K_D (160 n*M* for BAPTA) and the desired concentration of free calcium. The program then solves for the unspecified variable, in this instance total calcium. The uncertainty in the indicator's K_D is not a problem at this point because of the large excess of BAPTA. BAPTA's K_D, however, depends on salt concentration and temperature, both of which must be taken into account (**Note 5**). The set of calcium solutions described (**Table 1**) can be used to check successive batches of indicator or new indicators being tested. Although this might seem laborious (**Note 6**), one may wish to know, before completing a series of experiments, how responsive the calcium indicator is and whether or not its K_D is close to the advertised value (which may not be the case). These calibration solutions will also be used for in vivo "calcium clamp" experiments, where the indicator's behavior is determined inside living cells. One shortcut for checking new batches of indicator is to measure the indicator's fluorescence at two calcium levels: 0 calcium and saturating (10 mM) calcium. Although this provides only a dynamic range, it at least gives a sense of the indicator's performance with a particular imaging system and under actual experimental conditions.

3.2. "Cuvette" Calibrations: Fluorescence vs [Free Calcium]

Fluorescent calcium indicators are expensive, so it is desirable to work with relatively small volumes. An alternative is to work with a low concentration of indicator during the calibration, but this deviates from the experimental condition so we avoid it. From a 20X stock of the indicator (2 mM, in de-ionized water), 10 μL of indicator is taken and added to 190-μL aliquots of the different calcium solutions (**Table 1**) in small (1 mL) plastic conical tubes and mixed well. A Bio-Rad MRC 600 with an inverted Zeiss microscope (**Subheading 2.3.**) was used to collect the fluorescent measurements shown in this chapter, but the methods are directly applicable to other confocal systems. Our "cuvette" consists of a small well that is fashioned by cutting a 14 mm diameter hole in the center of a 35-mm plastic petri dish and then gluing a square glass coverslip onto the bottom of the dish, covering the hole (**Note 7**). For calibration, a small droplet (4–5 μL) from each solution in the calcium series is placed on the coverslip. About five droplets can be conveniently spaced out across the coverslip, allowing the fluorescence at five calcium levels to be quickly determined. The top of the well is then covered with a coverslip to slow evaporation. The fluorescence of each droplet is then measured and its fluorescence plotted as a function of calcium concentration (**Note 8**). Our calibration of fluo-3, shown in **Fig. 1**, used 10 calcium levels.

The indicator's in vitro or "cuvette" behavior can now be evaluated. The K_D is simply read from the plot of fluorescence vs calcium concentration: the concentration that produced one-half of the total fluorescence increase is the K_D. To more completely evaluate the behavior of the indicator, the predicted fluorescence is also plotted vs calcium concentration. The determination of predicted fluorescence is based on the observed K_D and dynamic range. The calcium calculation program is used as described previously, except that the indicator's observed K_D, rather than the estimated K_D, is used. Keeping the total concentration of indicator fixed at 100 μM, the amount of calcium-bound indicator is read from the program after entering each free calcium level (**Table 1**). Because fluorescence is directly proportional to the amount of calcium-bound indicator (after subtracting the 0-calcium fluorescence), the predicted fluorescence can be directly calculated. To standardize these calibration curves, the 0 calcium fluorescence (F_{min}) is normalized to a

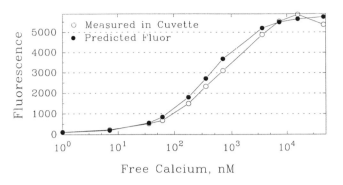

Fig. 1. Predicted vs observed fluorescence of fluo-3. The fluorescence of fluo-3 is measured in microliter-droplets placed on a glass coverslip. The droplets have varying concentrations of free calcium, as described in **Subheading 2.2.** Measurements are made with the MRC 600's aperture set at 1/3 open and with a 1.5 log unit neutral density filter. The gain of the photomultiplier tube was initially set at maximum for the zero calcium solution, but with increasing calcium (fluorescence) levels, was adjusted downward in 1 unit increments to accommodate the range of fluorescence values encountered. Fluorescence values are then scaled to the initial gain setting after determining gain correction factors based on the use of test droplets and test slides as fluorescence standards.

value of 100 and the maximum fluorescence (F_{max}) is then the maximal-fold change observed x 100, yielding a value of 5800 for the 58-fold dynamic range of fluo-3. For **Fig. 1** the predicted fluorescence curve is calculated from:

$$F = (100 - B) + 58 \cdot B \qquad \text{(Eq. 1)}$$

where B = the percentage of calcium-bound fluo-3 and is equal to the amount of fluo-3 bound, in μM, determined by the calculation program (**Table 1**).

In a perfect world the calculated and the observed curves would match exactly. As seen in **Fig. 1**, there was fair agreement between the predicted and the observed fluorescence using fluo-3's published K_D of 400 nM. One could attempt to better fit the predicted curve to the observed fluorescence by, e.g., varying the value for fluo-3's K_D in the program and recalculating the fraction bound and the predicted fluorescence. However, an indicator's intracellular behavior deviates so greatly from its in vitro behavior that small refinements in the cuvette value have little biological relevance.

3.3. Intracellular Calibration

The calcium binding and spectral properties of fluorescent calcium indicators are dramatically altered by the intracellular milieux. In muscle cells, perhaps 85% of the indicator is bound to proteins or other large molecules, resulting in an increase in the K_D of three- to fourfold (11,33). This appears to be true for most fluorescent calcium indicators, including fura-2, indo-1, fluo-3, and fura red (34–36). Also the indicators have a substantially reduced dynamic range intracellularly, versus their dynamic range in cuvettes (34,36,37). Furthermore, spectral shifts can result from indicator/protein interactions and other causes (34,38–42). Spectral shifts may be especially problematic for ratiometric indicators because a small shift in fluorescence spectra could have a much larger effect on ratio imaging than on single-wavelength imaging. Even greater problems are caused by compartmentalization of indicators into cytoplasmic organelles, such as the endoplasmic reticulum and mitochondria, which often contain high levels of free calcium. In such cases, the cytoplasmic fluorescence (or cytoplasmic ratio) is the sum of organellar plus cytosolic fluorescence, whereas the nuclear fluorescence (or ratio) has no organellar contribution and therefore will not equal the cytoplasmic fluorescence. This is quite a serious complication for the membrane permeant or AM form of indicators because AM-indicators may cross intracellular membranes just as readily as they cross the plasma membrane (25,26,28,29,43). For example, in cardiac ventricular cells, mitochondria were reported to contain roughly 50% of the indo-1 AM taken up by the cell (43,44). A final complication is the differential binding of calcium indicators to the assorted constituents of different subcellular regions (e.g., different proteins, DNA, etc.). In aggregate, these issues make it almost a foregone conclusion that the calibration curve will vary from one subcellular region to another. These complications apply to both single-wavelength and ratiometric calcium imaging (**Note 9**).

The standard approach to intracellular calibration of fluorescent calcium indicators is to bathe cells in a second class of membrane permeant compound—the calcium ionophores such as ionomycin and A23187. These ionophores facilitate equilibration of calcium levels across the membrane. This approach seems to work well on some cell types but less well on others—the ionophores may work very slowly and may not fully equilibrate calcium across the plasma membrane (27,45,46). Ionophores are especially difficult to employ when working with tissues such as brain slices. In addition, if there is compartmentalized indicator, the ionophores might also cause redistribution of calcium between different compartments and affect other ionic gradients and pH, all of which can alter the behavior of the indicator (26,38,47) and influence the calibration. It is thus perhaps not surprising that few detailed intracellular calibration curves have been published for fluorescent calcium indicators. Work by Blatter and Wier (43) and Schnetkamp et al. (48) provided detailed calibrations using batches of either permeabilized cells or rod outer segments, respectively, but we are aware of no subcellular calibration where a detailed series of calibration images, collected from a single cell, has been presented. The majority of calcium imaging studies either use in vitro calibration data or simply avoid estimating calcium concentrations altogether and report only fluorescence values. In this chapter, we will focus on alternative calibration approaches that rely on patch clamping. Although these approaches encounter some of the same limitations as the ionophore approach, they appear to have some advantages.

Two distinct electrophysiological approaches can be used to attempt intracellular calibration of calcium indicators: calcium pulsing and intracellular perfusion. Calcium pulsing involves patch-clamping cells (**Note 10**) and then using depolarizing voltage pulses to flood the cell with calcium. Intracellular perfusion also uses patch-clamping with the additional use of a fine tube that runs inside the patch-clamp electrode down to the tip; this tube is connected to a valve and allows perfusion of the pipette tip (and hence the cell) with any desired solution. Intracellular perfusion is substantially more difficult than calcium pulsing, but has clear advantages. In the context of these two calibration approaches, we will consider a particularly controversial topic: the existence of persistent calcium gradients that span the nuclear envelope. This ongoing controversy is deeply intertwined with the problem of subcellular calibration of fluorescent calcium indicators. It therefore provides a framework within which to discuss the interpretation of confocal calcium images.

3.3.1. Calibration via Intracellular Perfusion

In a typical experiment, the fluorescent indicator is dissolved in intracellular solution (**Subheading 2.1.**) and injected into a patch-clamp pipet. The patch-clamp pipet is then pushed against the cell membrane, forming a high-resistance seal. The small patch of membrane under the tip of the pipet is ruptured by gentle suction. This allows the contents of the pipette to diffuse into the cell. The intracellular solution in the pipette will gradually replace the cell's cytosol, i.e., the aqueous space of the cell outside of the organelles. This is why the solution is designed to mimic, to some extent, the intracellular milieux (**Note 3**). For all calibration experiments, calibration solution (**Subheading 2.1.1.**) is used; it has a high concentration of BAPTA to allow the setting of free calcium at a particular level. For intracellular perfusion, the patch electrode is initially filled with one solution (solution A) and after filling the cell to (or near) equilibrium with indicator (**Note 11**), the pipet tip is perfused with a second solution (solution B). The intracellular perfusion apparatus is set up by threading a fine piece of inflow tubing through a rubber O-ring into the patch-clamp electrode holder (**Note 12**). A fine quartz inner pipette fits snugly into this inflow tubing and extends very close to the tip of the patch pipet, which is filled with solution A. From the pipet, the inflow tubing runs to a reservoir filled with solution B. The electrode holder also has a port with a small tube running to an outflow chamber. This outflow line is clamped and held under a modest vacuum during the process of patch-clamping. At a given point in the experiment, the inflow line is unclamped and the vacuum from the outflow line then begins to pull solution B through the system. Once the reservoir is drained and the patch-electrode filled with solution B, the lines are re-clamped. By this point, solution B has begun diffusing into the cell.

Filling a cell with a 0 calcium solution (solution A) and subsequently perfusing it with a saturating calcium solution (solution B) provides the intracellular dynamic range of the indicator. That is, by taking an indicator from a calcium free to a calcium-saturated state, the maximal possible intracellular fluorescence change is obtained (**Note 13**). A key advantage of this style of calibration is that the only thing that is directly changed is the cell's free calcium concentration. This approach was illustrated for fluo 3 using cultured bullfrog sympathetic neurons (*see* Fig. 7A in **ref. 20**). In that study, the indicator showed a 24-fold fluorescence increase in the nucleus and an 18-fold increase in cytoplasm. This is the largest intracellular dynamic range demonstrated for fluo-3 in a calcium imaging study, which suggests that this method of controlling calcium may be as effective as other methods. In this calibration, the nucleus was initially twice as bright as the cytoplasm. This could be interpreted either as a simple fluorescence gradient or as a resting nuclear/cytosolic calcium gradient. Because intracellular perfusion produced roughly parallel fluorescence increases in both the nucleus and cytosol across a broad range of calcium levels, it seems much more likely that the resting fluorescence gradient is due to the behavior of the indicator, rather than to a nuclear/cytosolic calcium gradient that persists across all calcium levels (**Note 14**).

Confocal microscopy is a key tool for making such subcellular measurements accurately. Even in the case of the large-sized bullfrog neurons used here, good optical sectioning is needed to ensure that the nuclear signal is not contaminated by signal from the overlying cytoplasm. **Figure 2** shows two confocal views of a cell: a conventional "XY" image (**Fig. 2A**) and a vertical slice through the cell (referred to as an "XZ" image; **Fig. 2B**). The XY image shows an example of the nuclear and cytosolic regions from which our measurements are typically made. The XZ image is made by successively scanning a single line in the XY plane while focusing the objective at successive planes spanning the vertical (or "Z") extent of the cell. The XZ scan illustrates the optical sectioning achieved. In this instance the nucleolus (the bright spot in the center of the nucleus), which is only about 4 μm in diameter, is well-resolved. Few studies on nuclear calcium signals have so directly illustrated actual optical performance. This is unfortunate because XZ scans make it easy to graphically document said performance.

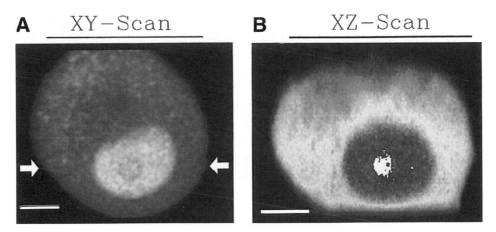

Fig. 2. Top and Side views of a cultured sympathetic neuron. In the conventional top view (**A**) the nucleus (bright region between the arrows) is easily distinguished because of its brighter fluorescence. This pattern of labeling always occurs when fluorescent calcium indicators are loaded into the aqueous space of nerve cells. A side view or "XZ"-scan (**B**) is generated by repetitively scanning a single line, in this instance, the line running between the two arrows in (A), while the lens is focused through the vertical or "Z" extent of the cell (lines were acquired at 0.4 μm intervals, with a total of 120 Z-steps). The scale bars are 10 μm; in (B) the scale bar is positioned at the level of the coverslip on which the cell is resting. All images in this chapter are linearly contrast enhanced except for the image in (B) which is a nonlinear representation provided only to highlight anatomical details such as the nucleolus—the small bright dot in the center of the nucleus (the nucleolus is 15% brighter than the nucleus which in turn is twice as fluorescent at the cytoplasm, the nucleus appears darker in this rendering because of the nonlinear contrast enhancement).

The combination of optical sectioning and intracellular perfusion thus confirms that the nucleus and cytoplasm show parallel fluorescence increases across a broad range of calcium levels. One might argue, however, that the calcium indicator was not saturated with calcium since it did not exhibit the full dynamic range observed in cuvettes. If so, then cytosolic calcium could, in theory, rest at one-half of the nuclear calcium level, show a 20-fold increase, and still be at a level far from saturating the indicator (thus preserving the idea that the nuclear signal is amplified relative to the cytosolic calcium signal). This is very unlikely because of the large size of the nuclear response, as well as the limited intracellular dynamic range of calcium indicators and the initial starting conditions (**Note 15**). Intracellular perfusion provides an even more compelling criticism of this idea. In Fig. 7b from **ref. 20**, a cell is initially filled with low calcium solution and is then perfused with 10 mM manganese, a concentration that is about 1,000,000-fold greater than fluo-3's K_D for manganese

(8 nM). This causes a rapid, parallel increase in both nuclear and cytoplasmic fluorescence (about 5.5-fold) that quickly saturates, virtually eliminating any plausible gradient hypothesis (in cuvettes, manganese produces an eightfold fluorescence increase when it binds to fluo-3; **ref. 49**). Again the nuclear/cytoplasmic ratio is about two, supporting the previous contention that calcium had saturated the indicator in both compartments of the cell. In addition, the time course required for manganese to saturate fluo-3 was similar, in both compartments, to that of calcium. In total, these findings provide powerful evidence that the nucleoplasm is in free communication with the cytosol (insofar as small molecules are concerned) and that nuclear and cytoplasmic fluorescence signals must be independently calibrated. These findings are supported by reports from other labs *(50–54)* and by earlier work from our lab *(17)*.

There are several factors that probably contribute to the resting nuclear/cytoplasmic fluorescence gradient. First, the cytoplasm is packed with membrane bound organelles, which will tend to exclude the highly charged potassium salt form of fluo-3 (it has a charge of -5). In the "nucleoplasm," there are no membrane bound organelles, so there is no "excluded" volume in the nucleus. In the cytoplasm, even with confocal calcium imaging, it is generally not possible to resolve the interstitial aqueous space (cytosol) from the organelles. Hence, the cytoplasmic signal is an integration of fluorescence coming from regions with fluo-3 (cytosol) and regions that are largely without fluo-3 (organelles). A second possible cause of nuclear/cytoplasmic differences is differential binding of fluo-3, including the possible intercalation of molecules such as fluo-3 into DNA *(55)*. Lastly, other aspects of the nuclear environment may enhance the fluorescence of fluo-3 *(54)*. Enhanced nuclear fluorescence is not restricted to calcium indicators because other dyes such as DAPI and lucifer yellow also show bright nuclei. Evaluating the competing explanations for the brighter nuclear signals is not trivial, so questions about these persistent subcellular fluorescence gradients are likely to persist.

With intracellular perfusion, the indicator's behavior was imaged in both compartments at essentially all calcium levels ranging from 10 nM to saturating calcium levels. Only 16 discrete time points (or "calcium levels") were sampled, but more could have been. This gives a comprehensive picture of how the indicator behaves in different subcellular regions across nearly the full range of calcium levels to which the indicator is responsive. This has not been directly illustrated (to our knowledge) in other calcium imaging studies. Those studies that take the effects of the intracellular environment into account typically provide a minimum fluorescence (F_{min}), a maximum fluorescence (F_{max}) plus one intermediate point for the K_D calculation. Indeed, we are not aware of the actual fluorescence images being shown for even these three points—in any study. With our calibration approach, the only factor that was directly changed inside the cell was the concentration of calcium in the aqueous space. One concern with a traditional calibration is how an amphipathic compound such as the ionophore will affect the stability of the plasma membrane. A related problem is that the fluorescence increase produced by high calcium plus A23187 or ionomycin is not always maximal *(27,45,56)*. Digitonin, a stronger permeabilizing agent, can yield a substantial further fluorescence increase of the calcium indicator, but also leads to leakage of indicator out of the cell. This seriously impedes calibration *(45,57)*. It would be instructive to compare calcium images generated via ionophoric calibrations to those produced by the electrophysiological approach so as to determine which protocol yields the largest and most systematic changes in subcellular fluorescence levels.

A particularly desirable feature of any calibration approach is reversibility. In the calibration mentioned *(20)*, the "electrical" health of the cell declined as intracellular calcium and manganese reached toxic levels. Although cells may look fine anatomically, a decline in their input resistance suggests a deteriorating plasma membrane. For smaller calcium changes, however, the effects of intracellular perfusion are reversible, as shown in a study of calcium-modulated potassium channels, where calcium was taken from 0 nM to 120 nM and back to 0 nM without any apparent ill effect on the potassium current or the health of the cell (*see* Fig. 2 in **ref. *18***). Such reversibility has generally not been evident in ionophoric calibrations. Intracellular perfusion, by virtue of its reversibility and noninvasiveness, is thus a potentially powerful technique for subcellular calibration of fluorescent indicators. Although technically challenging, this approach is also a general means to sequentially introduce impermeant compounds into cells at relatively defined concentrations during calcium imaging experiments. This approach may therefore have other applications.

3.3.2. Calibration via Calcium Pulsing

A technically less demanding calibration is to clamp free calcium at one concentration (by filling the cell via the patch-clamp electrode) and then take advantage of the cell's calcium channels to flood the cell with calcium, thereby saturating the calcium indicator. Nerve cells generally have a large complement of somatic calcium channels, so once a patch-clamp seal has been obtained and the cell filled with indicator, voltage-clamp can be used to repetitively deliver pulses of calcium (**Note 16**). To determine the dynamic range of the indicator, the cell is initially filled with 0-calcium solution and pulses of calcium are administered until the indicator is saturated (**Fig. 3**). One problem is that calcium channels tend to run down (inactivate) over time, limiting the ability to administer calcium. This makes it difficult to ensure that the indicator is saturated with calcium. The size of the calcium current, however, is continuously monitored during the calibration. If the voltage pulses are still

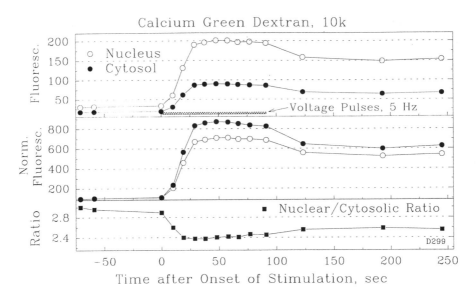

Fig. 3. Intracellular dynamic range of calcium green dextran (CGD). The neuron is loaded with a 0 calcium/10 mM BAPTA solution containing 100 μM CGD (10,000 Mr). The cell is voltage clamped at –70 mV and filled with CGD for approx 7 min prior to stimulation. At time 0, depolarizing voltage pulses (**Note 16**) are used to flood the cell with calcium. Fluorescence rapidly plateaus in both nucleus and cytosol, with a very similar time course. The fluorescence increases more in the cytosol than in the nucleus—a phenomenon that is usually observed with dextran-linked calcium indicators (the reverse is true with low molecular weight indicators). A slight recovery of the fluorescence toward baseline is observed after cessation of the voltage pulses, but cultured nerve cells seem to have difficulty in removing the very large quantities of calcium administered by this type of saturation protocol.

admitting appreciable quantities of calcium when the fluorescence reaches a plateau, which in this instance occurred about 25 s into the stimulation period, it suggests that intracellular calcium has surpassed the level required to saturate the indicator.

Another potential complication is that the calcium influx induced by the voltage pulses may be countered by an equal calcium efflux. Calcium extrusion mechanisms tend to be an order of magnitude or more slower than the peak influx mechanisms, so this seems unlikely. But, at high calcium loads, it is possible that Na^+/Ca^{2+} exchangers may produce a high efflux rate (**Note 17**). One control for this is to simply remove extracellular sodium (by, e.g., replacement of sodium with n-methyl glucamine). If the indicator is not already saturated, then halting sodium influx, which drives the largest component of calcium extrusion, should cause an immediate reduction in calcium efflux and a concomitant fluorescence increase. Alternatively, one could vary the extracellular concentration of calcium (normally 2 mM in our studies) between 0.5 and 10 mM. Any calcium level within this range of concentrations is sufficient to saturate the higher affinity calcium indicators discussed here (fluo-3, calcium green, oregon green-BAPTA ,and their dextrans), so the maximum obtainable fluorescence increase should not be greatly affected by changing extracellular calcium over this range once the indicator is saturated. If, however, calcium extrusion is preventing saturation, then voltage pulses at the higher concentrations of calcium should counteract extrusion, to some extent, and thus increase the observed dynamic range.

To estimate the indicator's intracellular K_D, calibration solutions are used to clamp calcium at intermediate calcium levels After setting intracellular calcium at each level, the indicator is subsequently saturated by calcium pulsing. The fluorescence responses determined for each starting calcium level allow independent estimates of the indicator's K_D to be made (**Note 18**). **Figure 4**, for example, shows a series of determinations of the intracellular K_D of calcium green dextran (CGD, 10,000 Mr), an indicator that provides robust, stable responses and has been quite useful for studying calcium dynamics in rat brain slices *(16)* and living zebrafish *(21,58)*. In this series, calcium is clamped at 3 intermediate levels and the maximal fluorescence response is determined for each starting level. For each calcium level, the experiment is repeated on several cells and the maximal fluorescence increase obtained is used to determine the K_D. There are trivial reasons for getting smaller than maximal fluorescence changes (such as failure of the calcium current or loss of cell integrity), so the largest observed change is taken as the true maximal fluorescence increase. For the three calcium set points used (10 nM, 107 nM, and 369 nM) the K_D determinations were, respectively: 100, 230, and 320 nM (**Note 19**). The low value (associated with the 10 nM set point) was partially discounted because of the greater error associated when correcting for autofluorescence and black level at very low starting fluorescence/calcium levels. The two higher values were closer to calcium green dextran's in vitro K_D of 250 nM. Taking a weighted average of the three determinations thus yields a value in agreement with the published K_D. This value was therefore used for our intracellular calcium measurements. One caveat is that while dextran-linked indicators show the least compartmentalization, and should be least influenced by the intracellular environment, they have been reported to undergo an up to twofold increase in K_D in muscle cells *(57)*. Given the uncertainty in our intracellular determinations, it would be most prudent to assume that CGD's K_D in neurons is somewhere between 100% and 200% of its in vitro value.

In cells lacking robust calcium channels, alternative schemes of flooding cells with calcium are possible, including different combinations of the follow-

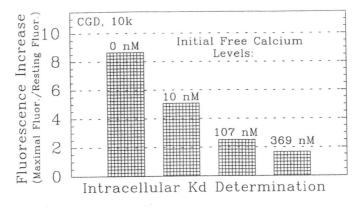

K_D Determination Parameters

Free Ca^{2+} nM	Fluoresc. Increase Max/Init.	Relative Starting Fluoresc.	Est. Ca^{2+}-Bound Fluo-3, %	Apparent K_D nM
0	8.65	100	0	-
10	5.07	171	9.3	100
107	2.52	343	31.8	230
369	1.66	521	55.0	320

Fig. 4. K_D determination. Cells are filled with calibration solution containing intermediate levels of free calcium (**Table 1**). The maximum fluorescence increase produced by calcium pulsing is shown for four different starting calcium levels. The value for the fluorescence increase (final fluorescence/resting fluorescence) is shown in the accompanying table along with a calculated "absolute" starting fluorescence that assumes the indicator is saturated after calcium pulsing (*see* text; this value is scaled to a 0 calcium value of 100). The amount of calcium-bound CGD corresponding to the calculated starting fluorescence is shown. The K_D is then determined iteratively for each intermediate calcium level using a calcium calculation program (*see* text for further details).

ing: high potassium, calcium-channel activators, blockers of the calcium pump or the Na^+/Ca^{2+}-exchanger, receptor agonists and releasers of intracellular calcium stores. The specific approach can be tailored towards the particular cell type being imaged. This calcium-clamp approach to determining intracellular K_D is equally applicable to the intracellular perfusion technique, where solution A is an intermediate calcium level and solution B a saturating level of calcium. The more robust nature of intracellular perfusion could make it better suited for calibrating higher K_D indicators, such as furaptra and calcium green 5N (*37,59–61*). Although this calibration approach is unorthodox, it appears to be least disruptive of the intracellular environment while providing a degree of control over intracellular calcium levels that approaches or, at least in some cell types, exceeds that provided by calcium ionophores.

3.3.3. Construction and Use of Intracellular Calibration Curves

The standard equation for converting fluorescence values to calcium concentrations (for single wavelength indicators) is:

$$[Ca^{2+}] = K_D \cdot (F - F_{min})/(F_{max} - F) \qquad \text{(Eq. 2)}$$

where F_{max} = maximum fluorescence (i.e., calcium-saturated fluorescence), F_{min} = minimum fluorescence (0 calcium fluorescence), and F = the fluorescence value to be converted to a calcium concentration (46). This equation can be used directly if one has F_{max} and F_{min} for each cell studied. In practice, it may be difficult or impossible to determine F_{max} and F_{min}. This is especially true for cells in brain slices or intact animals. In such cases, it is not known where the experimental fluorescence values fit within the overall dynamic range of the indicator (F_{min} to F_{max}), i.e., resting calcium is not known. In this situation, the simplest approach is to assume a resting value of calcium and equate the cell's resting fluorescence to it (**Note 20**). For CGD, using the estimated intracellular K_D of 250 nM and an arbitrary value for F_{min} of 100 (resulting in an F_{max} of 870 for CGD's 8.7-fold intracellularly determined dynamic range) the curve in **Fig. 5A** is generated (**Note 21**). If a resting calcium value of 100 nM is assumed, the standard curve yields a resting fluorescence (F value) of 320. Thus, 320 is the fluorescence value to which the experimental resting values will be normalized. A scaling factor is then determined to scale all of the experimental values to the standard curve. For example, if resting fluorescence in the cytoplasm is 32 arbitrary fluorescence units (after correcting for autofluorescence and/or black-level), then the scaling factor would be 10, i.e., all cytoplasmic fluorescence measurements are multiplied by 10 and substituted directly into equation 2 to calculate the corresponding calcium levels. The nuclear scaling factor would obviously be different. If resting nuclear fluorescence were 64, then the scaling factor for nuclear measurements would be 5. For this analysis, the 8.7-fold maximal fluorescence increase observed in the nucleus (**Fig. 3**) was used to set the dynamic range. For a more rigorous determination, a separate dynamic range would be used for nucleus and cytosol. For simplicity, just the nuclear calibration curve is used here.

This calibration curve can be replotted in an instructive form: calcium is plotted vs $\Delta F/F$ for a series of different resting calcium levels (**Fig. 5B**). For each resting calcium level to be plotted, the initial point on the curve is the resting calcium level (e.g., 50 nM) and the resting $\Delta F/F$ is equal to 0, because there is no ΔF yet. But there is a resting fluorescence level, which is read from the curve in **Fig. 5A** (or calculated using the equation). For example, a resting

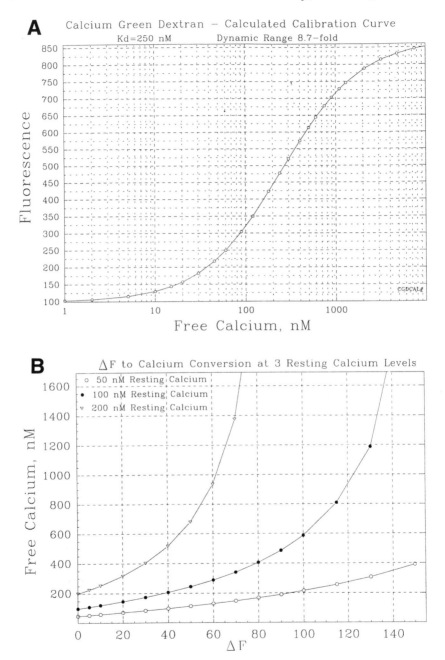

Fig. 5. Intracellular calibration curves. Based on the observed dynamic range and estimated intracellular K_D, fluorescence is plotted as a function of free calcium (**A**). This curve is generated by choosing different calcium levels and using the calculation program to determine "*B*", the percent of CGD with calcium bound. Fluorescence is

calcium of 50 nM has a fluorescence value of 228 in arbitrary units. All subsequent values for the 50 nM curve in **Fig. 5B** are calculated by incrementing the fluorescence by a specific ΔF, calculating the total fluorescence produced by that fluorescence increase ($F_{REST} + \Delta F$) and entering that total fluorescence into Eq. 2 to calculate free calcium. For example, on the 50 nM curve, a 20% $\Delta F = 0.2 \times 228 = 45.6$. The total $F = 228 + 45.6 = 273.6$ This total is entered into eq. 2, with $F_{max} = 870$ and $F_{min} = 100$, yielding a calcium level of 73 nM for a $\Delta F = 20$. Curves were generated for three different resting calcium values: 50, 100, and 200 nM. These values all fall within the current "range of uncertainty" of neuronal resting calcium levels. The utility of this type of graph is that one can enter the graph with any relative fluorescence increase ($\Delta F/F$) and directly read out the size of the calcium response. More importantly, one can do this for each resting calcium level that one wishes to consider. For example, if one assumes a resting calcium level of 50 nM, then a $\Delta F/F$ of 60% would equate to a 90 nM increase in free calcium. In contrast, if resting calcium is assumed to be 200 nM, then a 60% $\Delta F/F$ equates to a 700 nM increase in calcium.

Such calculations may seem esoteric, but this analysis had a significant impact on the interpretation of calcium signals generated by the low-threshold (LT) and high-threshold (HT) calcium channels found on thalamic relay cells in rat brain slices *(16)*. In that study, opposite conclusions were reached about the relative distribution of these two calcium channel subtypes, depending on whether a resting calcium of 50 nM or 100 nM was assumed. This ambiguity was due to a combination of the nonlinear relationship between calcium and fluorescence and the fact that different sizes of somatic fluorescence responses had been obtained in that particular experiment (**Note 22**). Although an exact answer on calcium channel distribution awaits further experimentation, had we not attempted to convert the raw fluorescence signals into calcium concentrations, we would have been unaware of the ambiguity inherent in the raw fluorescence data. In this particular instance, a simple change in protocol eliminates the source of ambiguity: the stimulus used to activate the channels is

----

then calculated from the equation $F = (100 - B) + 8.7B$. Alternatively, one could generate the curve by simply choosing fluorescence values and calculating free calcium directly from **Eq. 2**, where $F_{min} = 100$ and $F_{max} = 870$. **(B)** To illustrate the influence that resting free calcium level has on the interpretation of fluorescence responses, free calcium is plotted vs ΔF for several different resting calcium levels. To construct these curves, F is determined for a given resting calcium level from the calculation program or by trial and error usage of **Eq. 2**. The calcium values corresponding to different ΔF's are then calculated by calculating a total F value ($F_{REST} + \Delta F$) and then substituting that total F value into **Eq. 2**.

adjusted so that equal-sized somatic responses are produced for both channel types. With this modification, the relative direction in which the LT and HT calcium channel distributions are changing is clear regardless of the level of resting calcium. This approach thus provides the simplest starting point for the quantitative imaging of the distribution of functional calcium channels.

3.4. Indicator Comparison

Each calcium indicator seems to have its own unique "personality" owing perhaps to quirks in binding and compartmentalization. Rhod-2, e.g., exhibits a rather unique fluorescence pattern in that not only is the nucleus brighter than the cytoplasm, but the nucleolus is much brighter than the nucleus proper (**Fig. 6A**). It seems rather improbable that the nucleolar/nuclear fluorescence gradient reflects a standing calcium gradient. More likely, rhod-2 binds avidly to nucleolar constituents, or its fluorescence is enhanced by the nucleolar environment or some combination of both occurs. Calibration of rhod-2 by calcium pulsing shows that all three regions of the cell increase in parallel, with the nuclear and nucleolar signals showing relatively similar $\Delta F/F$ values, and the cytosolic signal reaching a much lower plateau (**Fig. 6B**). The large parallel increases observed in all compartments is far more easily explained as a quirk of the indicator rather than by a set of highly unusual calcium gradients maintained over a wide range of calcium levels. Indeed, the lack of a membrane around the nucleolus would seem to make it physically impossible for a nucleolar/nucleoplasmic calcium gradient to persist (**Note 23**), yet a nucleolar/nucleoplasmic fluorescence gradient was described as a calcium gradient in a study of cultured rat DRG neurons (*62*).

The maximum fluorescence increase shown by rhod-2 in vivo was much less than its response in cuvettes. This was true for all indicators whose intracellular dynamic ranges were determined in calcium pulsing experiments (**Fig. 7**; note that for several newer indicators, just the cuvette values were determined). It can be seen that the cuvette behavior approaches the intracellular behavior most closely for CGD, with fluo-3 being next closest and rhod-2 showing the most deviation. Compartmentalization of the indicators seems to play a role in the decreased dynamic range of both rhod-2 and fluo-3. For example, fluo-3 showed a 24-fold fluorescence increase in the nucleus and only an 18-fold increase in the cytosol (Fig. 7 in **ref. *20***). Such a result is consistent with a gradual internalization of the indicator into organelles which would leave some

---

Fig. 6. Intracellular behavior of Rhod-2. (**A**) After filling a cell with 80 µ*M* rhod-2 dissolved in 0 calcium solution (with 10 m*M* BAPTA; "low calcium"), the delivery of calcium pulses causes fluorescence to increase in the cytoplasm, nucleus and nucleo-

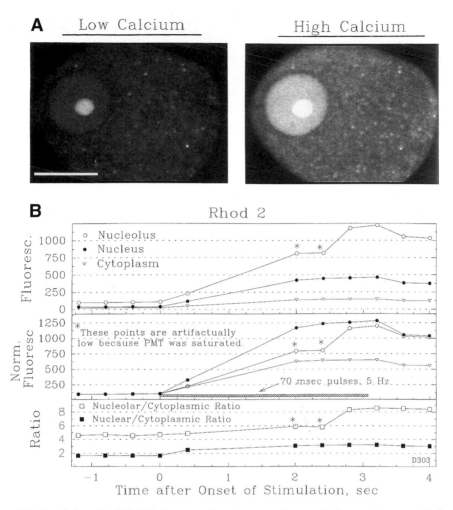

A   Low Calcium   High Calcium

B   Rhod 2

○ Nucleolus
● Nucleus
▽ Cytoplasm

These points are artifactually low because PMT was saturated.

70 msec pulses, 5 Hz

□ Nucleolar/Cytoplasmic Ratio
■ Nuclear/Cytoplasmic Ratio

D303

Time after Onset of Stimulation, sec

lus ("high calcium"). **(B)** All three regions increase in parallel, consistent with free diffusion of calcium throughout the aqueous space of the cell. As seen with fluo-3, the nuclear and nucleolar signals increase more than the cytoplasmic signal. Rhod-2 is unique, however, in that its nucleolar signal is twice as bright as the nuclear signal, even at the outset of the experiment (i.e., at a time when the cell is in excellent physiological condition). This nuclear/nucleolar fluorescence gradient is maintained across all calcium levels which argues against the idea of a "nuclear/nucleolar calcium gradient." Note that the nuclear/cytosolic fluorescence ratio reaches a higher value than is observed with fluo-3. This may be due to the reported accumulation of rhod-2 into mitochondria. Scale bar = 15 μm.

of the cytoplasmic indicator in a high-calcium (organellar) environment, where it would not respond to increasing cytosolic calcium levels. This would there-

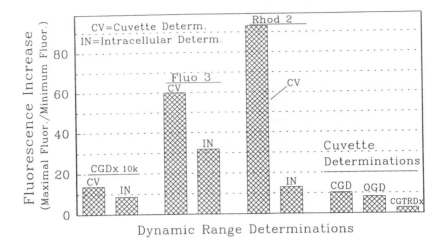

Fig. 7. Indicator Comparison: in vitro vs in vivo behavior. All three indicators tested intracellularly showed a decreased in vivo dynamic range, relative to their in vitro or cuvette value. This was most marked for rhod-2, whose intracellular dynamic range is reported for its nuclear response—its cytosolic dynamic range is even less. Fluo-3 showed the largest intracellular response but this can decline over time perhaps because of compartmentalization and gradual extrusion from the cell. The indicator manufacturers have not yet succeeded in producing a dextran-linked version of fluo-3, a compound that would constitute a major advance for calcium imaging. Cuvette determinations are shown for several newer indicators including Oregon green dextran BAPTA-488 (OGD) and a calcium green Texas red dextran conjugate (CGTRD). In preliminary in vivo studies, OGD was quite bright and yielded responses similar to CGD. In contrast, the cuvette behavior of CGTRD was disappointing and so it was not used inside cells. Our more recent batches of CGD (shown with the other cuvette determinations) have yielded somewhat smaller responses than previous batches, but it is not known whether this is due to the indicator itself or to other factors such as changes in the calibration solutions or imaging system.

fore decrease the cytoplasmic dynamic range relative to the nuclear dynamic range. Compartmentalization is not the entire explanation, however, as rhod-2's nuclear dynamic range is greatly reduced relative to its in vitro dynamic range, and there are no membrane bound organelles inside the nucleus. This reduction of intracellular dynamic ranges, relative to cuvette values, has held true for all fluorescent calcium indicators examined (35–37). Although the cause of the decrease is not certain, a quenching process associated with the binding of the indicators to intracellular constituents seems likely. What is certain is that the combination of an increased intracellular K_D and a decreased intracellular dynamic range creates in vivo calibration curves that bear little resemblance to the in vitro curves.

3.5. High Spatio-Temporal Resolution Imaging

To this point, the focus has been on the nature of the fluorescent calcium signals rather than confocal microscopy, although the Z-resolving power of the confocal, i.e. the ability to resolve signals to specific Z depths in the cell (**Fig. 2**), was mentioned. The power of the confocal for calcium imaging is best exemplified, however, by the use of linescans—rapid, 1-dimensional calcium images that have been used with great success in many applications. These include, for example, the imaging of calcium sparks in muscle *(63)*, calcium waves in starfish oocytes *(64)*, dendritic and somatic calcium signals in Purkinje cells and thalamic neurons *(16,60)*, and population activity in behaving zebrafish *(21,65)*. These linescans have an unparalleled combination of spatial resolution, temporal resolution and signal-to-noise ratio (**Note 24**). This is especially true of linescans acquired via 2-photon imaging, a remarkable tool that has provided insights into intracellular dynamics in stereocilia *(66)* and dendritic spines *(67,68)*. Although two-photon linescans are not technically "confocal," they exhibit remarkable Z-sectioning and are often acquired via a laser-scanning instrument in the same manner as confocal linescans. The beauty of linescans will be illustrated by their usage with two different calcium indicators as we return, for a last time, to the examination of persistent nuclear-cytoplasmic fluorescence gradients.

Figure 8A shows a linescan across a cultured bullfrog neuron loaded with the 70,000 mol wt version of CGD (CGD-70k). These experiments are done using a physiological solution (**Subheading 2.1.2.**) that contains 100 µM of the calcium indicator. Because there is no BAPTA, the indicator also serves as the primary calcium buffer. In this situation the cell controls its resting calcium level, while the indicator acts mainly as a "sensor" of calcium dynamics (but *see* **Note 25**). Depolarizing the cell for 200 ms admits a large bolus of calcium at the edges of this relatively large (50 µm diameter) cell. The spread of calcium across the cytoplasm, into the nucleus and across the nucleolus is clearly visualized. Fluorescence changes in the six specific regions indicated in **Fig. 8A** are plotted in **Fig. 8B**. Calcium rises fastest at the left and right edges of the cell. In the next two regions in from the edge (central cytoplasm; left edge of nucleus), calcium rises more slowly. Although these two inner regions are equidistant from the plasma membrane, calcium rises more slowly in the nucleus, most likely because diffusion into the nucleus is restricted to the small area encompassed by the nuclear pores *(69)*. The most central regions of the cell rise the slowest. The fluorescence values in **Fig. 8B** have all been normalized to the resting fluorescence in each region, which eliminates indicator concentration as a variable (**Note 26**). The resolution of even the nucleolar signal illustrates the detailed spatial and temporal resolution provided by confocal linescans.

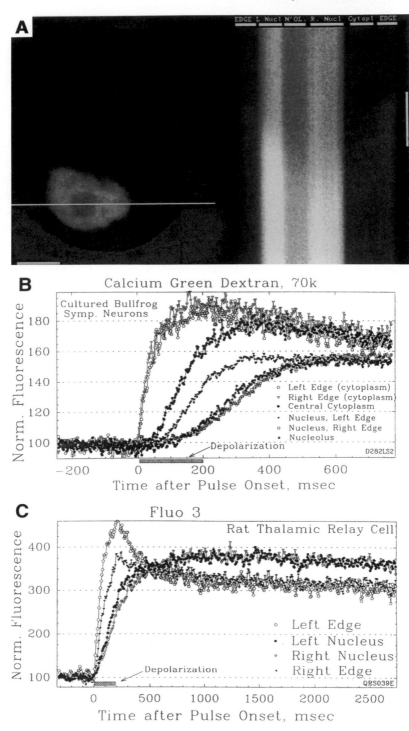

A

EDGE L Nucl N'OL. R. Nucl Cytopl EDGE

B Calcium Green Dextran, 70k

Cultured Bullfrog
Symp. Neurons

Norm. Fluorescence

180

160

140

120

100

○ Left Edge (cytoplasm)
▽ Right Edge (cytoplasm)
● Central Cytoplasm
▾ Nucleus, Left Edge
□ Nucleus, Right Edge
● Nucleolus

Depolarization

D282LS2

−200   0   200   400   600

Time after Pulse Onset, msec

C Fluo 3

Rat Thalamic Relay Cell

Norm. Fluorescence

400

300

200

100

○ Left Edge
● Left Nucleus
▾ Right Nucleus
▾ Right Edge

Depolarization

QRS039E

0   500   1000   1500   2000   2500

Time after Pulse Onset, msec

Using this indicator, the relative fluorescence change is higher in the cytoplasm than in the nucleus. Data of this sort have been used to argue that the nucleus is insulated from cytosolic calcium transients *(70)*. But it is not clear how calcium could, e.g., rapidly rise in the left nuclear region, for about 300 ms, and then abruptly level off even though the adjacent cytoplasm is, by this interpretation, much "higher" in calcium. Even more puzzling is how the deeper nucleoplasmic calcium levels rise for an additional 200 ms, i.e., well after the nuclear pores were ostensibly "shut" at a point about 300 ms into the experiment. The simpler explanation is that these different subcellular regions have independent calibration curves and show different relative fluorescence increases for the same increase in free calcium. Our conclusions are based on the intracellular perfusion and calcium pulsing approach to calibration where we observed free passage of calcium across the nuclear envelope at calcium levels well into the micromolar range. In contrast, Al Mohanna et al. *(70)* hypoosmotically shocked cells in distilled water to isolate the nuclei and then calibrated the fluorescence signals from the isolated nuclei. They concluded that the pores are effectively "shut" at cytosolic calcium levels above 300 nM.

Fig. 8. *(opposite page)* Nuclear and cytosolic calcium dynamics. (**A**) Fluorescence responses are measured in different regions of a cultured bullfrog sympathetic neuron filled with 100 μM 70K-CGD (and no BAPTA). The indicated line is scanned at 2 ms intervals; successive lines are plotted from top to bottom in the right-hand panel. The cell is depolarized for 200 ms, during the time indicated by the vertical bar. Calcium influx is evident at both edges of the cell beginning from the onset of the depolarization. *See* Color Plate III, following page 372. (**B**) The fluorescence responses in six different regions of this cell, normalized to their respective resting fluorescence levels, are plotted. Calcium is seen diffusing through the cytoplasm, into the nucleus, and across the nucleolus. A clear lag in the nuclear signal is seen by comparing the left edge of the nucleus (N LE) to the central cytoplasm (C Cyto); these regions are equidistant from the plasma membrane. The nucleolus and the right edge of the nucleus rise the slowest. The nuclear regions all show smaller relative fluorescence responses than the cytoplasm, as was also observed in calibration experiments (**Fig. 3**). Only the transient fluorescence gradients are believed to be calcium gradients. The persistent fluorescence gradients most likely reflect the behavior of the indicator. Calcium appears to have largely equilibrated across the cell at about 400 ms after the depolarization. (**C**) In a similar type of experiment, but now using a slice from rat brain, a thalamic neuron is loaded with 100 μM fluo-3. It shows fluorescence responses in the nucleus that are larger than those in the cytosol. This agrees with our results using fluo-3 in cultured neurons. The rapid flux of calcium across the nuclear envelope again suggests free communication (for small molecules) between nucleus and cytoplasm. (**Figure 8C** is courtesy of Dr. Qiang Zhou, Dept. of Pharmacology, UCSF, San Francisco, CA.)

Given the many and drastic ways that fluorescent calcium indicators are influenced by the intracellular environment, exposing nuclei to distilled water seems unlikely to provide a physiologically relevant calibration.

In contrast to the CGD results, with fluo-3-loaded bullfrog sympathetic neurons, the nuclear fluorescence responses are larger than the cytoplasmic responses. This is observed in other types of neurons including, e.g., thalamic relay cells in slices of rat brain (**Fig. 8C**). In this experiment, a relay cell was depolarized for 200 ms producing a rapid calcium increase at the edges of the cell. The two nuclear regions rise more slowly, but reach a higher plateau than the cytosolic regions. Data of this sort, including early data from our laboratory, have been widely interpreted to mean that the nucleus amplifies the cytosolic calcium signal *(19,62,71)*. This observation could alternatively be explained by compartmentalization of a small fraction of the cytoplasmic fluo-3 into calcium-containing organelles, as mentioned earlier, or by other differences in the nuclear and cytoplasmic environments. What is especially difficult for the proponents of persistent nuclear/cytosolic calcium gradients to explain is how the nucleus can insulate itself from cytoplasmic calcium gradients when loaded with calcium green dextran, but then generate amplified nuclear signals when loaded with fluo-3. As unlikely as this may seem, there are many active proponents of the idea that the nucleus actively regulates its free calcium levels *(72–76)* and so the mechanistic implications of the "persistent calcium gradient" interpretation should be discussed.

The nuclear envelope is actually a double membrane structure that forms a compartment (the lumen of the nuclear envelope) surrounding the nucleus. Thus, calcium ATPases in the outer membrane of the nuclear envelope will pump calcium into the lumen but would not pump calcium into the nucleus proper. The only structure that spans both the inner and outer membranes of the nuclear envelope are the nuclear pores. These pores are believed by most biologists to universally pass small molecules (**Note 27**). It is not clear how a gradient of free calcium could be created or maintained across the nuclear envelope given such cell biological data. In fact, if one believes both the amplification and the insulation theories, one must propose two sets of mechanisms—one pumping calcium across two membranes into the nucleus, the other pumping calcium across two membranes out of the nucleus. Our dynamic imaging data do not fit with this. Our observations show calcium rushing into the nucleus in every calcium imaging experiment we've done—including every indicator tested, every neuronal cell type studied, and in every preparation ranging from cultured bullfrog sympathetic neurons, to thalamic cells in rat brain slices, to hindbrain neurons in living zebrafish. It seems much simpler to attribute the fluorescence gradients to the known "eccentricities" of the various indicators (and to the vagaries of the loading mechanisms) rather than to a new and fantastic set of molecular machinery whose existence is based solely on persistent fluorescence gradients.

In spite of the foregoing arguments, it has not been "proven" that these persistent fluorescence gradients are *not* calcium gradients. This uncertainty reflects the core difficulty of the calibration problem: the difficulty in attaining absolute control over calcium in different subcellular regions while maintaining physiological conditions. This difficulty is at the center of our uncertainties about resting free calcium levels and the absolute size of physiological calcium signals. These limitations notwithstanding, it is clear that confocal calcium imaging provides excellent dynamic information about the location of calcium signals and the size of the fluorescence signals at different locations inside the cell. Indeed, it is the rapid dynamics of the fluorescence signals that gives us the best information about where calcium is changing and by how much. As optical methods continue to improve, and as indicators expand to chloride, sodium and more complex signaling molecules such as calmodulin, cAMP and gene products, confocal activity imaging should advance many areas of cellular and neurobiology. Such advances would be facilitated by vigorous efforts to better understand and judiciously interpret subcellular fluorescence gradients.

4. Notes

1. When fluorescent calcium indicators are loaded directly into the cytosol (via injection or patch-clamping) the nucleus is invariably brighter than the cytosol. This has held true for all cell types and calcium indicators examined, including the ratiometric indicators fura-2 and indo-1. In contrast, in AM-loaded cells, the nucleus is often darker than the cytosol (*62,77,78*), consistent with partitioning of an appreciable amount of indicator into either lipid bilayers or cytoplasmic organelles. These dark nuclei are "artifactual" in the sense that they do not reflect selective loading of indicator into the aqueous space of the cell (i.e., cytosol and nucleoplasm). This was confirmed by first loading cells with fluo-3 AM (which yielded dark nuclei) and subsequently perfusing the cells with the salt form of fluo-3 which then produced a bright nucleus. The aqueous labeling was much more responsive to calcium influx than the AM-loaded indicator (Fig. 8 in **ref. 20**). In other cases, AM loaded cells may show a less dark, equifluorescent or even a slightly bright nucleus, most likely reflecting varying degrees of compartmentalization of the indicator. This will cause varying degrees of nuclear/cytoplasmic fluorescence gradients (or ratio gradients) and absolutely precludes directly equating resting image gradients to resting calcium gradients (*20,26*). In addition, the subcellular calibration of calcium responses in AM-loaded cells will vary directly with the degree of compartmentalized indicator in each region. This seriously limits the utility of AM indicators in making absolute calcium determinations at different subcellular locations.

2. Although confocal imaging with the two main ratiometric indicators, fura-2 and indo-1, can be done with a UV laser *(79)*, confocal optics are not usually optimized for passing shorter wavelengths. Also, UV light is generally more harmful than visible light *(53,80)*. Because photodamage is a major concern in physiological experiments, our preference is to use visible wavelength indicators. There are several possibilities for visible-wavelength ratio imaging. One is to use a mixture of fura red and fluo-3 *(81)*, which has been used to demonstrate a rapid equilibration of calcium across the nuclear envelope in hamster oocytes *(53)*. Another potential option is a calcium green–Texas red dextran (CGTRD) conjugate. However, the batch of CGTRD purchased by us showed only a threefold in vitro dynamic range in the calcium green (fluorescein) channel, vs the 10- to 15-fold dynamic range that pure calcium-green dextran provides (other investigators have apparently experienced similar problems). We also attempted using a simple mixture of calcium green dextran and texas-red dextran, but found that Texas red dextran had a much faster bleach rate than the calcium green, which eliminated its utility for ratio imaging. In the fura red/fluo-3 study, the authors did not comment on the number of imaging trials over which the fura red/fluo-3 ratio was stable, but they did report better signal-to-noise ratio and less phototoxicity than with UV-ratio indicators *(53)*. A third option is to use two-photon (long-wavelength) excitation of a shorter wavelength ratio indicator, since this circumvents UV-induced photodamage. A pilot experiment that we performed in collaboration with the Webb group, using indo-1, suggests that this approach may be useful (*see* Fig. 8 in **ref. *82***).

   The fluorescent calcium indicators discussed in this paper were all purchased from Molecular Probes in Eugene, Oregon. Although some indicators may not have been as successful as hoped, it should be emphasized that all of the imaging work published from our laboratory (and many others) has relied exclusively on these indicators.

3. Our intracellular solution is specifically designed for recording calcium currents. Although a more "natural" intracellular solution would contain KCl, our solution contains cesium gluconate because gluconate is a slightly better anion than chloride for general preservation of the cell and because cesium reduces potassium currents that would interfere with our recordings of calcium currents. The compound tetraethylammonium (TEA) is also used to block potassium currents. The extracellular solution is a standard Ringer's, but contains tetrodotoxin (1 μM) to block sodium currents (*see* **ref. *20*** for more details). This simple formula allows the recording of robust calcium currents in cultured bullfrog sympathetic neurons. More complex formulae, e.g., containing an ATP-regenerating system, can be used *(16)*. A formulary providing many helpful details on the makeup of intracellular solutions is available *(83)*.

4. One calcium calculation program, called MAXC, is available via the internet at http://www-leland.stanford.edu/~cpatton, and was written by Chris Patton at the Hopkins Marine Station in Pacific Grove, CA. Bers et al. *(31)* discuss the use of this program which is currently available for IBM/Windows operating systems.

This program might be run on MacIntosh systems using "Virtual PC" (Connectix, San Mateo, CA). Our laboratory uses a similar type of program written by Alvaro Villaroel in the ASYST electrophysiological programming language.

5. For our experimental conditions, BAPTA's K_D would be about 210 nM, using the values of Harrison and Bers *(84)*. However, taking the original value of Tsien *(30)* and correcting it for osmolarity based on Harrison and Bers' results yielded a K_D of 160 nM at 23°C. This latter K_D provided a somewhat better fit between our observed and predicted fluorescence for fluo-3 and has been used for all calcium calculations in this chapter. A variety of factors should be taken into account in preparing calcium standards and calcium buffer solutions, including, e.g., purity of the buffer. These have been discussed in detail *(31,85)*.

It may seem problematic that the makeup of the calibration solutions is based, in part, on the K_D of the indicator, which is the unknown being determined. However, because there is a 100-fold greater concentration of BAPTA (10 mM) than indicator (100 μM), the error that results from uncertainty in the indicator's in vitro K_D is negligible. For example, increasing fluo-3's K_D from 400 nM to 800 nM would increase the free calcium in the "100 nM solution" to only 100.4 nM.

6. Calcium-sensitive electrodes are an alternate technique for measuring free calcium levels, but the most commonly available electrodes are not very accurate at low calcium levels, i.e., in the 50 nM to 300 nM range, where much calcium activity of interest occurs. A more sensitive calcium electrode has been developed, but its use and characterization have thus far been limited *(86)*.

7. A square coverslip is glued onto the bottom of a 35-mm Petri dish with General Electric RTV 615A adhesive (aquarium sealant has also been used). These chambers or "confocal cuvettes" are cleaned before each experiment. After they are cleaned and rinsed with distilled water, the excess water is shaken off and any remaining droplets of water in the chamber are removed by puffs of air from a pipette bulb fitted with a short plastic tube. This avoids residues left by Kim wipes or spray cleaners. We use either a conventional cove slip (no. 1–1/2) for cell culture or droplet-calibration studies, or a thinner cover slip (no. 1) when looking deeper into tissue, e.g., when looking 200 μm into the larval zebrafish hindbrain.

This "confocal cuvette" may be thought of as a virtual cuvette in the sense that it is delineated in the Z-dimension by the optical section thickness (principally a function of the microscope objective lens) and in the XY-dimension by the measuring box (a standard feature of confocal software) used to quantitate the fluorescence. With appropriate solution filtering and chamber cleaning, these droplets are fairly uniform in fluorescence when the "confocal cuvette" is fully enclosed in the droplet, i.e., positioned just high enough above the coverslip to obtain a maximal signal.

8. Obtaining reproducible, quantitative fluorescence measurements requires careful attention to the confocal instrument's settings: aperture, gain, neutral density filtering, black level, and acquisition mode (e.g., we use the "low signal" setting on the Bio-Rad MRC 600). Because our experiments are conducted in this mode,

we've constructed our calibration curves using this setting. With our settings, we find that turning the "autoblack" function off and manually setting the black level provides the most stable frame-to-frame baseline within imaging trials (such trials typically last about 1–10 s). However, turning off the autoblack function also reveals a slow but substantial drift in black level, especially while the scan box is warming up. We allow the scan box to warm up for 2 h or more prior to imaging. Thereafter, the black level shows modest slow fluctuations; it is periodically checked and adjusted as needed.

Ideally, all fluorescence measurements would be taken at the same instrument settings. However, with the 256 unit gray scale of the MRC 600 and the low signal setting, we found that the most linear fluorescence measurements were taken with average pixel intensities between about 30 and 200. With this limited range, it was not possible to record the fluorescence values of all calcium levels at a single photomultiplier (PMT) gain setting. Thus, it was necessary to decrease the gain when higher calcium concentrations were measured. By measuring test droplets and test slides at different gain settings, correction factors were generated and used to scale all the fluorescence measurements to a common scale *(20)*. Newer confocals with expanded (12-bit) gray scales may not require this, but the fluorescence linearities of all confocals should be checked under the settings used in the conduct of the calcium imaging experiments. Commercial fluorescence standards and serial dilution of fluorescent solutions can be used for this purpose.

9. Ratiometric calcium imaging is often believed to solve the "calibration" problem because it factors out indicator concentration and because a given ratio can, in theory, be directly equated to a free calcium concentration. In practice, the problem of intracellular alteration of indicator behavior is so great that ratio imaging has not resolved the conflict over resting free calcium in muscle cells, where this issue has been most intensively studied. Factors that alter the behavior of ratio indicators include spectral shifts caused by binding or compartmentalization of the indicator—processes that can vary over time and from cell to cell *(34,39–42)*. Other factors include changes in ratio with photobleaching (possibly oxidation) and, conversely, ratio changes with de-oxygenation, a technique that might be used to prevent photobleaching *(87–89)*. These indicators also show changes in calcium/magnesium selectivity *(90)* and, in the case of brain slices, changes in ratio with increasing depth in the slice (A. Konnerth, personal communication). These factors limit the accuracy with which the ratio method can establish the level of resting free calcium. Indeed, a set of calcium images documenting the subcellular, in vivo behavior of fura-2 or indo-1 across a range of known calcium levels remains to be published. If several ratio imaging studies can come to a consensus that a particular subcellular ratio of indo-1 or fura-2 represents a specific level of free calcium, then that would bolster the utility of ratio imaging for obtaining absolute calcium levels.

10. Electrophysiological techniques are described in several books *(91–93)* including a technical manual from Axon Instruments called The Axon Guide *(94)*. Standard patch-clamp techniques are used for these experiments, but nonelec-

trophysiologists would be best served by seeking an instructor or course to learn this skill. Our confocal configuration allows an electrophysiological recording station to sit beside the vibration isolation table that holds the confocal microscope. This minimizes the time required to set up an electrophysiological experiment. For our experiments, set up mainly involves placing the preparation on the stage of the inverted microscope, connecting the flow lines and attaching an electrode holder to a micromanipulator mounted on the stage. The bullfrog neurons are cultured on the cover-glass bottomed wells made from plastic Petri dishes (**Note 7**). These dishes are secured to a standard microscope stage (that accepts 35-mm Petri dishes) and connected to the inflow and outflow lines. The patch electrode is then inserted into its holder and positioned near the cell. The patch electrode is moved to within about 5 µm of the cell using a 20X objective lens, but the final seal is made only after switching to the 50X Leitz (water immersion, 1.0 NA). The 50X Leitz provides a very bright signal relative to many other objectives that we've tested. A newer 40X/0.75 Zeiss achroplan (infinity-corrected) lens was recently found to have comparable performance to the 50X Leitz, even though it is not coverslip corrected. Experiments on brain slices are more complicated and require use of an upright microscope as described in Zhou et al. *(16)*. In such experiments, the laser beam is redirected via a pair of mirrors to a Zeiss standard 16 upright microscope that is placed next to the inverted microscope. The slices are then imaged with a Zeiss 40X/0.75 dip lens. For studies on cultured cells, a minimum of one uninterrupted day per week of confocal time seems necessary to make even modest experimental progress; two or more contiguous days is desirable for brain slice or intracellular perfusion experiments.

11. For large bullfrog neurons (typically about 50 µm in diameter), the indicator fluorescence inside the cell did not reach a plateau even after 15–20 min of filling, although the rate of fluorescence increase slowed markedly by this point. This gradual fluorescence increase could be due to a variety of factors, including (1) slow binding or gradual compartmentalization, (2) depletion of indicator from the tip of the patch-electrode, (3) gradually rising calcium levels, (4) increased access resistance through the patch electrode, or (5) some combination of the above. However, after the initial rapid filling phase is complete, the minute-to-minute changes in fluorescence are quite modest in comparison to the rates of calcium increase produced by the experimental manipulations (calcium pulses and intracellular perfusion).

12. A detailed description of the intracellular perfusion technique is provided by Yu et al. *(18)* and Lopez *(95)*. Because the available time on a confocal microscope is usually limited, and because the setup and execution of intracellular perfusion experiments are somewhat intricate, this type of calibration is not easy. Proficiency at patch-clamping is essential. A useful approach is to first gain skill in intracellular perfusion on a nonimaging electrophysiology station, by e.g., changing the concentration of sodium in the patch pipet so as to reverse the cell's sodium gradient *(18,96)*.

The patch-clamp electrode is pulled with a very short shank to allow the closest possible apposition of the internal (perfusion) tube to the tip of the electrode. These electrodes are typically made to have a low resistance, about 1–4 megaohms. Colored solutions are helpful in working out the flow performance of the perfusion system. The "back end" of the inflow line rests in a small (typically 1 mL) reservoir or vial containing Solution B. The "front end" of this line is threaded though the O-ring seal in the electrode holder and connected to the quartz internal perfusion tube. The outflow line runs from a side port on the electrode holder to a small rubber-stoppered bottle, and is connected to one of two metal tubes inserted through the top of the bottle. A second line, connected to the second metal tube, runs from the outflow bottle to the electrophysiologist and is used to make the initial patch-clamp seal. By applying suction to and then clamping this second line, sufficient vacuum is created to pull solution B through the inflow line (after its valve is opened) and into the pipet tip. Solution A is simultaneously pulled through the outflow line and drops into the small outflow bottle. All solutions are carefully filtered with a 0.2 µm nylon filter; this is critical to prevent blockage of the small inflow tube inside the patch-clamp pipet.

13. Because binding of calcium to indicators is a hyperbolic process, indicators are not completely saturated even at very high calcium levels. However, attaining calcium levels reasonably near saturation, e.g., 95–97% saturated, is usually sufficient to provide an adequate description of the indicator's K_D and dynamic range—especially in the context of the other uncertainties faced. For example, whereas 99% saturation of fluo-3 in our calibration solution requires 39 µM free calcium, 97% requires only 13 µM and 95% saturation just 7.6 µM. For our purposes, 95% saturation would be acceptable because it would mean that we have 95% of the maximum dynamic range of the indicator. Given that free calcium in the pipet (when filled with high-calcium solution) is in excess of 100 µM, it seems likely that the cytoplasmic indicator is at least 95% saturated. By increasing total calcium to 12 mM (vs the 10 mM BAPTA in the solution), far higher (and more toxic) concentrations of free calcium can be employed.

14. That nuclear fluorescence exceeds cytosolic fluorescence is true for many fluorescent dyes such as Lucifer Yellow and for all non-AM fluorescent calcium indicators tested, including ratiometric indicators such as fura-2 and indo-1. Consideration of the alternative hypothesis, however, is instructive. Were the resting fluorescence gradient due entirely to a resting calcium gradient, than the large parallel increases during intracellular perfusion would mean that the nucleus is somehow able to sample cytosolic free calcium and, for every sized increase in cytosolic calcium, produce an increased but proportionately sized calcium response in the nucleus by a mechanism that spans the outer nuclear membrane, the lumen of the nuclear envelope, and the inner nuclear membrane. Such a remarkable mechanism would certainly be interesting. It is far simpler, however, to attribute the fluorescent gradient to the well-known binding of indicators to intracellular constituents or to the exclusion of highly charged indicators from cytoplasmic organelles, or perhaps, to some combination of these and other factors.

15. The "far from saturation" argument cannot explain the nuclear fluorescence increase if the resting fluorescence gradient is attributed to a calcium gradient. In these early experiments (using CsCl-based solutions) the cell was filled not with 0 calcium solution, but with a very low calcium solution (10 nM) to better preserve the health of the cell (this was not necessary in later experiments where cesium gluconate was used). A 10 nM free calcium concentration actually "uses up" a portion of the in vitro dynamic range, reducing the remaining available dynamic range to no more than about 35-fold. With this starting condition, if nuclear calcium were at twice the cytosolic calcium level, i.e., at 20 nM, than its observed 24-fold increase would not be possible. Even if one argues that only part of the resting fluorescence gradient is a calcium gradient (with the remainder being an indicator fluorescence gradient), the argument still encounters problems. Because the nuclear fluorescence begins "higher" on the calibration curve, it will begin to approach diminished fluorescence returns before the cytosolic calcium signal does. It should therefore show a smaller fluorescence increase than the cytosol. In fact, it does exactly the opposite—it increases 24-fold vs. the cytosol's 18-fold increase (Fig. 7 in **ref. 20**). This difference is most easily explained by a gradual internalization of a small amount of fluo-3 into cytoplasmic organelles, i.e., into a compartment where it is already resting at higher than cytosolic calcium levels and is therefore less sensitive to cytosolic calcium increases. This would thus diminish the cytoplasmic fluorescence increase relative to the nuclear increase (where none of the indicator is shielded inside membrane bound organelles). Another line of evidence in favor of this idea is the fall in the nuclear/cytoplasmic ratio when cells are initially filled with indicator (this was true for all calcium levels tested; *see* Fig. 5 in **ref. 20**). At the outset of filling there would be no compartmentalized fluo-3 (it has a -5 charge), but if it very gradually enters mitochondria, the endoplasmic reticulum (ER) or other compartments, it will selectively increase the cytoplasmic signal (especially if it is entering compartments that are high in calcium), thereby causing the nuclear/cytoplasmic ratio to fall.

16. Voltage clamp is described in *The Axon Guide* (**94**) and in Johnston and Wu (**93**), and involves injection of current (either positive or negative) to hold the cell at a specified voltage. From the holding potential, depolarizing voltage steps or pulses are used to administer calcium to the cell. From a membrane potential of –70 mV, depolarization to 0 mV maximally activates calcium currents in these cells. To flood cells with calcium, 70 ms long voltage pulses are given at 5 Hz until the cell's fluorescence reaches a steady plateau. This typically requires only a few tens of seconds or less.

17. Initially, the peak efflux rate may be quite substantial, as suggested by an initially rapid fall in dendritic calcium signals observed in thalamic relay cells (Fig. 3 in **ref. 16**). However, this falling phase involves not only efflux but also uptake of calcium into internal stores which have a limited capacity. In addition, the Na$^+$/Ca^{2+} exchanger's removal of calcium depends on the influx of sodium down its concentration gradient which may also be limited by a build up of sodium near

the plasma membrane during the large influx of calcium produced by repetitive voltage pulses *(2,48)*.

18. The K_D of an indicator is determined based on the maximum fluorescence increase it exhibits when starting from a given free calcium level. First, a normalized "starting fluorescence," F, is determined using the equation $F = (8.65/FI) \cdot 100$, where 8.65 was CGD's total dynamic range during this series of experiments and FI = the observed fluorescence increase (final fluorescence/pre-stimulation fluorescence). The percentage of indicator with bound calcium, B, is then directly calculated by entering F into the equation: $B = (F - 100)/7.65$. This value for B is the "observed" value for the percentage of indicator bound and is equal to the amount of indicator bound, in μM, as we use 100 μM total indicator. The calculation program is then entered with the starting free calcium level and the total concentration of calcium green dextran (100 μM) and is then used to obtain a "calculated" value for B. The K_D of calcium green dextran is now varied by trial and error until the calculated value for B is equal to the observed value for B—this is the intracellular K_D. Although this may not be the simplest approach for determining the intracellular K_D, it provides independent K_D's for each starting calcium level and is quite practical when using our calculation program.

19. There is obviously substantial variability in our determinations of CGD's intracellular K_D. Although this is far from ideal, we are not aware of any calcium imaging study where intracellular K_D determinations have been made from a range of different starting calcium levels. More work in this area is clearly needed. If nothing else, this determination provides, at a minimum, a check on the indicator's intracellular behavior and a sense of the reliability of the K_D used in one's calculations.

20. It may seem odd to assume a level for resting free calcium given the focus of this chapter on real calibrations. However, the absolute level of resting calcium remains controversial for muscle cells and even more uncertain for other cell types. Although ratio imaging might seem to ameliorate this problem, there are serious limitations to determining absolute calcium levels even with ratiometric indicators (**Note 9**). Regarding a ratiometric study reporting widely varying resting calcium levels among different cells in a population *(97)*, it seems that variability in compartmentalization of the indicator is at least as tenable an explanation for the different resting ratios as variability in resting calcium (perhaps even more tenable because the calcium pumps that set resting calcium would be expected to be identical from cell to cell). Until the issue of resting calcium is resolved, it seems that assuming a resting free calcium level (and testing the effects of this assumption) is as satisfactory as other methods for estimating the magnitude of intracellular calcium signals.

21. **Equation 2** can be used to construct the curves in two ways. The simplest is to just insert a range of F values into the equation and directly calculate free calcium. Alternatively, one can start with a range of calcium concentrations and solve for F. This is essentially what the iterative calculation programs do, although our program actually yields the amount of indicator-bound calcium from

which we calculate the fluorescence, F, using the CGD version of **Eq. 1**: $F = (100 - B) + 8.7B$, where $B = \%$ of indicator bound to calcium. If one is lacking a calculation program, it is also possible to choose a free calcium level and solve **Eq. 2** for F by trial and error.

22. The crux of the problem was that the somatic high-threshold (HT) calcium channel response was substantially larger than the somatic LT calcium channel response. Responses for both channel types increased from the soma to the dendrites, but the relative increase for the low-threshold channels was greater than for the HT channels (Fig. 3 in **ref. *16***). When 100 nM resting calcium was assumed, the dendritic fluorescence response of the HT channels, even though proportionately smaller than the LT response, pushed the HT response into the near-plateau region of the fluorescence vs calcium curve, creating a calcium increase that was proportionately larger than the LT calcium response. In contrast, with the 50 nM resting calcium assumption, the responses were in a more linear region of the curve and the relatively larger fluorescence increase observed with the LT channels would be consistent with the LT calcium signals having increased more from soma to dendrites than did the HT calcium signals.

23. A possible misconception is that calcium buffering might account for persistent calcium gradients, such as the proposed nucleolar/nucleoplasmic gradient. Although buffer concentration shapes the dynamics of calcium transients and affects the magnitude of the increase in free calcium produced by a bolus of calcium (***98,99***), it does not set steady-state levels of calcium. The resting level of free calcium (and therefore the gradient across the cell's membranes) is set and maintained actively by pumps and exchangers in the plasma membrane and organellar membranes. It is only by expenditure of energy that a persistent nucleolar/nucleoplasmic calcium gradient could, in theory, be generated. Because there are no pumps or exchangers at the nucleolar border, and no membrane at the nucleolar border to maintain any gradient that was generated, it is difficult to imagine how this persistent fluorescence gradient could be due to a calcium gradient. When a "finding," such as this, seems to violate physical laws, such as the first and second laws of thermodynamics, one should carefully scrutinize the underlying assumptions— in this case the assumption that the indicator is behaving the same in different nuclear regions.

Another observation of nucleolar calcium signals raises further problems for subcellular calibration. In cells filled with fluo-3 or calcium green dextran, the nucleoli are typically dimmer than (or equifluorescent with) the nucleus proper (see **Fig. 2A** and **Fig. 8**). However, when cultured neurons were in declining health, the nucleoli of fluo-3-loaded cells usually became brighter than the nucleus proper, as shown in **Fig. 2B**, where an XZ scan of a cell at the end of an experiment showed a prominent nucleolus. Although this cell looked fine morphologically, its input resistance had declined significantly indicating possible damage to the plasma membrane. In every fluo-3-loaded cell that showed nucleolar changes, the pattern was always one of dark nucleoli becoming bright (relative to the nucleus), indicating a change in state of the nucleolus that affected the binding or fluorescence of fluo-3. A more deliberate means of inducing an irreversible subcellular fluorescence gradient is to repetitively scan a single line at a higher laser intensity. This rapidly induces a permanent increase in the fluorescence of that line (see Fig. 3 in **ref. *100***). Such permanent alterations in fluorescence obviously pose considerable problems to calibration efforts, and should be considered if unusual fluorescence gradients are encountered.

24. Single lines can be scanned at 2 ms intervals on the Bio-Rad MRC 600 and even faster on newer confocal microscopes. Although these images are spatially "one-dimensional," they provide detailed information about calcium dynamics at different locations either within a cell or across multiple cells. The information sought in a "two-dimensional" image can, in many cases, be acquired by placing a scan line across the regions of interest. Although the MRC 600 provides only horizontal linescans, some newer confocal imaging systems allow the scan line to be oriented in any direction. One caution is that such off-axis (nonhorizontal) linescans may not be truly linear, but can be slightly elliptical, thereby skirting a target if the target is small. The fidelity of the oblique linescan can be checked by allowing a fluorescent solution to evaporate in a glass-bottomed Petri dish, creating a thin film of dye. Repetitively scanning a line at one specific orientation will then bleach the film and allow the fidelity of the linescan to be evaluated (E. Lumpkin, personal communication). Although the rate for acquiring two-dimensional images is becoming faster, the pixel dwell time must be drastically reduced in order to achieve the same scan rate as a linescan. It is not clear that a signal to noise ratio comparable to that of linescans can be achieved with such short pixel dwell times.

25. This is not strictly true—buffering of calcium by fluorescent indicators can alter the "shape" of cellular calcium signals, including the rate that calcium diffuses across the cell and the height of calcium transients *(59,98,99)*. The exact manner in which indicators perturb the "normal" calcium transient depends on their K_D, concentration and molecular weight. Whereas small molecular weight indicators might collapse or dissipate gradients *(98)*, large molecular weight indicators, such as the 70K dextran used in **Fig. 8A**, may act as a barrier, slowing the diffusion of calcium across the cell. The effect of indicator concentration depends on the extent to which it overwhelms the endogenous buffering capacity of the cell. Endogenous buffers, including both calcium binding proteins and negatively charged phospholipids, are thought to constitute a total buffering capacity in the range of 50–500 μM *(101–103)*, with some classes of neuron having large amounts of specific calcium binding proteins *(3,104)*. Our use of a 100 μM concentration of indicator might, therefore, constitute a considerable fraction of total cellular buffering and would change the shape and time course of calcium dynamics. Nonetheless, we found that the pattern of calcium influx is similar at both 100 μM and 20 μM fluo-3 (unpublished observations). The lower concentration would be expected to have only minimal effects on the native transients, so we expect that, in general, our experiments reflect fairly accurately the pattern of influx, which is important for such issues as localization of calcium channels

subtypes. Lower indicator concentrations would be expected to require more intense laser illumination, thereby increasing phototoxicity. However, use of 10 μM indicator has recently been reported to not only yield larger relative fluorescence responses, but also, to produce a more stable, long-lasting signal (Kovalchuk and Konnerth, pers. comm.). At the opposite end of the spectrum, in cells loaded with 10 mM BAPTA and 100 μM fluo-3, the BAPTA completely suppressed the fluorescence transients that would normally have occurred after a long calcium pulse (200 ms or more; *see*, e.g., Fig. 5b in **ref. *20***). Thus the buffering capacity of the indicators, while not problematic for some questions, should be considered in cases where the precise time-course and absolute magnitude of the subcellular calcium transients are essential to interpretation of the experiment.

26. As noted previously, ratiometric imaging eliminates indicator concentration as a variable. The analysis used in **Fig. 8B** and **Fig. 8C**, taking a ratio of the evoked fluorescence increase over the resting fluorescence, is analogous to ratiometric imaging in that indicator concentration is again factored out of the equation. As such, this approach should (as a first approximation) give the relative calcium changes in different subcellular regions. This assumes, however, that free calcium is uniform across the cell at rest. This seems likely to be the case in the soma, at least, because calcium can equilibrate across the cell fairly quickly (dendrites and axons are a separate matter). But if the resting fluorescence gradients represented resting calcium gradients, then this approach provides less information. The higher resting nuclear fluorescence observed is not unique to single wavelength indicators, because the main ratiometric indicators, indo-1 and fura-2, also show a bright nucleus at both wavelengths used for the ratio imaging.

27. In every study where molecules such as fluorescent dextrans have been injected into the cytosol, they have been shown to freely enter the nucleus (almost certainly via the nuclear pores) so long as their molecular weight was 20,000 or less *(105,106)*. Indeed, in our cells, even the 70,000 M_r CGD entered the nucleus almost immediately after formation of a whole-cell patch-clamp recording. Although compounds of molecular weights over 70,000 (up to 500,000) traverse the pores by an active mechanism that can be regulated, there is a large body of data showing that compounds under 20,000 are never prevented from diffusing into or out of the nucleus *(107)*. Given this physical situation, it is difficult to imagine how the nuclear pores can be completely open to calcium one instant, and then closed to calcium the next instant, which would seem necessary if one is to interpret the persistent fluorescence gradients in **Fig. 8** as persistent "calcium" gradients. If one does not argue for complete closure of the pores, then one must construct an elaborate transenvelope pumping system that would maintain the gradients (and in fact create the gradient for the amplification scheme).

References

1. Krause, K. H. (1991) Calcium-storage organelles. *FEBS Lett* **285,** 225–229.
2. White, R. J. and Reynolds, I. J. (1995) Mitochondria and Na^+/Ca^{2+} exchange

buffer glutamate-induced calcium loads in cultured cortical neurons. *J. Neurosci.* **15,** 1318–1328.

3. Baimbridge, K. G., Celio, M. R., and Rogers, J. H. (1992) Calcium-binding proteins in the nervous system. Trends Neurosci. 15, 303–308.

4. Takei, K., Stukenbrok, H., Metcalf, A., Mignery, G. A., Sudhof, T. C., Volpe, P., and De Camilli, P. (1992) Ca^{2+} stores in purkinje neurons, endoplasmic reticulum subcompartments demonstrated by the heterogenous distribution of the $InsP_3$ receptor, Ca^{2+}-ATPase, and calsequestrin. *J. Neurosci.* **12,** 489–505.

5. Andressen, C., Blumcke, I., and Celio, M. R. (1993) Calcium binding proteins, selective markers of nerve cells. *Cell Tiss. Res.* **271,** 181–208.

6. Pozzan, T., Rizzuto, R., Volpe, P., and Meldolesi, J. (1994) Molecular and cellular physiology of intracellular calcium stores. *Physiol. Rev.* **74,** 595–630.

7. Miller, R. J. (1988) Calcium signaling in neurons. *Trends Neurosci.* **11,** 415–419.

8. Gnegy, M. E. (1993) Calmodulin in neurotransmitter and hormone action. *Annu. Rev. Pharmacol. Toxicol.* **32,** 45–70.

9. Kasai, H. and Petersen, O. H. (1994) Spatial dynamics of second messengers, IP3 and cAMP as long-range and associative messengers. *Trends Neurosci.* **17,** 95–101.

10. Ghosh, A. and Greenberg, M. E. (1995) Calcium signaling in neurons, molecular mechanisms and cellular consequences. *Science* **268,** 239–247.

11. Baylor, S. M. and Hollingworth, S. (1988) Fura-2 calcium transients in frog skeletal muscle fibres. *J. Physiol.* **403,** 151–192.

12. Llinas, R., Sugimori, M., and Silver, R. B. (1995) The concept of calcium concentration microdomains in synaptic transmission. *Neuropharm.* **34,** 1443–1451.

13. Robinson, I. M., Finnegan, J. M., Monck, J. R., Wightman, R. M., and Fernandez, J. M. (1995) Colocalization of calcium entry and exocytotic release sites in adrenal chromaffin cells. *Proc. Natl. Acad. Sci.* **92,** 2474–2478.

14. Grynkiewicz, G., Poenie, M., and Tsien, R. Y. (1985) A new generation of calcium indicators with greatly improved fluorescence properties. *J. Biol. Chem.* **260,** 3440–3450.

15. Tsien, R. Y. (1989) Fluorescent probes of cell signaling. *Ann. Rev. Neurosci.* **12,** 227–253.

16. Zhou, Q., Godwin, D. W., O'Malley, D. M., and Adams, P. R. (1997) Visualization of calcium influx through channels that shape the burst and tonic firing modes of thalamic neurons. *J. Neurophys.* **77,** 2816–2825.

17. Marrion, N. V. and Adams,P. R. (1992) Release of intracellular calcium and modulation of membrane currents by caffeine in bull-frog sympathetic neurones. *J. Physiol.* **445,** 515–535.

18. Yu, S. P., O'Malley, D. M., and Adams, P. R. (1994) M-current regulation by intracellular calcium in bullfrog sympathetic neurons. *J. Neurosci.* **14,** 3487–3499.

19. Hernandez-Cruz, A., Sala, F., and Adams, P. R. (1990) Subcellular calcium transients visualized by confocal microscopy in a voltage-clamped vertebrate neuron. *Science* **247,** 858–862.

20. O'Malley, D. M. (1994) Calcium permeability of the neuronal nuclear envelope, evaluation using confocal volumes and intracellular perfusion. *J. Neurosci.* **14,** 5741–5758.

21. O'Malley, D. M., Kao, Y.-H., and Fetcho, J. R. (1996) Imaging the functional organization of zebrafish hindbrain segments. *Neuron* **17,** 1145–1155.

22. Westerblad, H. and Allen, D. G. (1994) Methods for calibration of fluorescent calcium indicators in skeletal muscle fibers. *Biophys. J.* **66,** 926,927.

23. Baylor, S. M., Harkins, A. B., and Kurebayashi, N. (1994) Response to Westerblad and Allen. *Biophys. J.* **66,** 927,928.

24. Almers, W. and Neher, E. (1985) The calcium signal from fura-2 loaded mast cells depends strongly on the method of dye-loading. *FEBS Lett.* **192,** 13–18.

25. Glennon, M. C., Bird, G. S. J., Takemura, H., Thastrup, O., Leslie, B. A., and Putney, J. W. (1992) In situ imaging of agonist-sensitive calcium pools in AR4-2J pancreatoma cells. *J. Biol. Chem.* **267,** 25,568–25,575.

26. Connor, J. A. (1993) Intracellular calcium mobilization by inositol 1,4,5-trisphosphate, intracellular movements and compartmentalization. *Cell Calcium* **14,** 185–200.

27. Maltsev, V. A., Wolff, B., Hess, J., and Werner, G. (1994) Calcium signalling in individual T-cells measured by confocal microscopy. *Immun. Lett.* **42,** 41–47.

28. Trollinger, D. R., Cascio, W. E., and Lemasters, J. J. (1997) Selective loading of rhod-2 into mitochondria shows mitochondrial calcium transients during the contractile cycle in adult rabbit cardiac myocytes. *Biochem. Biophys. Res. Comm.* **236,** 738–742.

29. Zhao, M., Hollingworth, S., and Baylor, S. M. (1997) AM-loading of fluorescent calcium indicators into intact single fibers of frog muscle. *Biophys. J.* **72,** 2736–2747.

30. Tsien, R. Y. (1980) New calcium indicators and buffers with high selectivity against magnesium and protons, design, synthesis, and properties of prototype structures. *Biochem.* **19,** 2396–2404.

31. Bers, D. M., Patton, C. W., and Nuccitelli, R. (1994) A practical guide to the preparation of calcium buffers. *Methods in Cell Biology,* **40,** 4–29.

32. Fabiato, A., Fabiato, F. (1979) Calculator programs for computing the composition of the solutions containing multiple metals and ligands used for experiments in skinned muscle cells. *J. Physiol.* (Paris) **75,** 463–505.

33. Konishi, M., Olson, A., Hollingworth, S., and Baylor, S. M. (1988) Myoplasmic binding of fura-2 investigated by steady state fluorescence and absorbance measurements. *Biophys. J.* **54,** 1089–1104.

34. Hove-Madsen, L. and Bers, D. M. (1992) Indo-1 binding to protein in permeabilized ventricular myocytes alters its spectral and calcium binding properties. *Biophys. J.* **63,** 89–97.

35. Harkins, A. B., Kurebayashi, N., and Baylor, S. M. (1993) Resting myoplasmic free calcium in frog skeletal muscle fibers estimated with fluo 3. *Biophys. J.* **65,** 865–881.

36. Kurebayashi, N., Harkins, A. B., and Baylor, S. M. (1993) Use of fura red as an intracellular calcium indicator in frog skeletal muscle fibers. *Biophys. J.* **64,** 1934–1960.

37. Zhao, M., Hollingworth, S., and Baylor, S. M. (1996) Properties of Tri- and Tetra-carboxylate calcium indicators in frog skeletal muscle fibers. *Biophys. J.* **70,** 896–916.
38. Reers, M., Kelly, R. A., and Smith, T. W. (1989) Calcium and proton activities in rat cardiac mitochondria. *Biochem. J.* **257,** 131–142
39. Owen, C. S. (1991) Spectra of intracellular fura-2. *Cell Calcium* **12,** 385–393.
40. Bancel, F., Salmon, J.-M., Vigo, J., and Viallet, P. (1992) Microspectrophotometry as a tool for investigation of non-calcium interactions of Indo-1. *Cell Calcium* **13,** 59–68.
41. Bancel, F., Salmon, J.-M., Vigo, J., Vo-Dinh, T., and Viallet, .P (1992) Investigation of non-calcium interactions of fura-2 by classical and synchronous fluorescence spectroscopy. *Anal. Biochem.* **204,** 231–238.
42. Baker, A. J., Brandes, R., Schreur, J. H. M., Camacho, A., and Weiner, M. W. (1994) Protein and acidosis alter calcium binding and fluorescence spectra of the calcium indicator Indo-1. *Biophys. J.* **67,** 1646–1654.
43. Blatter, L. A. and Wier, W. G. (1990) Intracellular diffusion, binding and compartmentalization of the fluorescent calcium indicators indo-1 and fura-2. *Biophys. J.* **58,** 1491–1499.
44. Spurgeon, H. A., Stern, M. D., Baartz, G., Raffaeli, S., Hansford, R. G., Talo, A., Lakatta, E. G., and Capogrossi, M. C. (1990) Simultaneous measurement of calcium, contraction and potential in cardiac myocytes. *Am. J. Physiol.* H574–H586.
45. Erdahl, W. L., Chapman, C. J., Taylor, R. W., and Pfeiffer, D. R. (1994) Calcium transport properties of ionophores A23187, ionomycin and 4-BrA23187 in a well defined model system. *Biophys. J.* **66,** 1678–1693.
46. Kao, J. P. Y. (1994) Practical aspects of measuring calcium with fluorescent indicators. *Methods in Cell Biology* **40,** 155–181.
47. Mason, M. J. and Grinstein, S. (1993) Ionomycin activates electrogenic calcium influx in rat thymic lymphocytes. *Biochem. J.* **296,** 33–39.
48. Schnetkamp, P. P. M., Li, X.-B., Basu, D. K., and Szerencsei, R. T. (1991) Regulation of free cytosolic calcium concentration in the outer segments of bovine retinal rods by Na-Ca-K exchange measured with fluo 3. *J. Biol. Chem.* **266,** 2275–2292.
49. Kao, J. P. Y., Harootunian, A. T., and Tsien, R. Y. (1989) Photochemically generated cytosolic calcium pulses and their detection by fluo-3. *J. Biol. Chem.* **264,** 8179–8184.
50. Giovannardi, S., Cesare, P., and Peres, A. (1994) Rapid synchrony of nuclear and cytosolic calcium signals activated by muscarinic stimulation in the human tumour line TE571/RD. *Cell Calcium* **16,** 491–499.
51. Lin, C., Hajnoczky, G., and Thomas, A. P. (1994) Propagation of cytosolic calcium waves into the nuclei of hepatocytes. *Cell Calcium* **16,** 247–258.
52. Allbritton, N. L., Kuhn, O. E., and Meyer, T. (1994) Source of nuclear calcium signals. *Proc. Natl. Acad. Sci.* **9,** 12,458–12,462.
53. Shirakawa, H. and Miyazaki, S. (1996) Spatiotemporal analysis of calcium dynamics in the nucleus of hamster oocytes. *J. Physiol.* **494,** 29–40.

54. Perez-Terzic, C., Stehno-Bittel, L., and Clapham, D. E.(1997) Nucleoplasmic and cytoplasmic differences in the fluorescence properties of the calcium indicator Fluo-3. *Cell Calcium* **21**, 275–282

55. Neidle, S. and Abraham, Z. (1984) Structural and sequence-dependent aspects of drug intercalation into nucleic acids. *CRC Crit. Rev. Biochem.* **17**, 73–121.

56. Merritt, J. E., McCarthy, S. A., Davies, M. P., and Moores, K. E. (1990) Use of fluo-3 to measure cytosolic calcium in platelets and neutrophils. *Biochem. J.* **269**, 513–519.

57. Konishi, M. and Watanabe, M. (1995) Resting cytoplasmic free calcium concentration in frog skeletal muscle measured with fura-2 conjugated to high molecular weight dextran. *J. Gen. Physiol.* **106**, 1123–1150.

58. Fetcho, J. R. and O'Malley, D. M. (1995) Visualization of active neural circuitry in the spinal cord of intact zebrafish. *J. Neurophys.* **73**, 399–406.

59. Berlin, J. R. and Konishi, M. (1993) Calcium transients in cardiac myocytes measured with high and low affinity calcium indicators. *Biophys. J.* **65**, 1632–1647.

60. Eilers, J., Callewaert, G., Armstrong, C., and Konnerth, A. (1995) Calcium signaling in a narrow somatic submembrane shell during synaptic activity in cerebellar Purkinje neurons. *Proc. Natl. Acad. Sci.* **92**, 10,272–10,276.

61. Regehr, W. G. and Atluri, P. P. (1995) Calcium transients in cerebellar granule cell presynaptic terminals. *Biophys. J.* **68**, 2156–2170.

62. Birch, B. D., Eng, D. L., and Kocsis, J. D. (1992) Intranuclear calcium transients during neurite regeneration of an adult mammalian neuron. *Proc. Natl. Acad. Sci.* **89**, 7978–7982.

63. Cheng, H., Lederer, W. J., and Cannell, M. B. (1993) Calcium sparks, elementary events underlying excitation-contraction coupling in heart muscle. *Science* **262**, 740–744.

64. Stricker, S. A., Centonze, V. E., and Melendez, R. F. (1994) Calcium dynamics during starfish oocyte maturation and fertilization. *Dev. Biol.* **166**, 34–58.

65. Fetcho, J. R. and O'Malley, D. M. (1997) Imaging neuronal networks in behaving animals. *Curr. Opin. Neurobio.* **7**, 832–838.

66. Denk, W., Holt, J. R., Shepherd, G. M., and Corey, D. P. (1995) Calcium imaging of single stereocilia in hair cells, localization of transduction channels at both ends of tip links. *Neuron* **15**, 1311–1321.

67. Yuste, R. and Denk, W. (1995) Dendritic spines as basic functional units of neuronal integration. *Nature* **375**, 682–684.

68. Svoboda, K., Tank, D. W., and Denk, W. (1996) Direct measurement of coupling between dendritic spines and shafts. *Science* **272**, 716–719.

69. Berg, H. (1993) Random walks in biology. Princeton University Press, Princeton, NJ, pp. 34,35.

70. Al-Mohanna, F. A., Caddy, K. W. T., and Bolsover, S. R. (1994) The nucleus is insulated from large cytosolic calcium ion changes. *Nature* **367**, 754–750.

71. Przywara, D. A., Bhave, S. V., Bhave, A., Wakade, T. D., and Wakade, A. R. (1991) Stimulated rise in neuronal calcium is faster and greater in the nucleus than the cytosol. *FASEB J* **5**, 217–222.

72. Davis, M. A., Chang, S. H., and Trump, B. F. (1995) IP$_3$-mediated cytosolic and nuclear calcium elevation in NRK-52E cells using 'caged' GPIP$_2$. *Cell Calcium* **17,** 453–458.

73. Badminton, M. N., Campbell, A. K., and Rembold, C. M. (1996) Differential regulation of nuclear and cytosolic calcium in HeLa cells. *J. Biol. Chem.* **271,** 31,210–31,214.

74. Bkaily, G., Gros-Louis, N., Naik, R., Jaalouk, D., and Pothier, P. (1996) Implication of the nucleus in excitation contraction coupling of heart cells. *Molec. Cell. Biochem.* **154,** 113–121.

75. Ferrier, J. and Yu, H. (1996) Nuclear versus perinuclear and cytoplasmic calcium in osteoclasts. *Cell Calcium* **20,** 381–388.

76. Jovanovic, A., Lopez, J. R., and Terzic, A. (1996) Cytosolic calcium domain-dependent protective action of adenosine in cardiomyocytes. *Eur. J. Pharm.* **298,** 63–69.

77. Dani, J. W., Chernjavsky, A., and Smith, S. J. (1992) Neuronal activity triggers calcium waves in hippocampal astrocyte networks. *Neuron* **8,** 429–440.

78. Segal, M. and Manor, D. (1992) Confocal microscopic imaging of calcium in cultured rat hippocampal neurons following exposure to N-methyl-d-aspartate. *J. Physiol.* **448,** 655–676.

79. Kuba, K., Hua, S.-Y., and Hayashi, T. (1994) A UV laser-scanning confocal microscope for the measurement of intracellular calcium. *Cell Calcium* **16,** 205–218.

80. Grapengiesser, E. (1993) Cell photodamage, a potential hazard when measuring cytoplasmic calcium with fura 2. *Cell Struc. Func.* **18,** 13–17.

81. Lipp, P. and Niggli, E. (1993) Ratiometric confocal calcium-measurements with visible wavelength indicators in isolated cardiac myocytes. *Cell Calcium* **14,** 359–372.

82. Williams, R. M., Piston, D. W., and Webb, W. W. (1994) Two-photon molecular excitation provides intrinsic 3-dimensional resolution for laser-based microscopy and microphotochemistry. *FASEB J.* **8,** 804–813.

83. Kay, A. R. (1992) An intracellular medium formulary. *J. Neurosci. Meth.* **44,** 91–100.

84. Harrison, S. M. and Bers, D. M. (1987) The effect of temperature and ionic strength on the apparent Ca-affinity of EGTA and the analogous Ca-chelators BAPTA and dibromo-BAPTA. *Biochim. Biophys. Acta.* **925,** 133–143.

85. Williams, D. A. and Fay, F. S. (1990) Intracellular calibration of the fluorescent calcium indicator fura-2. *Cell Calcium* **11,** 75–83.

86. Baudet, S,. Hove-Madsen, L., and Bers, D. M. (1994) How to make and to use calcium-specific mini- and micro-electrodes. *Methods in Cell Biology* **40,** 94–113.

87. Becker, P. L. and Fay, F. S. (1987) Photobleaching of fura-2 and its effect on determination of calcium concentrations. *Am. J. Physiol.* **253,** C613–C618.

88. Moore, E. D. W., Becker, P. L., Fogarty, K. E., Williams, D. A., and Fay, F. S. (1990) Calcium imaging in single living cells, theoretical and practical issues. *Cell Calcium* **11,** 157–179.

89. Stevens, T., Fouty, B., Cornfield, D., and Rodman, D. M. (1994) Reduced PO$_2$ alters the behavior of fura-2 and indo-1 in bovine pulmonary artery endothelial cells. *Cell Calcium* **16,** 404–412.

90. Lattanzio, F. A. and Bartschat, D. K. (1991) The effect of pH on rate constants, ion selectivity and thermodynamic properties of fluorescent calcium and magnesium indicators. *Biochem. Biophys. Res. Comm.* **177,** 184–191.
91. Matthews, G. G. (1991) Cellular physiology of nerve and muscle. Blackwell Scientific Publications, Cambridge, MA.
92. Hille, B. (1992) Ionic channels of excitable membranes. Sinauer Associates, Sunderland, MA.
93. Johnston, D .and Wu, S. S.-M. (1995) Foundations of cellular neurophysiology. MIT Press, Cambridge, MA.
94. Sherman-Gold, R. (Ed.) (1993) The Axon Guide. Axon Instruments, Foster City, CA.
95. Lopez ,H. S. (1992) Kinetics of G protein-mediated modulation of the potassium M-current in bullfrog sympathetic neurons. *Neuron* **8,** 725–736.
96. Jones, S. W. (1987) Sodium currents in dissociated bull-frog sympathetic neurones. *J. Physiol.* **389,** 605–627.
97. Toescu, E. C., Lawrie, A. M., Gallacher, D. V., and Petersen, O. H. (1993) The pattern of agonist-evoked cytosolic calcium oscillations depends on the resting intracellular calcium concentration. *J. Biol. Chem.* **268,** 18,654–18,658.
98. Sala, F. and Hernandez-Cruz, A. (1990) Calcium diffusion modeling in a spherical neuron, relevance of buffering properties. *Biophys. J.* **57,** 313–324.
99. Nowycky, M C. and Pinter, M. J. (1993) Time course of calcium and calcium-bound buffers following calcium influx in a model cell. *Biophys. J.* **64,** 77–91.
100. Sobierajski, L., Avila, R., O'Malley, D. M., Wang, S., and Kaufman, A. (1995) Visualization of calcium activity in nerve cells. *Computer Graphics Appl.* **15,** 55–61.
101. Berlin, J. R., Bassani, J. W. M., and Bers, D. M. (1994) Intrinsic cytosolic calcium buffering properties of single rat cardiac myocytes. *Biophys. J.* **67,** 1775–1787.
102. Zhou, Z. and Neher, E. (1993) Mobile and immobile calcium buffers in bovine adrenal chromaffin cells. *J. Physiol.* **469,** 245–273.
103. Neher, E. (1995) The use of fura-2 for estimating calcium buffers and calcium fluxes. *Neuropharm.* **34,** 1423–1442.
104. Mize, R. R., Luo, Q., Butler, G., Jeon, C. J., and Nabors, B. (1992) The calcium binding proteins parvalbumin and calbindin-D 28K form complementary patterns in the cat superior colliculus. *J. Comp. Neurol.* **320,** 243–256.
105. Peters, R. (1984) Nucleocytoplasmic flux and intracellular mobility in single hepatocytes measured by fluorescence microphotolysis. *EMBO J.* **3,** 1831–1836.
106. Peters, R. (1986) Fluorescence microphotolysis to measure nucleocytoplasmic transport and intracellular mobility. *Biochim. Biophys. Acta.* **864,** 305–359.
107. Csermely, P., Schnaider, T., and Szanto, I. (1995) Signalling and transport through the nuclear membrane. *Biochim. Biophys. Acta* **1241,** 425–452.

17

Intracellular pH and pCa Measurement

Michal Opas and Ewa Dziak

1. Introduction

Recent improvements in confocal technology permit the use of a confocal microscope as an effective tool to both measure concentration and visualize distribution of several ions. Effective discrimination against out-of-focus information in the confocal microscope permits for mapping of ion distribution at the subcellular level *(1)*. To generate such a map with any degree of accuracy it is necessary to compensate for inherent cellular inhomogeneities in optical path, probe distribution, etc. To do so ratiometric measurements are usually employed *(2,3)*. Fluorescence ratiometry takes advantage of a differential spectral sensitivity of a ratio probe to the measured parameter (e.g., ion concentration). Whereas fluorescent properties of the ratio probe at one wavelength are parameter sensitive in one manner, the emission (or excitation) of the probe at a distinct, well-separated wavelength must be either parameter insensitive, inversely sensitive, or exhibit a different sensitivity profile. Emission intensities for the two wavelengths are then divided by each other and thus the resulting ratio becomes normalized for inhomogeneities in probe distribution and concentration and in the system geometry. To obtain meaningful data the ratio values must be calibrated against a standard, which usually is a graded change in the parameter. Modern day confocal microscopy depends on lasers as light sources which causes a severe limitation in number of wavelengths available for fluorochrome excitation. Several H^+-sensitive fluorochromes are excited at wavelengths close to those emitted by standard lasers, however, this is not the case with Ca^{2+}-sensitive fluorochromes *(4)*.

This chapter provides protocols for measurement of intracellular pH (pH_i) and intracellular pCa (pCa_i) with a typical confocal microscope of the present day. Such a microscope consists of a scanner capable of simultaneous dual

From: *Methods in Molecular Biology, vol. 122: Confocal Microscopy Methods and Protocols*
Edited by: S. Paddock © Humana Press Inc., Totowa, NJ

excitation as well as simultaneous collection of dual emissions. The scan time for a full frame is about 1 second in a 'standard' mode or 2 s in a "slow" mode where the pixel dwell time is greater and, consequently, the signal-to-noise ratio is higher. The simultaneous dual excitation/dual emission mode of operation requires a FITC/Texas Red-type filter module in which the two emission wavelengths are separated at 560 nm with a dichroic mirror. The light source is likely to be a mixed gas krypton/argon laser. The fluorochromes, use of which is described herein, are seminaphthofluorescein-calcein acetoxylmethyl ester (SNAFL-calcein AM) for pH$_i$ measurement (*5,6* and *see* **Note 1**) and, for pCa$_i$ measurement, a mixture of Calcium Green and Fura-Red based on a mixture of fluo-3 and fura-red, first described by Lipp and Niggli (*7*). SNAFL-calcein seems to be a good choice for pH$_i$ measurements with the confocal microscope as it loads easily, is well retained in cells, and can be used without any modification to the currently most popular hardware. The Calcium Green/Fluo-3 and Fura-Red dye mixture is a necessity due to the lack of commercially available indicators that exhibit a spectral shift upon Ca^{2+} binding (necessary for ratioing) and are excited with visible light.

2. Materials

1. SNAFL-calcein AM, Calcium Green AM, Fura-Red AM, calcium calibration Kit 1
2. Pluronic F-127 can be purchased from Molecular Probes (Eugene, OR)
3. Ionomycin from Calbiochem (La Jolla, CA).
4. Dimethylsulfoxide (DMSO) and nigericin from Sigma Chemical Co. (St. Louis, MO).
5. Trypsin/EDTA.
6. α-minimal essential medium (α-MEM).
7. α-MEM with 20 m*M* HEPES (MEM-HEPES, pH 7.36)
8. Heat-inactivated fetal bovine serum (FBS) is from Gibco (Canadian Life Technologies, Burlington, ON, Canada). All MEM-HEPES media should be NaHCO$_3^-$ free.
9. Amino acid-free MEM-HEPES or a buffers shown in **Tables 1** and **2** are used for dye loading.
10. Nigericin-containing calibration solutions: Five solutions with pH values of 6.4, 6.7, 7.0, 7.3, and 7.6. made of K$^+$-MEM-HEPES supplemented with 10 μ*M* of nigericin.

3. Methods

3.1. Measurement of pH$_i$

These conditions have been optimized for near-confluent MDCK cells. **Conditions for each cell type have to be established experimentally**.

1. Load cells with 5 μ*M* SNAFL-calcein AM in serum-free and amino-acid-free medium for 30–40 min. Higher dye concentrations and/or longer incubation times will cause compartmentalization of the dye, which may cause measurement

Table 1
Composition of the Media Used for Dye Loading (A), pH$_i$ Measurement (B), and Calibration (C)

	A	B[a]	C[b]
NaCl, mM	118	118	0
KCl, mM	5.3	5.3	123.3
CaCl$_2$, mM	1.8	1.8	1.8
MgSO$_4$, mM	0.8	0.8	0.8
NaH$_2$PO$_4$·2H$_2$O, mM	1	1	1
HEPES, mM	20	20	20
D-Glucose, mM	10	10	10
Phenol red, mM	0.03	0.03	0.03
Lipoic acid, mM	1	1	1
Sodium pyruvate, mM	1	1	1
Set of amino acids	–	+	+
Set of vitamins	+	+	+
Fetal bovine serum, %	0	3	0
Nigericin, µM	0	0	10

[a]Composition of the medium for pH$_i$ measurement is essentially the same as that of the culture medium (α-MEM), in which HEPES was replaced by sodium bicarbonate and 10% FBS was added.
[b]pH of the calibration medium was adjusted with KOH or HCl.
The + or – represents the presence or absence of a component in the medium, respectively.

Table 2
Composition of the Buffers Used for Dye Loading (A), pCa$_i$ Measurement (B), and Calibration (C)

	A	B	C
NaCl, mM	140	140	–
KCL, mM	4	4	100
CaCl$_2$, mM	1	1	–
MgCl$_2$, mM	1	1	–
HEPES, mM	10	10	–
Glucose, mM	10	10	–
Ionomycin, µM	–	–	2
pH	7.2	7.2	7.2
MOPS, mM	–	–	10
EGTA/CaEGTA, mM	–	–	0–10
Pluronic F = 127, 0.02%	+	–	–

The – represents the absence of a component in the medium.

errors. (*see* **Notes 2** and **3**). We have never had any need to use Pluronic F-127 to aid loading of SNAFL-calcein. This may, however, be necessary with some cell types. For loading of calcium probes we use 0.02% Pluronic F-127 (*see* **Table 2**).

2. Incubate cells for 15–20 min in α -MEM with 20 mM HEPES containing 2% serum to deestrify the probe and let cells recover. After this incubation is completed rinse cells twice with K$^+$-MEM-HEPES (*see* **Note 4**).

3. Place the cell under the microscope. An inverted microscope is the configuration of choice as it is possible to use one of many types of cell culture chambers designed for microscopy. We use a Bio-Rad MRC 600 on an upright microscope, therefore the coverslips with cells are placed in MEM-HEPES containing 2% FBS in a 60-mm culture dish. Images of cells (20 cells per image) are collected with a Zeiss Plan-Neofluar objective (40x, multiimmersion). Measurements can be made at either room temperature or 37°C.

4. Using **simultaneously** 488 **and** 568 nm laser bands, the FITC/Texas Red-type filter set (e.g., K1 and K2 in a Bio-Rad MRC 600) and simultaneous dual-excitation **and** simultaneous dual-emission recording feature of a confocal microscope collect intracellular emission of SNAFL-calcein from the green channel (excitation: 488 nm, emission: 525 nm) and from the red channel (excitation: 568 nm, emission: 615 nm). Be sure that the gain ranges of both channels cover the expected dynamic range of both emissions, i.e., while the weakest emission is still recorded, the strongest emission does not saturate the photomultiplier. This is done on a Bio-Rad MRC 600 by manual adjustment of gain control ("auto" gain control in OFF position) and confocal aperture (the pinhole).

   Specifically, for a Bio-Rad MRC 600, we use the following settings for pH$_i$ imaging (for **both** *photomultiplier tube 1* and *photomultiplier tube 2*): (**1**) *confocal pinhole diameter*, fully open, (**2**) *enhancement control*, off, (**3**) *gain control*, manual, (**4**) *black level control*, manual, (**5**) *neutral density filter* no. 2 (neutral density = 1.5; 3% transmission), (**6**) *scanning speed* 1 s per frame.

   The measurements must be internally calibrated.

5. Calibrate pH$_i$ immediately after the end of each experiment under the same microscopic working conditions as those for pH$_i$ measurements. Use a fresh batch of cells loaded with SNAFL-calcein AM, equilibrated and rinsed twice with K$^+$-MEM- HEPES. Flush the dish with nigericin-containing K$^+$-MEM-HEPES of desired pH value and incubate for 3–5 min before taking the calibration measurements. Repeat for all the pH values.

   We normally take a series of images from **different** areas of the cell monolayer in the nigericin-containing K$^+$-MEM-HEPES with pH values of 6.4, 6.7, 7.0, 7.3, and 7.6. At least five **different** image pairs are collected for pH calibration points. We use acidic solutions first to minimize the cell blebbing. Between each calibration measurement, samples are washed twice in the buffer for the next-point calibration and allowed 3–5 min equilibration. (*see* **Note 5**).

6. Ratio the green and red emissions by performing an arithmetic division of one image by the other. This is usually followed by a multiplication by a constant factor. Any software that performs arithmetic operations on image files will do the job. The calibration data are used to generate a standard pH curve for each experiment.

7. Be aware that useful pH range of SNAFL-calcein AM is 6. 2–7. 8, as the dye is difficult to calibrate outside this pH range according to Zhou et al. (*6* and *see* **Notes 6–8**).

3.2. Measurement of pCa$_i$

Conditions optimized for sparse retinal pigment epithelial cells of the chick embryo. **Conditions for each cell type have to be established experimentally.**

1. Load cells with a mixture of acetoxylmethyl esters of Calcium Green (8 μM) and Fura-Red (5 μM) in the loading buffer in presence of 0.02% Pluronic F-127 (1 μL of 20% stock solution per milliliter of stain solution) for 30–40 min. Higher dye concentrations and/or longer incubation times will cause compartmentalization of the dye, which may cause measurement errors (*see* **Note 9**).
2. Incubate cells for 15–20 min in α-MEM with 10 mM HEPES containing 2% serum to deestrify the probe and let cells recover.
3. Place the cell under the microscope. An inverted microscope is the configuration of choice as it is possible to use one of many types of cell culture chambers designed for microscopy. We use a Bio-Rad MRC 600 on an upright microscope, therefore, the coverslips with cells are placed in measurement buffer (*see* **Table 2**) in a 60-mm culture dish. Images of cells (20 cells per image) are collected with a Zeiss Plan-Neofluar objective (40x, multiimmersion). Measurements can be made at either room temperature or 37°C.
4. Using **only** the 488-nm laser band, the FITC/Texas Red-type filter set (e.g., K1 and K2 in a Bio-Rad MRC 600) and simultaneous dual-emission recording feature of a confocal microscope collect intracellular emission of Calcium Green from the green channel (excitation: 488 nm, emission: 525 nm) and of Fura-Red from the red channel (excitation: 568 nm, emission: 615 nm). Be sure that the gain ranges of both channels cover the expected dynamic range of both emissions, i.e., while the weakest emission is still recorded, the strongest emission does not saturate the photomultiplier. **This has also to be taken into consideration while testing the proportion of the dyes in the mixture.** The system calibration is done on a Bio-Rad MRC 600 by manual adjustment of gain control ("auto" gain control in OFF position) and confocal aperture (the pinhole).
5. Calibrate pCa$_i$ immediately after the end of each experiment under the same microscopic working conditions as those for pCa$_i$ measurements. Use a fresh batch of cells loaded with the dye mixture, equilibrated and rinsed twice with appropriate Ca^{2+} buffer (*see* **Note 10**). Flush the dish with a series of solutions of different Ca^{2+} concentrations containing a Ca^{2+} ionophore, ionomycin. A series of measurements is taken of the dye-loaded Ca^{2+}-permeable cells flushed with the graded [Ca^{2+}] solutions to generate a standard pCa$_i$ curve for each experiment. Repeat for all the Ca concentrations.

We normally take at least five **different** image pairs of images from **different** areas of the cell monolayer for each pCa$_i$ calibration point. Between each calibration measurement, samples are washed twice in the buffer for the next-point calibration and allowed 3–5 min equilibration.

6. Ratio the green and red emissions by performing an arithmetic division of one image by the other. This is usually followed by a multiplication by a constant factor. Any software that performs arithmetic operations on image files will do the job. The calibration data are used to generate a standard pCa$_i$ curve for each experiment (also *see* **Notes 11–14**).

4. Notes

1. The acidic form of SNAFL-calcein AM (Molecular Probes cat. no. S-3052) has spectral properties similar to fluorescein isothiocyanate (FITC) (excitation: Ç490 nm, emission: 535 nm) while its basic form behaves like a hybrid of TRITC and Texas Red (excitation: 540 nm, emission: 620 nm).
2. Before loading the coverslips are rinsed twice with serum free α-MEM-HEPES. They are then incubated at room temperature in amino-acid free α-MEM-HEPES containing 5 μM of SNAFL-calcein AM from a stock solution in anhydrous DMSO, which was either prepared immediately before use or used from aliquot stored at –20°C. Aliquots are good for 1 mo in DMSO. The final concentration of DMSO in the culture medium should never exceed 0.1%.
3. For subconfluent cells time and the temperature of loading is very important. To prevent dye compartmentalization, with SNAFL-calcein concentration at 5 μM, loading at room temperature should take no longer than 40 min. Cells could be loaded at 37°C as well, but the time of loading should then be shortened to 30–35 min. For very packed cells loading times should be slightly extended.
4. Never use antibiotics for any pH or pCa$_i$ measurements or calibrations. NaCl is replaced with KCl in the medium for calibration only.
5. Some cell lines are very sensitive to any change in conditions of their medium (such as a difference in pH) and do not survive the complete calibration procedure very well. It might be advisable to load a few smaller coverslips under exactly the same conditions (we were doing it in one Petri dish) and use one or two of them for each pH point calibration instead of using one coverslip for the calibration of all pH points. In our experience, the entire experiment, including calibration, should be completed within ca. 3 h, otherwise fresh batch of cells must be loaded.
6. The quality and the time resolution of pH$_i$ images are restricted by the microscopic working conditions. Samples are localized and brought into focus with fast scanning modes (1/4 s per frame) to minimize photobleaching. We routinely record images with the laser scanning speed of 1 sec per frame (512 lines), the confocal pinholes of both channels fully open and manually adjusted gain control. With the scanner *gain* on, the background noise levels of the 525–614 nm emission images and the 615 nm ones were quite different even with the same manual settings for the black level control. This is due to the gain device automatically increasing the photon/output voltage ratio *(8,9)*.
7. This calibration of pH$_i$ is performed according to the high [K$^+$]/nigericin method of Thomas et al. *(10)*.

8. Calcium Green has FITC-like spectral properties (excitation at 506 nm, emission at 530 nm) and its fluorescence intensity increases with an increase in Ca^{2+} concentration. Calcium Green has the same spectral properties as Fluo-3; however, Molecular Probes indicate that Calcium Green is five times brighter than Fluo-3 and that it bleaches more slowly. Fura-Red is a ratiometric dye that is excited at 440 and at 490 nm and emits with a huge Stokes shift at 660 nm. It is unusual in that its emission intensity decreases with an increase in Ca^{2+} concentration.

9. Conditions of loading for each cell type should be established experimentally in such a manner that proportion of the dyes will give good quality images in whole expected spectrum of pCa_i. It is important to keep in mind that Fura-Red images will get **darker** with higher pCa_i, while Calcium Green images will get **brighter** with higher pCa_i. Reported dye proportions for loading the mixture of acetoxylmethyl esters vary from 3:4 μM, to 1:10 μM, to 50:100 μM to 100:300 μM for fluo-3:Fura-Red; we load Calcium Green:Fura-Red at an 8.3:4.6 μM ratio.

10. The calibration procedure based on Grynkiewicz's equation *(11)* is not applicable to cells loaded with a mixture of acetoxylmethyl esters of the two dyes. Even after modification of Grynkiewicz's equation to accommodate the two different Ca^{2+} K_ds of the dyes, its applicability remains highly problematic *(12)*. Therefore, a calibration procedure similar to that used for pH_i measurements should be implemented. For calibration measurements several buffers of increasing Ca^{2+} concentrations were prepared according to Molecular Probes protocol included in their calibration kit. Molecular Probes sells two Ca buffer kits (cat. no. C-3008 and C-3009). For each pCa point we use a fresh coverslip with cells loaded under the same conditions. The coverslip is rinsed with appropriate Ca^{2+} buffer and placed in the same buffer containing 2 μM of ionomycin. Several different image pairs should be collected for a given calibration point.

11. Because a single and uniform optical section can be imaged by a confocal microscope, very successful attempts have been made at measuring pCa_i in thick samples using non ratio fluorochromes *(13,14)*.

12. Maximal recording speed ("time resolution") is limited by the scanning speed (often: 512 lines/s). While ionic changes comprise extremely fast mode of cell signaling, a standard confocal microscope takes a very long time to collect a full-field image of a decent quality. The time resolution can be further increased by reducing the image frame size which in some applications can consist of just one line. The confocal microscope in a single line scanning mode can visualize very fast (millisecond range) changes in an ion concentration along a rapidly scanned single line. Slit scanning confocal microscopes, provide an option of video-rate imaging of ionic changes.

13. Physiologically meaningful data can be collected only when the probe is where we think it is. Therefore extreme caution has to be exercised while loading cells with a fluorescent probes. Probe concentration, cell density, time of loading, temperature, the presence or absence of serum and particular amino acids in the medium, not to mention all sorts of spells, will affect cell loading with AM forms

of dyes. Overloaded probes will compartmentalize immediately, and most probes loaded into the cytosol will compartmentalize eventually, meaning that they end up in various cell compartments such as the mitochondria, endoplasmic reticulum, and all types of vesicles. This will lead to measurement errors. Unfortunately there is no ready recipe for cell loading. Dye loading conditions must be worked out for each experimental setup. The effect of presence or absence of serum on (1%) efficacy of loading and (2%) compartmentalization of loaded dye should be checked first. Next temperature of loading should be experimented with. Whereas some cell types will load quickly and well in 37°C others may need more time at lower temperature to load uniformly. To ascertain a uniform diffusion of dyes into thicker specimens (such as very packed multilayered epithelia or tissue sections) it is not unheard to use a refrigerator as an incubator! However, the presence or absence of extracellular esterases will also be a factor dictating how much time and what concentration of dye can be used to load the specimen. If an incubation in temperature lower than room temperature is required, it will be necessary to warm up the specimen for 20–30 min at 37°C to allow the dye deestrification.

14. Pluronic F-127 in the range of 1–4 µL/mL of stain solution has been often used to aid loading of AM forms of dyes. Finally, cells attempt to expel the dye by pumping it out. It has been reported that, in particular case of Fluo-3, an addition of 2.5 mM probenecid (Sigma) inhibits the expulsion of the dye *(15)*.

Acknowledgments

Our research has been supported by the Ontario Heart and Stroke Foundation Grant T-2922 and by the Medical Research Council of Canada Grant MT-9713 to Michal Opas.

References

1. Opas, M. (1997) Measurement of intracellular pH and pCa with a confocal microscope. *Trends Cell Biol.* **7,** 75-80.
2. Bright, G. R., Fisher, G. W., Rogowska, J., and Taylor, D. L.(1989) Fluorescence ratio imaging microscopy. *Methods Cell Biol.* **30,** 157–192.
3. Dunn, K. W., Mayor, S., Myers, J. N., and Maxfield, F. R. (1994) Applications of ratio fluorescence microscopy in the study of cell physiology. *FASEB J.* **8,** 573–582.
4. Schild, D. (1996) Laser scanning microscopy and calcium imaging. *Cell Calcium* **19,** 281–296.
5. Whitaker, J. E., Haugland, R. P.and Prendergast, F. G.(1991) Spectral and photophysical studies of benzo[c]xanthene dyes: dual emission pH sensors. *Anal. Biochem.* **194,** 330–344.
6. Zhou, Y., Marcus, E. M., Haugland, R. P., and Opas, M.(1995) Use of a new fluorescent probe, seminaphthofluorescein-calcein, for determination of intracellular pH by simultaneous dual-emission imaging laser scanning confocal microscopy. *J. Cell. Physiol.* **164,** 9–16.

7. Lipp, P. and Niggli, E. (1993) Ratiometric confocal Ca^{2+}-measurements with visible wavelength indicators in isolated cardiac myocytes. *Cell Calcium* **14,** 359–372.

8. Webb, R. H. and Dorey, C. K. (1995) The pixelated image, in *Handbook of Biological Confocal Microscopy* (Pawley, J. B., ed.), Plenum Press, New York, pp. 55–67.

9. Sandison, D. R., Williams, R. M., Wells, K. S., Strickler, J., and Webb, W. W. (1995) Quantitative fluorescence confocal laser scanning microscopy, in *Handbook of Biological Confocal Microscopy* (Pawley, J. B., ed.), Plenum Press, New York, pp. 39–53.

10. Thomas, J. A., Buchsbaum, R. N., Zimniak, A., and Racker, E. (1979) Intracellular pH measurements in Ehrlich ascites tumor cells utilizing spectrscopic probes generated in situ. *Biochemistry* **18,** 2210–2218.

11. Grynkiewicz, G., Poenie, M., and Tsien, R. Y. (1985) A new generation of Ca^{2+} indicators with greatly improved fluorescence properties. *J. Biol. Chem.* **260,** 3440–3450.

12. Floto, R. A., Mahaut-Smith, M. P., Somasundaram, B., and Allen, J. M. (1995) IgG-induced Ca^{2+} oscillations in differentiated U937 cells, a study using laser scanning confocal microscopy and co-loaded fluo-3 and fura-red fluorescent probes. *Cell Calcium* **18,** 377–389.

13. Tsugorka, A., Rios, E., and Blatter, L. A. (1995) Imaging elementary events of calcium release in skeletal muscle cells. *Science* **269,** 1723–1726.

14. Robb-Gaspers, L. D. and Thomas, A. P. (1995) Coordination of Ca^{2+} signaling by intercellular propagation of Ca2+ waves in the intact liver. *J. Biol. Chem.* **270,** 8102–8107.

15. Merritt, J. E., McCarthy, S. A., Davies, M. P. A., and Moores, K. E. (1990) Use of fluo-3 to measure cytosolic Ca^{2+} in platteletes and neutrophils. *Biochem. J.* **269,** 513–519.

18

Measuring Dynamic Cell Volume In Situ by Confocal Microscopy

Rachel J. Errington and Nick S. White

1. Introduction

Regulation of cell volume is a fundamental homeostatic mechanism in the face of osmotic stress *(1,2)*. One approach to understanding aspects of cell volume regulation involves the removal of cells from the matrix and manipulation in culture *(3)*. Measurements are often also needed from cells within intact tissue that are operating in their correct physiological context. These include the effects of cell-cell interactions and the mechanical, ionic, and physiological effects of the extracellular matrix (ECM). The procedure described here is an *in situ* approach to volume measurement using an organ culture system, which maintains tissue integrity and hence the spatial organisation of cells in the ECM *(4)*. We present a comprehensive protocol for investigating volume regulatory behaviour using confocal laser scanning microscopy (CLSM). Although we focus on articular and fetal growth plate cartilage tissues, the protocols can be applied to other intact animal, plant, and fungal tissues *(5)*. As we show below, many factors must be taken into account in the choice of experimental parameters, there are no "best settings" that work for all tissues, in all cases.

1.1. Aims

The protocol comprises a series of key methods and associated optimization or correction steps that are necessary to extract and view cell volume data (**Fig. 1**). We describe a microscopy regime, optimizing signal-to-noise ratio and cell viability (**Fig. 2**). As physical interactions between the specimen and light significantly affect the accuracy of the confocal optical probe, we present practical tools to assess tissue-induced attenuation and axial dis-

From: *Methods in Molecular Biology, vol. 122: Confocal Microscopy Methods and Protocols*
Edited by: S. Paddock © Humana Press Inc., Totowa, NJ

MEASURING *IN SITU* CELL VOLUME

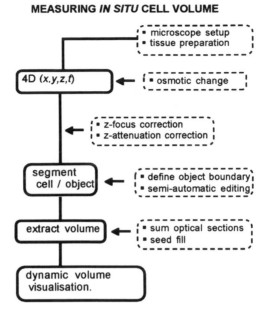

Fig. 1. Schematic summarizing the steps involved in dynamic volume measurements by CLSM.

MICROSCOPE AND TISSUE SETUP

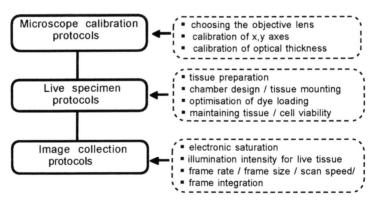

Fig. 2. Schematic summarizing the steps involved in setting up the microscope and optimizing the live specimen conditions.

tortion. Two methods for extracting cell volumes are then given and finally we describe our 4D image visualization methods for the display of dynamic changes in cell volume.

2. Materials

2.1. Biological Material

1. Metacarpal joints of 8-mo-old pigs (used within 4–5 h of slaughter).

2.2. Medium/Buffers

1. Standard: HEPES-buffered DMEM (280 mOsM, pH 7.4)
2. Hypotonic: 1:2 dilution of standard medium (140 mosM, pH 7.4)
3. Hypertonic: standard medium plus 50 mM NaCl (380 mosM, pH 7.4)

2.3. Fluorescent Dyes

1. Vital dye: 5-chloromethyl fluorescein diacetate (CMFDA), 2.25 mM stock (Molecular Probes)
2. Calibration dye: 50 µM aqueous fluorescein (Sigma Chemicals)

2.4. Preparation of Biological and Calibration Microscopy Specimens

1. Tissue-cover glass adhesive such as: (a) "super glue"; (b) poly-L-lysine; (c) low percentage agarose; (d) medical glue e.g. TissueTAK™; (e) physical method (wire mesh, etc.)
2. Plastic electrical tape to make test chambers
3. Glass slides
4. Cover glass (no 1.5 thickness) (BDH)
5. Tissue fixative: 4% paraformaldehyde in phosphate-buffered saline (PBS).
6. Fluorescein-labeled latex beads (approx the size of cells), e.g., 7 ± 0.3 µm (Polysciences Ltd.)

3. Methods

3.1. Calibration of the Confocal Microscopy Imaging Conditions (Fig. 2)

3.1.1. Choosing the Objective Lens

When measuring cell volume deep within tissue, a major limitation is the working distance of the lens. Data should not be collected from the cut surface, where cells are damaged or undergoing a wound healing response. The finest confocal sectioning requires the maximum possible numerical aperture (NA), (i.e., up to 1.33 for aqueous samples), but a reduced NA of approx 0.8–1.0 may be needed for an acceptable working distance. Although a high NA objective lens provides optimum sectioning near the surface of a preparation, the image becomes less sharp and attenuated when imaging deeper. This occurs often when oil immersion is used and/or with small confocal apertures, further reducing the working distance. The choice of immersion objective lens is not always simple. It is important to use a lens whose prescribed immersion medium matches the refractive index of the bulk of the specimen you will focus through. Most tissue is hydrated, but the ECM constituents of cartilage (e.g., collagen, fibronectin and proteoglycans) scatter and refract light. Therefore, all objective lenses should be calibrated for measurements (**Subheadings 3.1. and 8.7.**). For tissue experiments we used a 25x, 0.8 NA Plan Neofluar objective with a variable correction allowing for oil, glycerol or water immersion.

3.1.2. Pixel Size with Different Objective Lenses: Calibration of Image (X,Y) Axes

Since the objective magnification may not be exactly what is inscribed on the lens and the microscope system may have added optical components, the (X,Y) pixel size calibration must be checked.

1. Image a microscope graticule in transmission or reflection mode.
2. Measure the length of the largest number of divisions that will fit the field of view (in pixels). The pixel size (in micrometers) is the total length of the graticule units (in micrometers) divided by the number of pixels.
3. Check this over the image size, zoom and scan speed range you will later use for measurements.

3.1.3. Confocal Pinhole or Iris Setting: Calibrating the Optical Section Thickness

The optical section thickness depends on the out-of-focus fluorescence in the final image (according to the size of the confocal aperture) and the size of the illuminated and detected spots (determined by the NA, wavelength, sample properties, etc). Accurate calculation is difficult for practical cases of high NA, variable confocal aperture and biological specimens. A good compromise is to measure a test sample under the experimental conditions. The axial intensity profile through a simple planar object (much thinner than the axial resolution) gives the plane response function (PRF) from which the section thickness is derived, and *see* **Note 1**.

3.1.4. Axial Response Test Specimen: Reflection Contrast

The PRF is measured from a confocal reflection optical section through a flat mirror (an infinitely thin boundary) deposited on a cover glass using aluminum, gold, or silver as used for SEM specimens (**Fig. 3**).

1. Sputter coat a cleaned no. 1.5 cover glass with the reflective material
2. Mount on a microscope slide with super glue and the mirror surface facing the slide (**Fig. 3A**), and *see* **Note 2**.

3.1.5. Axial Response Test Specimen: Fluorescence

It is difficult to make a planar fluorescent test sample truly subresolution and much easier to derive a PRF from an edge-response sample where fluorescent medium meets an optical boundary (**Fig. 4A**).

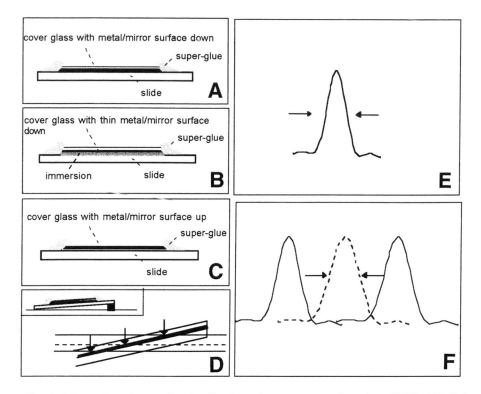

Fig. 3. Measuring the confocal reflection plane response function (PRF). (A) Side view of 100% reflecting mirror sputter coated onto cover glass and mounted face down on slide. (B) Partially reflecting mirror mounted on slide in immersion medium. (C) As in (A) but for objectives corrected for use without cover glass. (D) Tilted version of test specimen (inset). Arrows point to intersection of horizontal optical sections with tilted mirror surface (solid band). Calibration sections at two focus positions (solid horizontal lines) and section to measure axial response in centre of field (dotted horizontal line). (E) Reflection PRF extracted from vertical (X,Z) section through (A). Arrows indicate the axial width of the peak between the 50% intensity points. (F) Reflection PRF plots extracted from corresponding horizontal (X,Y) sections in (D). Arrows indicate width, between 50% intensity points, of central peak corresponding to experimentally determined reflection PRF measurement.

1. Dissolve dye (e.g., fluorescein) in mountant/medium used for biological specimens (50–100 μM).
2. Stick a spacer (~ 200 μm × 25 mm × 25 mm) of foil or electrical adhesive tape onto a slide.
3. Make a shallow chamber by cutting out the center (~10 mm × 10 mm), and *see* **Note 3**.
4. Add 10–20 μL of the fluorescent solution. Avoid overfilling the chamber.

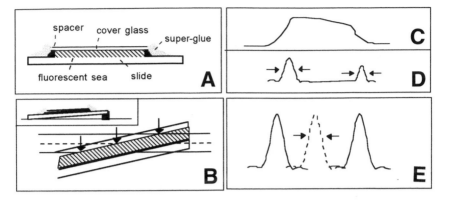

Fig. 4. Measuring the confocal fluorescence plane response function (PRF). (**A**) Fluorescent medium mounted between cover glass andslide. (**B**) Tilted version of test specimen (inset). Arrows point to intersection of horizontal optical sections with cover glass surface and fluorescent medium (hatched band). Calibration sections at two focus positions (solid horizontal lines) and section to measure axial response in centre of field (dotted horizontal line). (**C**) Fluorescence axial response extracted from vertical (X,Z) section through (A). (**D**) Fluorescence PRFs derived by differentiation of the axial response in (C). Arrows indicate the width between 50% intensity points of peaks corresponding to cover glass/medium (left) and medium/slide (right) interfaces. (**E**) Fluorescence PRF plots extracted from corresponding horizontal (X,Y) sections in (C). Arrows indicate width, between 50% intensity points, of central peak corresponding to experimentally determined fluorescence PRF measurement.

5. Slowly lower a cover glass onto the chamber, avoiding air bubbles.
6. Seal with super glue.

3.1.6. Collecting the Axial Response Using the Motorized Focus

The confocal microscope focus is usually controlled by a motor in steps of approx 50–100 nm. The accuracy (of the mechanical stage movement), when transmitted through the microscope to the specimen, is always worse than this (even with temperature control) but estimates of the axial performance can be made with this method (**Figs. 3A** and **4A**).

1. Set up the required microscopy conditions, ensuring the system is correctly aligned.
2. Mount the reflection or fluorescence test specimen securely on the microscope stage.
3. Collect a vertical (X,Z) section or a 3D (X,Y,Z) z-series across the boundary. For low NA lenses or non-ideal conditions several micrometers on each side should be scanned initially to capture the full axial response, and *see* **Note 4**.
4. Measure an intensity z-profile across the boundary (averaged in X/Y to reduce noise) (**Figs. 3E** and **4C**).

3.1.7. Extracting the PRF from the Fluorescence Axial Step-Response

The PRF is approximated by the first differential (gradient) of the fluorescence Z step-response (**Fig. 4C,D**).

1. Export the boundary profile into your spreadsheet package.
2. Generate the gradient plot (e.g., the difference between neighbouring points).

3.1.8. Collecting the Axial-Response with a Single (X,Y) Section

When the Z-response must be assessed repeatedly, or more accurately than the focus steps allow, (X,Y) scanning is used. By slightly tilting the test specimen the boundary plane passes through the focus across the field of view (**Figs. 3D** and **4B**). An (X,Y) section through the boundary gives the z-response, and *see* **Notes 5** and **6**.

1. Set up the required imaging conditions, ensuring the system is correctly aligned.
2. Attach a spacer (~1 mm) at one end of the test slide.
3. Calibration using the known tilt : mm (Z) per pixel (X,Y) = tilt/(X,Y) pixel size in mm.
4. Calibration using two (X,Y) sections with a focus difference (df): mm (Z) per pixel (X,Y) = df (μm)/distance between boundary centers (**Figs. 3D,F** and **4B,E**).
5. Store the calibration as the image x-increment and use to convert later measurements.
6. Collect an (X,Y) section with the boundary in the centre of the field. Measure an intensity X-profile across the boundary (averaged in y to reduce noise) (**Figs. 3D,F** and **4B,E**).

3.1.9. Extracting the Optical Section Thickness from the PRF Data

The complete optical section is the distance between the first minima on either side of the PRF central peak. A better practical measure is the width between the 50% intensity points (**Figs 3E,F** and **4D,E**).

3.1.10. Setting the Optimum Confocal Aperture Iris Setting

Three things happen to a confocal image as the aperture closes: First, out of focus blur steadily reduces, giving more contrast. Second, near the optimum contrast (under ideal conditions) resolution should increase to a final value, and will not then change if the iris is closed further. The third effect is a pronounced loss of intensity. Sectioning and resolution are balanced against the signal-to-noise ratio and must be determined empirically. About 15–20 sections per cell (~10 μm) are needed to measure cell volumes in cartilage tissue and focus steps should ideally be half the section thickness, (e.g., 0.5 μm steps and 1.0 μm sections). The confocal aperture is key as it influences (1) section thickness, (2) signal-to-noise, (3) resolution/contrast, and (4) depth attenuation. Test specimens are a starting point for optimizing the biological experiment.

3.2. Cartilage Tissue Preparation

1. Remove 20 mm × 10 mm × 3 mm cartilage explants from the ridge of porcine meta-carpal joints, and *see* **Note 7**.
2. Cut approx 1 mm slices longitudinally and suspend in standard HEPES-buffered DMEM.

3.3. Vital Dye Loading

1. Incubate the tissue in standard medium containing 4.5 μM CMFDA at 37°C for 15 min, and *see* **Note 8**.
2. Wash in fresh medium
3. Leave for a further 10 min to ensure all of the dye-ester is cleaved by intracellular enzymes.

3.4. Mounting the Tissue in the Observation Chamber

1. Place the tissue in a heated chamber (37°C) with perfusion (2 mL/min).
2. Mount the chamber on the microscope (we used a Bio-Rad MRC 600, with a Nikon Diaphot)
3. For perfusion, use a chamber inlet supplying fresh medium close to the cover glass and aspirate the spent medium from the top (*see* **refs.** *7,8* for details of some available chambers).

3.5. Optimizing Image Collection

3.5.1. Fluorescence Image Statistics: Avoiding Detection Saturation

For quantitative images, all signal must be accomodated within the detection limits (e.g., 8 bits, 0–255).

1. Focus into the cover glass of the chamber and scan single (X,Y) planes, with no averaging.
2. Plot a histogram of the pixel intensity distribution after each scan.
3. Adjust the black level so that all intensities are recorded above a gray-level of 15.
4. Adjust the gain and/or laser power so the maximum histogram intensity is below 240.
5. Alternatively, use a range check display LUT indicating these limits, e.g., green (0–15), gray scale (16–239) and red (240–255). Ensure no pixels in a single (nonaveraged scan) are colored.

3.5.2. Image Collection Conditions

1. Ideally, control the laser power (directly or by neutral density [ND] filters) at the specimen using a power meter (typically 20–100 μW for physiological specimens).
2. An iterative process should be used to establish the instrument settings: Our aim is to collect 4D images where neither the tissue nor fluorophore are disrupted. Use PI as a viability marker and mean intensity/variance of a homogenous region to assess signal-to-noise ratio.

3. Set the frame rate by independently adjusting frame size and scan speed/pixel dwell time.
4. CLSM offers many different time-lapse modes. Select either 3D (X,Y,Z) "Z-series" imaging and auto/manually repeat for different time points or 4D (X,Y,Z,T) imaging (if your system supports this).
5. Set the frame averaging to balance noise reduction (s/n increases by the square root of the no. of frames), against photobleaching and lower temporal resolution, and *see* **Notes 9** and **10**.

3.6. Experimental Regime: Chondrocyte Volume Regulation In Situ

The following protocol is for collecting 4D images through chondrocytes *in situ* under anisotonic conditions *(4)*. From osmotic rest, we apply a short swelling stimulus followed by hypertonic shrinking and then follow the regulatory volume increase.

1. Mount the cartilage as described previously.
2. Set up the microscope for fluorescence, ensuring the gain and black level are correctly set for the specimen region you will image (*see* optimizing the imaging conditions above).
3. Perfuse with standard DMEM (280 mos$M$) and collect the initial 3D image of the 4D series.
4. Replace the medium with a hypotonic solution: the swelling stimulus (140 mos$M$)
5. Collect a series of 3D images at 5-min time intervals for 10 minutes.
6. Replace the medium with a hypertonic solution: the shrinking stimulus (380 mOsM).
7. Collect a series of 3D images at 2-min intervals for 15 min.
8. To ensure the preparation is physiologically sound, collect a final 3D image 30 min later.
9. The result is a 4D image, composed of a series of 10–12 3D images.

3.7. How Does the Specimen Affect the Imaging? Calibrating the Sample Induced Errors

*3.7.1. Calibration of Axial Attenuation (**Fig. 5**)*

It is important to record exactly how the live data was collected, to replicate the conditions, so that the calibration can be meaningfully applied to 4D data. Features that effect axial attenuation include: (1) objective lens (plus correction collar setting if available); (2) pinhole/iris setting; (3) excitation/emission spectra; and (4) explant tissue or segments.

It is best to use the same piece of tissue for calibration (e.g., at the end of each experiment), but accurate attenuation assessment requires penetration of the cartilage with fluorescein, which can take 48 h. The explant site must therefore be accurately logged to locate an equivalent region for calibration.

1. Fix the tissue in 4% paraformaldehyde in PBS (24 h for cartilage, less for other tissues).

Z-ATTENUATION CORRECTION

- fix tissue
- permeabilise tissue

- infiltrate fluorochrome

- collect xz section

- measure intensity
 values at varying depth

- subtract background

- normalise to start of
  tissue

- fit with quadratic
  function

- inverted function
 = z-attenuation correction

- apply to raw data

Fig. 5. Schematic summarizing the steps involved in axial attenuation correction.

2. Incubate in aqueous 50 μM fluorescein for 48 h to ensure dye penetration, and *see* **Note 11**.
3. Mount the tissue in the chamber surrounded by fresh fluorescein solution (**Fig. 6A**).
4. Set up the microscope ensuring the gain and black level are correctly set (*see* **Subheading 3.5.1.**).
5. Collect a 2D (X,Z) section or 3D (X,Y,Z) image of the tissue over the experiment Z-range.
6. Measure the image black level as the average intensity collected with no laser illumination.
7. Plot a vertical intensity profile (averaged across a homogenous region in the X dimension) through the vertical section and export this to your spread sheet package.
8. Subtract the average black level from all the values in the intensity profile.

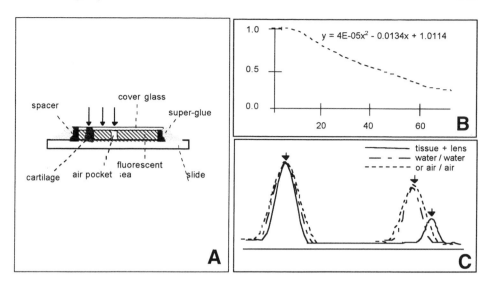

Fig. 6. Calibrating the specimen-induced attenuation and axial focus distortion. (**A**) Fluorescence test specimen (as **Fig. 4**) with the addition of biological material and reference (fluorescence sea or air pocket). (**B**) Measured axial fluorescence attenuation response through cartilage impregnated with fluorescein. Intensities are normalised to value at top of biological material. Quadratic parameters fitted in Microsoft Excel are shown. (**C**) Axial reflection PRF plots through the biological material (**solid line**) and reference regions (**dotted lines**). Arrows point to the peaks corresponding to cover glass/medium (**left**) and medium/slide (**right**) interfaces.

9. Normalize the intensities to the first point in the tissue, this then gives the relative axial intensity change owing to the combined optical and tissue-dependent attenuation (**Fig. 6B**).
10. Parameterise the normalized data by fitting a trend line (e.g., quadratic: $Y = aX^2 + bX + c$) (**Fig. 6B**).
11. Subtract the black level from all images and divide by the attenuation for each Z-position (**Fig. 7**).

3.7.2. Calibration of Z-Focus Distortion

Refractive and scattering boundaries in biological specimens result in significant aberrations in confocal microscopy. There is an axial focus shift and attenuation when the spherical wavefront coming from the objective, or specimen, passes between immersion and mountant with unmatched refractive indices. Specimens with horizontal planar or other continuous refractive index boundaries may also show this effect. Distances measured in the sample must be corrected by a factor that approximates to (sample refractive index)/(immersion refractive index). For some combinations of immersion and

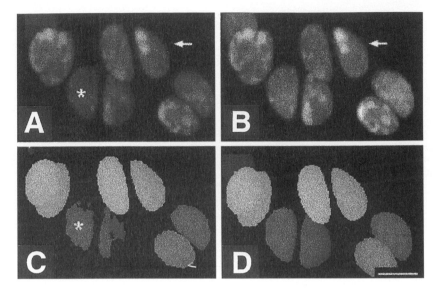

Fig. 7. Correction of sample-induced axial attenuation for in situ chondrocytes in cartilage explant. (**A**) Maximum projection of a single uncorrected 3D time point. Asterisk shows a cell with a lower apparent fluorescence intensity and deeper into the tissue volume than the neighboring cells. Arrow shows brighter cell nearer the surface. (**B**) As in (A) after correction of axial attenuation. Cell nearer tissue surface (**arrow**) and deeper cell now have similar fluorescence intensity. (**C**) Height coded projection of data in (A). Asterisk shows cell deeper into the tissue incorrectly visualized owing to the axial attenuation. (**D**) As in (C) after correction of axial attenuation.

medium or sample the factor may be up to 5% different from this ratio in confocal microscopy or the refractive indices may be unknown. Under these circumstances a value is determined empirically (**Fig. 8**).

1. Prepare the fluorescence test specimen chamber as described previously but thick enough to just accommodate a 100 μm thick piece of tissue (just a few microns of medium above and below).
2. Place a representative specimen into one half of the chamber. Fill with medium and seal.
3. Collect a reflection PRF profile (described above) through the cover glass, tissue, and into the slide (below the tissue). Use the objective you normally use for physiological experiments (**Fig. 6C**).
4. Repeat **step 3** with a cover glass corrected water immersion lens through the clear part of the chamber, and *see* **Note 12**.
5. Measure the distance between the two medium/glass boundaries for the tissue and reference.
6. Measure the width of the optical sectioning as indicated by the PRF above the specimen and below, and *see* **Note 13**.

Z-FOCUS CORRECTION

- x,z reflection image through test specimen and tissue
- x,z reflection image through reference 'sea'

- measure peak-peak distance for sample and

- ratio distances = correction factor

- apply correction to raw data

Fig. 8. Schematic summarizing the steps involved in axial focus correction.

7. The ratio of the distances in **step 5** gives the Z-focus error (a factor to apply to all Z-distances)
8. The difference between the estimates in **step 6** gives an estimate of sectioning distortion.

3.7.3. Confocality, Optical Section Thickness, and Object Dimensions

Our measurements of volume require the optical section thickness to be much smaller than the size of the object. The confocal microscopy image can be described as each fluorescing point in the object, replaced, or imaged as a volume element with the dimensions of the microscope resolution, i.e., a few hundred nm in (X,Y), (depending on the NA and confocal aperture), and the optical section thickness in Z. Because the section thickness is always greater than the (X,Y) resolution by at least a factor of three, the major error is an inflation of the Z-dimension of objects by at least one optical section thickness. The precise effect at the boundary of an object depends on the radius of curvature of the edge compared with the microscope resolution *(9)*. For low NA lenses, or with large confocal apertures this can lead to overestimation of volumes. The only simple solution is to use a higher NA lens and/or smaller aperture. When this is not possible, i.e., for depths of many hundreds of microns into tissue, two more involved options may have to be considered. The first is to use computational deblurring (e.g., deconvolution) software to partially restore the correct dimensions of the object. This subject is fully discussed elsewhere *(10)*. The second, more pragmatic approach, is to measure a test object of the same shape and dimensions as your cells using the imaging protocol in question. A rough indication of the size inflation can then be calibrated by ratioing the measured size with an independent estimate by, e.g., light scattering, Coulter Counter etc. A representative but thinner preparation could also be used for this calibration and imaged with a high NA objective.

3.8. Segmentation: Objective Determination of the Cell Boundary

To accurately measure the volume of labelled cells (positive or negative), an objective estimate of the cell boundary position must be made. The threshold used to segment an object profoundly affects the apparent volume *(4,5)*. The position of this boundary for positive labelling is somewhere between the maximum intensity (definitely within the cell) and the minimum intensity (definitely within the extracellular matrix).

3.8.1. Defining the Boundary Threshold Rule Using Fluorescent Beads

1. Find the modal value of fluorescent bead diameters (~ 7 µm) by Coulter Counter (**Fig. 9A**).
2. Obtain 3D images of the beads adhered to a cover glass coated with 0.1% (w/vol) poly-L-lysine.
3. Select a central section through the beads.
4. Measure an average intensity within a bead and for a background region.
5. Obtain a series of intensity thresholds as a percentage of the maximum intensity above background, i.e.:

$$[X \cdot (I_{cell} - I_{background})] + I_{background} \qquad \text{e.g., for } X = 0.3 \text{ to } 0.7$$

6. Generate a maximum Z-projection (see 3D visualization below) of the image data.
7. Binarise the projected image by setting all pixel values below X to 0 and all values above X to 255.
8. Obtain estimates of bead diameter (area of cross-section/3.1416). Plot a histogram (**Fig. 9**).
9. Compare the modal values from the confocal microscopy and Coulter Counter distributions.

Diameters of 7 µm beads by Coulter Counter and confocal microscopy (50% threshold) correlate well, and *see* **Note 14**.

3.8.2. Segmenting the Chondrocyte Data

1. Segment the CMFDA-stained cells from the cartilage images (as for the bead calibration previously described). Determine background and cell intensities on a cell by cell basis, avoiding errors from uneven staining.
2. Obtain confidence limits for the boundary thresholds at 50% ± 1 SD of the cell intensity. For example, for a 50% threshold

(i) $\qquad\qquad 0.5 \cdot (I_{cell} - I_{background}) + I_{background} = (I_{cell} + I_{background})/2$

(ii) $\qquad\qquad (I_{cell} + I_{background} + I_{std})/2$

(iii) $\qquad\qquad (I_{cell} + I_{background} - I_{std})/2$

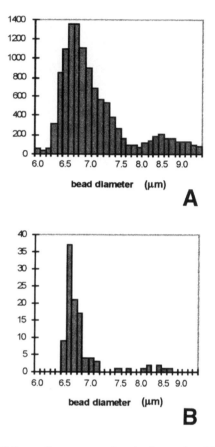

Fig. 9. Diameters of 7 μm fluorescent beads determined by Coulter Counter and confocal microscopy. Histograms of bead diameter distributions. (A) Coulter Counter (mode = 6.64mm). (B) Confocal fluorescence microscopy (mode = 6.6 mm).

3.9. Two Methods for Extracting Cell Volume

You can now use the segmentation thresholds to measure volumes in 3D time points or 4D image series.

3.9.1. Summing Cross-section Areas

For each cell in the volume (**Fig. 10A**):

1. Analyze the (X,Y) optical section through the cell.
2. Determine the cross-sectional area above the threshold intensity in pixels or micrometers (**Fig. 10B**)
3. Determine the cross-sectional area for the two confidence limits.
4. Sum the areas for all sections through the cell.

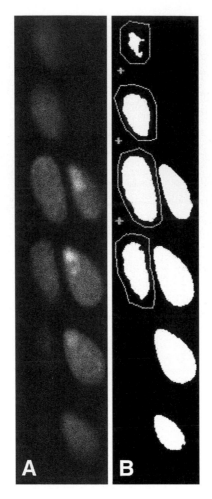

Fig. 10. Measurement of *in situ* chondrocyte volumes by summing optical sections. Six optical sections through two cells fluorescently stained with CMFDA. **(A)** Raw image data. **(B)** Data in (A) after binarisation based on 50% intensity threshold between background and mean cell intensity.

5. Convert the sum into volumes for each cell (and confidence limits) by multiplying a pixel area by the product of X, Y, and Z pixel spacings) or multiplying an area in mm by the corrected Z spacing, and *see* **Note 15**.

3.9.2. 3D Segmentation

Some software packages can automatically segment 3D objects without processing each section separately *(11)*. These use a threshold or other boundary-condition algorithm, searching for image voxels that are connected to each

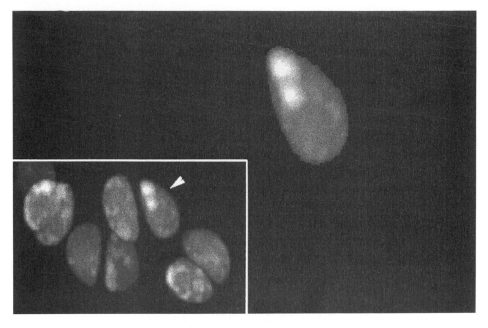

Fig. 11. Measurement of *in situ* chondrocyte volumes by seed filling. Maximum projections of 3D confocal fluorescence images. A single cell (**arrow**) extracted from one time point (**inset**) by voxel seed filling.

other. Different versions allow for various degrees of connectivity between adjacent voxels to include in the object. The seed fill algorithm starts inside the object at a seed position determined manually or (semi-)automatically, filling out the object to the boundary and reporting the volume (**Fig. 11**).

3.10. Visualizing Volume Data

The amount of data collected in a typical 4D confocal microscopy image (often 10s or 100s of Mb per experiment) results in many logistical problems for analysis, data manipulation and visualization. It is important to consider multidimensional visualization as selective data reduction. A 3D data set, or sub volume from a 4D series, must be reduced to a 2D view for display on a monitor, printout, slide etc. Some display systems, such as computer or video screens, allow sequences of 2D views to be presented and most can use 3D effects to aid the interpretation of the presented data. Visualization requires the projection of 3D/4D data onto one or more 2D planes, the data intensities being rendered into display 'views' in such a way as to retain some important features. The result is then presented to an audience, using additional effects or processes to partially restore the "3D/4D nature" of the original data. We strive to make the most efficient use of the available display resources using the simplest projections and rendering to maximize the audiences ability to understand what they see with the minimum of misinterpretation *(12)*.

A typical confocal data visualization session would normally progress from simple section animations to data projections of increasing sophistication including stereo and other display techniques.

3.10.1. Displaying 2D Sections Through the 4D Image Data

1. Ensure the images are correctly scaled to fill the range of the display device (0–255 gray levels, for 8 bit data). Measure the minimum and maximum intensities (of the entire image) and then either rescale all the values or simply using the intensity limits to control the display look-up-table (LUT), and *see* **Note 16**.
2. Ensure the dimensions (X, Y, and Z pixel sizes in μm) are correctly represented in the file format and/or the display software, otherwise images displayed outside of the microscope software or computer system may not retain the correct proportions. In particular ensure that the axis calibrations have had any necessary corrections.
3. Use the section display/animation tools to show a sequence of views as if focussing along each of the three main axes of the data. Some display packages may not be able to do this sort of animation rapidly for anything other than standard Z series of (X,Y) sections, Y series of (X,Z) sections, etc. Many will allow general sections at any angles to be precalculated for later animation, and *see* **Note 17**.
4. Use the section animation tool to view a time series at a single Z-focus position (if your software will allow this). If this capability is not available in your software, you should be able to rearrange the data so that , e.g., it is organised as a series of files containing timelapsed (X,Y), (X,Z), etc. sections, each file corresponding to a particular focus level, rather than each time point in a separate file.

3.10.2. Simple 3D Projections

Only a single value is required for each pixel in a 2D view. During a 3D projection, voxels are traced along lines cast through the volume, in the direction of view, and processed to yield a single value at each view pixel (e.g., **Figs. 7** and **11**). Processing rules'for simple ray cast projections include the maximum brightness (e.g., for surface reflections, sparse fluorescence staining, etc.) or the mean value (a good unbiassed method for an initial view). It may be necessary after projection to rescale the view for optimum contrast.

1. Use the simple projection functions in your visualisation package to make a single 2D view of each 3D time point of the 4D data. Store these views in a single file and then animate the time series, and *see* **Note 18**.
2. Now use the projection tool to make several 2D views of just one 3D time point but from a range of viewing directions (e.g., a series with tilt and/or rotation steps).

3.10.3. Voxel Surface Visualization: Height or Depth Views

To help restore depth in the views, some algorithms render the first or front voxel nearest the viewer. This gives a surface effect (but no connection is necessarily assumed between adjacent voxels). These surface voxels are represented either by their original intensity or by a brightness coresponding to their depth, height, or Z-coordinate in the data volume. Surface projection modes have the effect of hiding the data voxels behind and/or within the structure, conveying a sense of solid character to the view. Always use the section animation tools previously described to review the entire data volume before making solid reconstructions.

1. Recall the objective estimates of boundary thresholds for the 4D images of CMFDA-stained cells.
2. Use these values as the surface threshold in a surface or height projection, and *see* **Note 19**.
3. Use the visualization tools to make a surface view at each 3D time point. Animate these to see the changing morphology of the cells over time (**Fig. 12**).
4. The display colour now represents the height of a voxel, looking into the volume. Use a colour display LUT to highlight cells near the front of the volume, middle or further back, etc. Find a pre-set LUT like the standard geographical colours in an atlas (blue/black for valleys, red/grey for peaks etc.) to show topographical features. Alternatively, use a single narrow color band (e.g., a 10 level-wide red band) and move it over the grey scale to contor different topographical features of the surface view.

3.10.4. More Advanced Visualization Tools

Alpha blending *(12)* is a tool very similar to the projection techniques described above that is available in some visualisation packages. Ray casting is used and the data is traversed either away from or towards the viewer. Each pixel of the 2D view is determined by first multiplying voxels along the cast ray by an opacity or alpha value that simulates masking of voxels behind it in the volume. High opacity voxels will appear as solid features near the front. Bright voxels towards the back of the data volume will tend to show through more transparent data near the front. Alpha blending requires careful control of the alpha function used to map data values to opacity/transparency attributes which confer an artificial material character to the raw data. The alpha function must be inspected with the rendered views, otherwise you cannot unambiguously separate, e.g., dimmer voxels from bright regions masked by overlying opaque structures.

We used an alternative approach to visualize chondrocyte volume data by a combination of average projection (conserving original fluorescence intensity values) and surface visualization (show the cells at the front of the volume

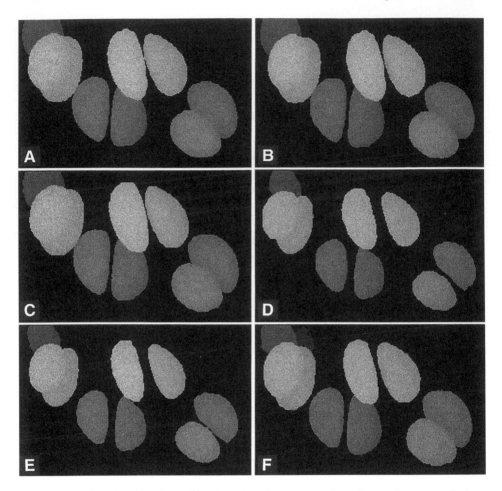

Fig. 12. 4D visualization of in situ chondrocytes undergoing volume regulation. Six time points **(A–F)** during a one-hour physiological experiment manipulating osmotic load. A 3D confocal fluorescence image collected at each time point is visualized as a height coded projection. (A) 0 min (280 mosM). (B,C) 5–15 min (140 mosM hypotonic swelling stimulus). (D) 20 min (380 mosM hypertonic shrinking stimulus). (E,F) 25–60 min (380 mosM). Cells respond by regulatory volume increase.

occluding the cells behind. We have also devised a second type of 4D reconstruction that uses color to represent relative changes in cell volume, hence contrasting individual and population volume responses. This process is not fully implemented in the standard commercial CLSM packages, but it is possible with many systems to produce output as presented below. We describe the process, step by step below, using a combination of standard components, with a details of the new techniques we have added.

1. Use the boundary thresholds defined above to generate height coded views of each time point.
2. On a pixel-by-pixel basis, use the height image to define a reference surface about which to apply a limited or 'local' average intensity projection (i.e. ,from a few voxels in front to a few voxels behind the surface). In this way a layer around the outer region of the cell is rendered, showing the mean fluorescence intensity. By increasing the distance projected, the cell becomes more transparent, and *see* **Note 20**.
3. Animate these reconstructions (in rotation, tilt, etc., or over time depending on how you have made them) to see the staining pattern of the cells and their arrangement in the tissue.

3.10.5. Visualizing Cell Volume Using a Depth Sensitive Color Display

1. Seed fill each cell in turn (as described above) and render the voxels of each cell, using the two pass method described above, into the view.
2. Maintain a height (Z-coordinate) view of the result as it builds up.
3. Convert the volume of each cell to a relative volume by normalising to the first time point.
4. Generate a height (Z-coordinate) view of each cell.
5. As each cell is rendered into the view (in **ref. *1***), the height view (in **ref. *3***) is compared with that of the final result and only the front-most voxels are rendered onto the display. In this way, cells at the front of the volume obscure cells behind.
6. Generate the rendered view as a 24-bit color image, each voxel contributing an intensity corresponding to its brightness in the projection proportioned between the three channels according to an RGB colour ratio. Obtain this ratio by passing the relative cell volume through a topographic colour LUT to provide the necessary red, green, and blue components.
7. Now observe a single visualized 3D time point to see the relative volume of each cell in the volume.
8. Animate a rotation series of one time-point to see the detailed 3D disposition of the cells and any correlation with their relative volume.
9. Animate a series of volume-coded reconstructions made at each time point to see the dynamic volume changes as transitions through different colors. You should be able to see subtle volume changes that are not apparent by the small changes in the linear dimensions observed in a monochrome view.
10. Finally, make a montage of each visualized time-point together with a graphical plot of the relative volumes during the physiological experiment (**Fig. 13**).

4. Notes

1. The PRF is often confused with the point spread function (PSF) obtained by serially sectioning sub resolution fluorescent beads. Profiles through the PSF give the Z- and X/Y- resolution, but the total intensity in each plane gives the axial PRF. Brownian motion of low intensity beads makes these measurements unreliable in aqueous samples.

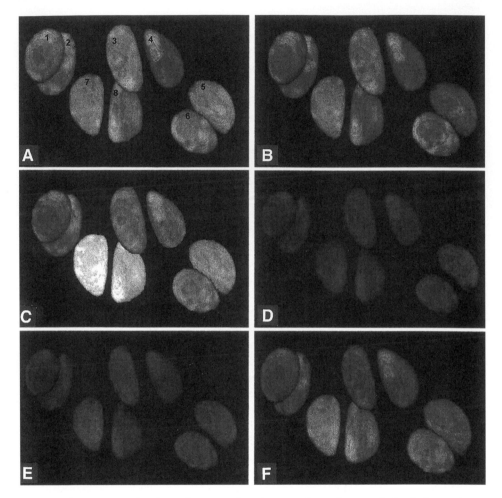

Fig. 13. 4D visualization of *in situ* chondrocytes by colour coded relative volume changes. Six time points (**A–F**) (as in **Fig. 12**). (A) Cells at initial time point have relative volume = 1 (green, shown here in mid gray). (B,C) Cells swell to a relative volume > 1 (yellow-red, shown here in light gray) asynchronously and to different extents (0.89–1.1). (D) Cells shrink to a relative volume < 1 (blue, shown here in dark gray) coordinately (to 0.56–0.66). (E,F) Cells slowly increase their relative volume (cyan–green) achieving a mean recovery of 0.92 (0.82–0.92). (*See* Color Plate I.)

2. If the metal layer passes light through to the slide, this may reflect back causing interference fringes. Thin metal layers should be mounted towards the slide in immersion oil sealed with super glue, or solid mountant with the refractive index of glass (**Fig. 3B**). For an NCG (no cover glass) lens the mirror is mounted toward the objective lens (**Fig. 3C**).

3. See dye data sheets for solubilities (e.g., DMSO may be needed for some dyes in water). Rhodamine 6-G, Ex 514 nm, Em > 590 nm, (Eastman Kodak) dissolves in water, glycerol, and oil.

4. With immersion objectives, focus so that force applied to the specimen is towards the stage (i.e., stage and lens approaching on upright microscopes or moving apart on inverted). This limits axial errors if the sample is not compressible or extendable. Sample distances must be adjusted by the ratio of medium/immersion refractive indices.

5. Use the sharpness of the peak in this reflection Z-response to fine tune the alignment of a confocal microscope.

6. The second method is more accurate if the stage is not perfectly orthogonal to the optical axis.

7. The tissue is placed into a physiological medium that must maintain the cells at the desired osmolarity. Removing explant tissue often causes swelling, so sucrose and sodium chloride are used to increase medium osmolarity. HEPES eliminates the need to supply CO_2 during the experiment, and buffers pH so that the tissue is not exposed to acid load.

8. Other dyes (e.g., BCECF*, calcein) may be used to measure volume but during cell swelling and recovery dye leaks and/or is pumped from the cell. CMFDA is retained better as it covalently binds to intracellular thiols. To avoid disruption of cell function at high dye concentrations, ascertain the upper limit of loading consistent with normal physiology before optimizing the microscopy conditions. This requires objective markers of function in loaded and unloaded cells, for example; membrane potential or integrity [e.g., propidium iodide or PI (6)], cytoplasmic streaming, metabolic activity, growth rate, etc.). The ECM excludes large molecules, has highly charged proteoglycans and cells may leak enzymes that cleave dye-esters. Loading may be enhanced by adding enzyme (e.g., protease) inhibitors.

9. It is not always clear whether the best signal-to-noise vs cell viability regime is single slow scans or faster scans with integration, as this usually depends on the fluorochrome relaxation time and is heavily influenced by the environment.

10. For many of our 4D physiological cartilage experiments on a MRC 600, we used a 25× 0.8 NA objective, image size = 768 pixels × 180 lines × 40 frames (0.21 × 0.21 × 1.0 μm voxels), scanning at three frames/s with two-frame integration. This 3D image was repeated every 2–5 min for one hour. Laser excitation (at the sample) was ~ 20 mW, intracellular dye (CMFDA) concentration ~ 5 mM with a confocal aperture ~2 mm.

11. Fluorescein was chosen, because its fluorescence is similar to CMFDA but, unlike CMFDA, it stains fixed cells. Strictly, the attenuation profile starts at the cover glass/medium boundary and should include the medium above the specimen. If you normalise to the top of the specimen itself you should collect the experimental data with the tissue at the same distance from the cover glass. For a water immersion or low NA objective the error is small.

---

*2',7'–Bis (2–carboxyethyl–5,6–carboxyfluorescein.

12. If you do not have a CC water immersion lens, you will have to make this reference section through an air pocket trapped in the medium when you make the specimen and image with an air lens (**Fig. 6C**).

13. The integrated intensity of the PRF responses above and below the specimen should correspond to the attenuation curve determined earlier. Any differences may reflect small changes due to fixation.

14. Take care when using beads to calibrate axial dimensions, including volumes, for CLSM. Some beads have a significantly higher refractive index than water (often over 1.6). These act like lenses, focussing the light with the effect of lengthening apparent axial dimensions measured in the bead and asymmetric axial blurring.

15. In some software packages, semi- or fully-automatic procedures may be available to segment these cross sectional areas. It may, alternatively, be necessary to work with (X,Z) sections if the object is particularly flattened in the axial direction (to avoid significant errors at the top and bottom of the structure).

16. For 8-bit images, software packages allow the image values to be represented on the screen in several ways. The simplest is to use a grey scale (equal amounts of red, green, and blue for each value) directly related to the voxel value. Data values are mapped through the display LUT and the output red, green, and blue colors are displayed on the screen. To enhance areas of similar intensity a nonlinear color LUT is used to map brightness to particular hues (combinations of red, green, and blue). Such techniques are widely used to display subtle changes in ion concentrations stored as the emission ratio of dual wavelength ion-sensitive dyes. This way of coloring 8-bit data is usually called 'ndexed color. When multiple-channel images (e.g., from several confocal detectors) or the output from a color camera are displayed, the data usually exist as three 8-bit values, each representing a separate level of red, green, or blue. The final color displayed is not determined by mapping values through a LUT but by altering the actual value stored in each color plane. This is normally called 24-bit or true color. Because of the increased amount of data, and the need to change stored values to change the display colors, 24-bit color display, e.g., for animation, is usually much slower than 8-bit indexed color displays. Some packages convert 24-bits to 8-bit indexed color for animation.

17. Animation of sections at 10–20 frames/s or faster will give an impression of smooth progression through the data. The viewer will start to benefit from persistance of vision where the retina carries the previous view as the next is seen. The brain, too, will aid this process by subconsiously linking cross-sections of objects seen in neighboring sections to give the impression of a continuous structure running through the volume *(12)*. Features that cannot be seen clearly in a single section or montage can be more readily understood. It is important to ensure good lighting and a high contrast, good quality display to benefit from these visual perception processes.

18. When you play a smooth animation of this sequence the reconstruction will apear to rotate on the screen. Again, your eye and brain fills in the gaps between the

presented views, to enhance features that appear to move quickly round the periphery of the data volume, or more slowly toward the center of the data volume.
19. The purpose of the surface visualizations is to see the boundaries of all the cells and the way they are arranged in the tissue. This threshold technique uses a single value to pick out several objects in a volume and so you may have to use an average threshold for a group of cells. More sophisticated gradient boundary operators are offered by some packages that do not require a single threshold and can highlight both bright and darker objects in the same field. All the volume measurements presented here are done with thresholds on a cell by cell basis.
20. This first part of the visualisation uses one of the two-pass visualization modes of the Bio-Rad Lasersharp software. The distance on each side of the surface is entered as a use variable (typically 5–10 voxels each side).

Acknowledgments

This work was funded by The Wellcome Trust (R. J. E.). N. S. W. is a Royal Society Industry Fellow.

References

1. Hoffmann, E. K. and Simonsen, L. O. (1989) Membrane mechanisms In volume and pH regulation in vertebrate cells. *Physiol. Rev.* **69**, 315–382.
2. Hoffmann, E. K. and Dunham, P. B. (1995) Membrane mechanisms and intracellular signalling in cell volume regulation. *Int. Rev. Cytol.* **161**, 173–262.
3. Raat, N. J. H., de Smet, P., Van Driessche, W., Bindels, R. J. M. and van Os, C. H (1996). Measuring volume perturbation of proximal tubular cells in primary culture with three different methods. *Am. J. Physiol.* **271**, C235–C241.
4. Errington, R.J., Fricker, M.D., Wood, J.L., Hall, A.C. and White, N.S. (1997) Four-dimensional imaging of living chondrocytes in cartilage using confocal microscopy: a pragmatic approach. *Am. J. Physiol.* **272**, C1040–C1051.
5. White, N.S., Errington, R.J., Fricker, M.D., and Wood, J.L. (1996) Aberration control in quantitative imaging of botanical specimens by multi-dimensional fluorescence microscopy. *J. Microsc.* **181**, 99–116.
6. Beletsky, I. P. and Umansky, S. R. (1990) A new assay for cell death. *J. Immunol. Meth.* **134**, 201–205.
7. Ince, C., Ypey, D. L., Dieselhoff-den Dulk, M. C. C., Visser, J. A. M., De Vos, A. and van Furth, R. (1983) Micro-CO_2-incubator for use on a microscope. *J. Immunol. Methods* **60**, 269–275.
8. Terasaki, M. and Dailey, M.E. (1995). Confocal microscopy of living cells, in *Handbook of Biological Confocal Microscopy*, second edition (Pawley, J. B., ed.), pp. 327–344.
9. van Vliet, L. J. (1993) Grey-scale measurements in multi-dimensional digitised images. PhD thesis Delft University Press.

10. Shaw, P. J. (1998) Computational deblurring of fluorescence microscope images, in *Cell Biology A Laboratory Handbook*, second edition (Celis, J. E., ed.), pp. 206–218.
11. Guilak, F. (1993) Volume and surface area measurement of viable chondrocytes in situ using geometric modelling of serial confocal sections. *J. Microsc.* **173,** 245–256.
12. White, N. S. (1995) Visualisation systems for multidimensional CLSM images, in *Handbook of Biological Confocal Microscopy*, second edition (Pawley, J. B., ed.), pp. 211–254.

19

Imaging Thick Tissues with Confocal Microscopy

Delphine Imbert, Janet Hoogstraate, Emmeline Marttin, and Christopher Cullander

1. Introduction

Confocal laser scanning microscopy (CLSM) can be used to obtain optical sections of thick tissues that are relatively free of interfering autofluorescence, and that do not strongly scatter or absorb either the excitation or emission light. This chapter provides protocols used to examine three such tissues: the cornea of the eye, the buccal mucosa (which lines the inner cheek), and the nasal respiratory epithelium. Although in each case our overall motivation was to study the transport of drugs or model compounds across the particular epithelium, the approaches taken were quite different.

1.1. Viability Assay for an Ex Vivo Tissue (Cornea)

Determining the viability of an ex vivo tissue sample is important in any experiment where the physiological state of the tissue may influence the results obtained. In particular, excised cornea viability is of interest for in vitro diffusion study design and ocular toxicity risk assessment *(1)*. Simultaneous vital staining by calcein AM (CAM) and ethidium homodimer-1 (EH-1), as "live" and "dead" probes, respectively, has been used for viability determination in monolayer cultures, but has not found wide application with thick tissue sections *(2)*. Using the CAM/EH-1 probe pair, we developed a confocal laser scanning microscopy (CLSM) assay to determine corneal epithelial and endothelial viability as well as cornea thickness. The assay described for cornea here can readily be adapted for other thick tissues.

1.2. Dynamic Visualization of Diffusion (Buccal Mucosa)

Diffusion cells are often used to measure permeation rates of compounds through a tissue sample. The tissue is mounted between two chambers, each of

From: *Methods in Molecular Biology, vol. 122: Confocal Microscopy Methods and Protocols*
Edited by: S. Paddock © Humana Press Inc., Totowa, NJ

which typically contains a buffered salt solution. The compound of interest is present in the "donor" chamber, and its flux through the tissue is determined by its rate of appearance in the "acceptor" chamber (or its rate of disappearance from the donor). Although this provides important information about bulk flux, it does not show how and where the compound traversed the tissue, that is, how the distribution of the marker in the tissue changed as a function of time. To accomplish this, we developed an in situ model (essentially a miniature diffusion cell on a microscope stage) that made it possible to use CLSM to continuously monitor and measure changes in the distribution and transport pathways taken by a fluorescent marker as it permeated across a tissue, and thus to localize any regions of lower flux that indicated the presence of barriers to transport. Although our tissue of interest was the buccal mucosa (the lining of the cheek), this in situ cell can be used for other tissues. The flow probe used was fluorescein insothiocyatnate (FITC), but other probes may be used as appropriate.

An important feature of the in situ cell was that the tissue was imaged by making optical sections parallel to the plane of a mechanical cross-section and below the region where cells had been damaged by the razor blade. There is sufficient scattering and absorption of light in most tissues, as well as index of refraction mismatch between the oil immersion objective lenses and the aqueous sample, so that there is significant loss of fluorescence intensity and spatial resolution when imaging deep sections parallel to the tissue surface. By viewing the tissue "edge-on" we could visualize the penetration of the fluoroprobe across all cell layers simultaneously, and compare relative intensity between regions.

1.3. Imaging Transport Across Fragile Tissues (Nasal Epithelium)

In contrast to corneal and buccal tissue, the nasal respiratory epithelium consists of only two cell layers, with a thickness of approximately 20 μm, and is very fragile. Its viability and integrity are difficult to maintain in vitro, and it is also easy to damage the tissue during removal. The approach that we developed was to administer the fluorescent probe intranasally in vivo, and to fix the tissue in situ. Transport of the fluorescent probe thus occurs under exactly the same circumstances as in vivo drug absorption studies (3). The animal was then sacrificed, and the tissue removed from the nasal cavity, post-fixed, and mounted for examination with CLSM (4). The essential features of this approach are the use of a fixable probe and a rapidly acting in vivo fixation procedure. The fixative must be able to immobilize the compound rapidly during the initial fixation, and remain stable during storage and imaging.

2. Materials

In all cases, fluorescent probes are kept protected from light throughout the experiment until the tissue is imaged. It is also assumed that the tissue being used does not have unacceptably high autofluorescence.

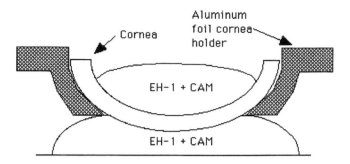

Fig. 1. Exposure of excised cornea to EH-1/CAM solution. About 100 mL of the EH-1/CAM solution is pipetted onto a glass slide and the cornea in the aluminum holder is gently floated onto the liquid, epithelium down. Another 100 mL of probe solution is then pipetted onto the concave endothelial side.

2.1. Viability Assay for an Ex Vivo Tissue (Cornea)

1. Microscope used: upright CLSM.
2. Media optimized for tissue storage. For the cornea, we used Optisol™ (Chiron Vision, Irvine, CA) (*see* **Note 1**).
3. Tissue collection (cornea) requires an 11-mm trephine blade and holder (Solan Ophthalmic Products, Sacramento, CA), a pair of surgical scissors, and 2 pairs of microdissecting forceps (*see* **Note 2**).
4. The buffer solution used was bicarbonate Ringer (BR) solution pH 7.5 containing 115 mM NaCl, 5.5 mM KCl, 1.8 mM CaCl$_2$, 0.8 mM MgCl$_2$, 35.0 mM NaHCO$_3$, and 5.5 mM glucose was prepared fresh, and equilibrated with carbogen (a mixture of 95% O$_2$ and 5% CO$_2$) until pH stabilized (*see* **Note 3**).
5. Fluorescent probes used: Ethidium homodimer-1 (EH-1) and calcein AM (CAM) (Molecular Probes, Eugene, OR; *see* **Note 4**).
6. Stock solutions are prepared as follows: For the CAM stock solution, dissolve CAM powder in anhydrous dimethyl sulfoxide (DMSO) to form a 10 mM solution (*see* **Note 5**). For the EH-1 stock solution, use the 2 mM solution in DMSO/H$_2$O 1:4 (v/v) that is commercially available. Both stock solutions should be stored sealed, frozen (–20°C), protected from light, and desiccated (*see* **Note 6**).
7. For each cornea, 200 µL of EH-1/CAM working solution in BR is prepared immediately before use in a conical centrifuge tube by mixing 2.5 µL of the EH-1 stock solution and 1 µL of the CAM stock solution (the final concentrations are 25 µM EH-1 and 50 µM CAM). The working solution should be vortex-mixed and then degassed by vacuum for 1 min, keeping it protected from light at all times.
8. Cornea holder for probe exposure (**Fig. 1**). Make a simple holder out of aluminum foil to suspend a cornea in the probe solution (*see* **Note 7**).
9. Lucite sample holder for microscopy (**Fig. 2**). You will need a sample holder similar to one that we made for the rapid examination of thick tissue specimens in an

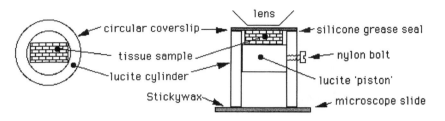

Fig. 2. Piston sample holder for CLSM viability assay.

upright microscope (*see* **Note 8**). Also have ready a clean 25-mm diameter circular glass no. 1 coverslip, silicone or vacuum grease, two pairs of tweezers, a wide glass microscope slide (75 × 38 mm), and either double-sided adhesive tape (Scotch, 3M, Minneapolis, MN) or a small piece of Stiki-wax (*see* **Note 9**).

2.2. Dynamic Visualization of Diffusion (Buccal Mucosa)

Because this protocol is focused on the construction of an in situ cell, only minimal information regarding the particular tissue and flow probe we used in our study is provided here. For more detail, *see* **ref. 5**.

1. Microscope used: inverted CLSM (*see* **Note 10**).
2. Fresh tissue should be stored chilled in an appropriate storage buffer (*see* **Note 11**). For the buccal mucosa, we used isotonic pH 7.5 Krebs buffer (*see* **Note 12**).
3. A flat, clean plastic cutting surface (e.g., a smooth nylon plate), and new de-oiled single-edged razor blades (*see* **Note 13**) to prepare the sample for the cell.
4. The fluorescent probe (the donor solution) should be prepared fresh in an appropriate buffer, and kept from light until use (*see* **Note 14**). We used 0.5 mM FITC in pH 7.5 Krebs buffer (*see* **Note 15**).
5. The acceptor solution, which is usually the same buffer used to prepare the probe (pH 7.5 Krebs buffer in this case).
6. Several round wooden toothpicks; two small (20-cc) hypodermic syringes, each with a no. 19 needle.
7. Sample holder for in-situ cell: A 1.5 inch × 3 inch × 0.25 inch piece of aluminum plate (approximately the length and width of a wide glass slide) is made with a (right circular) conical well bored through it at center (**Fig. 3**). This well provides clearance for the microscope objective lenses as the turret swings them into position. A 25.5 mm diameter cylindrical well is then bored at center from the other side, so that it intersects the conical well. The hole at the intersection of the well and the cone should be approx 15 mm in diameter. The sharp edge of the hole is removed, leaving a circular lip about 5 mm wide at the bottom of the cylindrical well. This lip supports the in situ cell described below. The completed cell holder should be anodized or painted black.
8. Preparation of coverslips for in situ cell: One such coverslip is needed to construct each in situ cell.

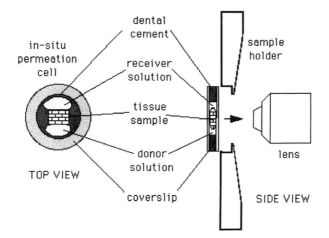

Fig. 3. In situ cell and sample holder. Cell sits in well in center of holder, and holder sits on stage of inverted microscope.

    a. Thoroughly clean and dry a 25 mm diameter circular glass no. 1 coverslip

    b. Dip the coverslip into a 0.5% (w/w) solution of Pioloform (polyvinyl-formaldehyde, Agar, The Netherlands) in chloroform.

    c. Slowly lift the coverslip vertically out of the Pioloform solution

    d. Dry it at ambient temperature.

The dried coverslip will be covered with a very thin hydrophobic film, which prevents "wicking" (capillarity) of the probe solution between the tissue surface and the coverslip. Other hydrophobic materials can also be used. One additional 25 mm circular coverslip (clean, but not prepared as above) is also required to construct the in situ cell.

   9. Preparation of in-situ cell wall material: Two-component silicone dental clay is prepared by mixing Provil® L Base with Provil® L Catalyst (Bayer Dental, Leverkusen, Germany) in a ratio of 1:1. This adhesive elastomer (*6*) hardens in about 5 min, thus this step should be done just before it is to be used. Other substances may be used to form the in situ cell walls, provided that they are not fluorescent.

2.3. Imaging Transport Across Fragile Tissues (Nasal Epithelium)

   1. Microscope used: either upright or inverted CLSM.

   2. Preparation of fixative: For 100 mL of Bouin's fixative, add 75 mL of saturated picric acid (for microscopy, Fluka A.G. Buchs, Germany), 25 mL of formaldehyde solution (extra pure, Merck, Darmstadt, Germany), and 5 mL 98% of acetic acid (Baker Deventer, The Netherlands). Make up the fixative freshly before use (*see* **Note 16**).

   3. Preparation of postfixative: After fixation in Bouin, tissue samples are postfixed and stored in a solution of 25 mL of formaldehyde solution (extra pure, Merck,

Darmstadt, Germany) and 75 mL of phosphate-buffered saline (PBS) (pH 8.0) (137 mM NaCl, 2.5 mM KCl, 8.8 mM Na$_2$HPO$_4$·2H$_2$O, 0.45 mM KH$_2$PO$_4$). Make up the solution fresh before use.

4. PBS pH 8.0
5. Fluorescent probes:
    a. 0.5% (w/v) FITC-dextran solution is prepared by dissolving 1 mg of aldehyde fixable FITC-dextran (mol wt 3000 or 10,000, Molecular Probes, Eugene, OR) in 200 µL of physiological saline (0.9% NaCl in distilled water; w/v; *see* **Note 17**).
    b. 1% (w/v) of Evans blue solution is prepared by dissolving 10 mg Evans Blue (ICN Biomedicals Inc., Aurora, OH) in 1000 µL of physiological saline. Filter the solution through a Millipore filter.
    c. 0.5% (w/v) of DiIC18(5) solution is made by dissolving 5 mg of DiIC18(5) oil (Molecular Probes, Eugene, OR) in 500 µL ethanol. Add 375 mL of propylene glycol and 125 mL of PBS (pH 8.0).
    The fluorescent probe solutions should be stored at –20°C, protected from light.
6. Apparatus
    a. 5-cm silicone cannula.
    b. 20-cm silicone cannula.
    c. Small mirror.
    d. 100-µL syringe, with stub Luer fitting and 2 cm of polyvinyl chloride (PVC) tubing on it.
    e. Syringe for fixative.

3. Methods

For all the methods below, it is assumed that the CLSM has been turned on and allowed to warm up for at least 15 min prior to imaging, that the appropriate laser lines and filters have been selected, and that the optics of the system are aligned.

3.1. Viability Assay for an Ex Vivo Tissue (Cornea)

3.1.1. Tissue Collection and Preparation

1. As soon as possible after animal sacrifice, both corneas are excised (*see* **Note 2**) and immediately placed into either storage medium or test medium, depending upon the experimental purpose.
2. Immediately prepare the CAM/EH-1 working solution as described previously.
3. Rinse cornea with BR, position it in the cornea holder (**Fig. 1**), and expose it to the fluorescent probe solution for 120 min, away from light (*see* **Note 7**).
4. During this time, set up the microscope for 488 and 568 nm excitation if this has not already been done (*see* **Note 18**).
5. Rinse the corneas with BR and use the forceps and surgical scissors to isolate the circular area that was exposed to the fluorescent probes from the rest of the tissue to create a small, roughly circular "button" about 7 mm in diameter.

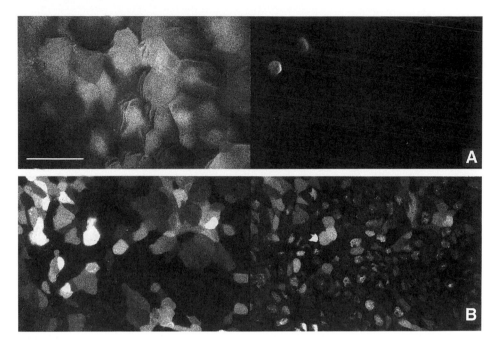

Fig. 4. Epithelium viability of freshly excised corneas stored at room temperature in PBS. Calcein fluorescence is shown on the **left**, EH-1 fluorescence is shown on the **right**. (**A**) Healthy control cornea; (**B**) After 4 h in PBS. All images are at the same magnification. Scale bar = 50 μm.

6. Mount this corneal button with the epithelium side up in the lucite sample holder (**Fig. 2**) without further treatment (*see* **Note 19**).

3.1.2. Imaging

1. Scan the sample: Use a 40× or 60× lens to survey the sample with a fast scan (typically F3 on MRC 600), identifying any heterogeneous regions and/or representative areas. Be careful to stay away from the edges of the sample.
2. Imaging the epithelium: Collect full screen averaged images of areas of interest at normal speed. Begin with the calcein image (calcein is more susceptible to photobleaching than the EH-1; *see* **Notes 20** and **21**). Typical calcein and EH-1 signals observed on control cornea samples are shown in **Fig. 4A.** After 4 hours at room temperature in PBS (**Fig. 4B**), severe damage to the epithelium can be observed.
3. Imaging the endothelium: Remove the coverslip from the sample holder, rinse the corneal button in BR, and place it back in the sample holder with the endothelium side up (*see* **Note 22**). Endothelium is imaged as described previously for the epithelium. Healthy New Zealand white rabbit endothelium is shown on **Fig. 5A.** After 6 h in PBS, massive endothelial cellular death is seen (**Fig. 5B**).

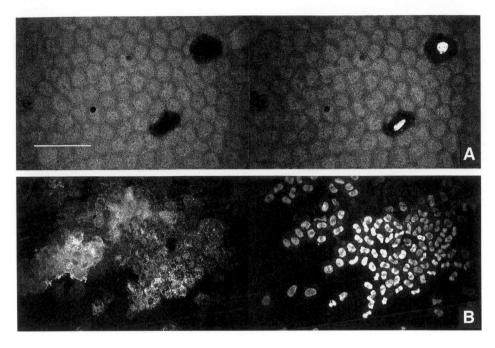

Fig. 5. Endothelium viability of freshly-excised corneas stored at room temperature in PBS, as determined by CLSM. Calcein fluorescence is shown on the **left**, EH-1 fluorescence is shown on the **right**. (A) Healthy control cornea; (B) After 6 h in PBS. All images are at the same magnification. Scale bar = 50 μm.

Confocal Z-sections through the entire sample of cornea can provide a measure of relative thickness, and therefore of corneal swelling *(7)*.

3.1.3. Display of Dual-Label Images

Simultaneous display of both calcein and EH-1 signals on a single color-coded image can be accomplished with the confocal software or with NIH Image (*see* **Note 23**), or any other image processing software.

3.2. Dynamic Visualization of Diffusion (Buccal Mucosa)

3.2.1. Preparation of Tissue for Cell

1. If the tissue sample is not already thin (0.5–2.0 mm) and flat, then it should be sectioned so that it is (*see* **Note 24**). This tissue slice will also establish which surfaces will face the donor and receptor chambers in the cell (*see* **Note 12**).
2. Use a de-oiled razor blade to make the slice, balancing the pressure evenly across the razor's edge; i.e., push straight down rather than rocking or sliding the blade (which may tear the tissue) (*see* **Note 25**). Be careful to maintain the orientation of the sample (*see* **Note 26**). The thickness of the tissue slice is "Th."

3. Now orient this tissue slice so that one side is flat against the cutting surface (*see* **Note 25**). It's helpful if you can now cut out a square of tissue with sides slightly shorter than the width of the razor.
4. Carefully make a clean cross-section cut (pushing straight down as in **step 1** above) parallel to one edge of the tissue.
5. Move the blade over 1–2 mm from the new edge and cut again to make a strip with parallel sides about 1–2 mm wide and as long as the blade.
6. Make several such strips, and select the best cut. A 6 mm long portion of the best strip is your sample. Again, be careful to maintain the orientation of the sample throughout.

3.2.2. Construction of In Situ Cell

1. Position the sample along the diameter of one hydrophobic coverslip so that one "Th" edge is against the surface of the coverslip (**Fig. 3**) (*see* **Note 27**).
2. Surround the tissue slice with the freshly mixed dental clay so that a crude Greek "theta" (θ) is formed. The tissue is the bar across the center of the letter, and the open areas above and below the tissue are the donor and receptor chambers of the cell (**Fig. 3**). Use a toothpick to move the clay around (*see* **Note 28**). The top surface of the clay should be about even or just below the height of the tissue strip (1–2 mm).
3. Cover the top side of the θ loosely with the second coverslip to prevent the tissue sample from drying out, but do not press it down (*see* **Note 29**).
4. Insert this coverslip/tissue/coverslip sandwich (the in situ cell) into the sample holder so that the center bar of the θ is along the line of either X or Y stage motion.

3.2.3. Imaging

1. Mount the sample holder with the cell assembly on the stage of the inverted microscope, and check the sample for autofluorescence using the laser and confocal settings that will be employed during the experiment (*see* **Note 30**).
2. Position the sample holder so that the field of view includes the area through which transport is expected to occur, and determine the Z-location of the tissue/coverslip interface (*see* **Note 31**). Make a note of this position as Ztc.
3. Move the focal place about 30 μm away from the interface, deeper into the tissue and away from the damaged cells at the cut edge. Note the Z-location position here as well, will be Zim.
4. Fill one 20-cc hypodermic with the probe (donor) solution, and the other with the buffer (acceptor) solution.
5. At $t = 0$ (the start of the experiment), use the appropriate syringe to fill the acceptor chamber with a drop (about 10 μL) of Krebs buffer, and the donor chamber with a drop of probe solution.
6. After filling the chambers, quickly move the focal plane to Ztc, and check to see is there is any leakage of the probe (seen as a bright fluorescent film) along this interface.
7. If leakage is observed, discard the in situ cell. If not, continue with the experiment, returning to Zim for image acquisition (*see* **Note 32**).

8. Images are then acquired at set time intervals from the same plane in the sample, without changing any settings, and the time/date stamped images stored without image enhancement (*see* **Note 33**).
9. Successive images will show the gradual spread of fluorescence through the tissue as the probe molecule moves through it *(5)*.

3.3. Imaging Transport Across Fragile Tissues (Nasal Epithelium)

3.3.1. In Vivo Intranasal Administration of Fluorescent Dextran Solution

1. Anesthetize the animal and place it on its back.
2. When the rat is sedated, insert a 5-cm silicone cannula into the trachea via tracheotomy to enable the animal to breathe.
3. Insert a second 20-mm silicone cannula via an incision in the esophagus into the posterior end of the nasal cavity (*see* **Note 34**). This cannula will be used to flush the nasal cavity with fixative.
4. Administer the fluorescent probe solution (FITC-dextran in our study) to both nostrils with a 100 µL syringe with a PVC tube attached to it. Insert the PVC tube at least 0.5 cm into each nostril, and administer at least 50 µL of the probe solution.
5. Keep the animals supine during exposure to the probe solution (15 min in our experiments).

3.3.2. Fixation and Postfixation

1. Give an intravenous overdose of anaesthetic 30 s before flushing the nasal cavity with fixative to ensure the complete sedation of the animal.
2. Perfuse 5 mL of Bouin's fixative through the nasal cavity via the esophageal cannula. The fixative will flush from the nasal cavity through the nostrils.
3. Remove the nasal septum, divide it into three equal parts (*see* **Note 35**).
4. Post-fix the tissue by immersing the parts in Bouin's fixative for another 2 h.
5. Store the fixed tissue in a solution of 10% formaldehyde at pH 8.0 (*see* **Note 36**).

3.3. Counterstaining the Tissue

Identification of cell types in the tissue is aided by counterstaining with Evans blue, which binds to proteins, or dioctadecyl indodicarbocyanine (DiIC18(5)), which stains lipids (*see* **Note 37**) 24 h after the completion of the in vivo portion of the procedure.

1. Counterstain the tissue by immersing the septum parts in 1 mL of Evans Blue solution for 15 min, or in 1 mL of DiIC18(5) solution for 60 min.
2. Rinse the tissue after staining in PBS pH 8.0.
3. Carefully remove the fixed epithelium from both sides of the septum with a sharp scalpel.
4. Arrange the epithelium with the apical side facing the objective lens (*see* **Note 38**), using either of the sample holders described previously.
5. Examine the fixed tissue.

4. Notes

1. Optisol (Chiron Vision, Irvine, CA) is a medium-term storage medium for human cornea. It is no longer commercially available, but can be replaced by Optisol-GS, also from Chiron Vision. These transplant media, however, although very elaborate, are also rather expensive. In most instances, simpler media can be prepared in the laboratory or purchased from cell culture facilities. The most commonly used of such cost-effective media include glutathione bicarbonate Ringer solutions [e.g., *(7,8)*] and medium 199 *(8)*. PBS should be avoided whenever possible as it has been shown to alter excised cornea integrity within a few hours *(1,9)*.

2. Cornea collection and storage: A clear and illustrated protocol for the excision of cornea from the enucleated ocular globe can be found in **ref. *11***. A similar procedure can be followed without prior enucleation of the globe. In our study, New Zealand white rabbit corneas were excised immediately after sacrifice, sans scleral rim, without prior enucleation, using the trephine blade and surgical scissors.

3. BR can also be stored at 4°C overnight. If so, after equilibration at room temperature, pH should be checked and readjusted if necessary with carbogen.

4. CAM is a nonfluorescent cell-permeant dye (mol wt 994.8) that is cleaved by intracellular esterases to fluorescent calcein (mol wt 666.5, λ_{max} absorption = 490-510 nm, λ_{max} emission = 515-535 nm). EH-1 (mol wt 856.7, λ_{max} absorption = 528 nm, λ_{max} emission = 617 nm) passes through damaged cell membranes to bind DNA and undergo a 40-fold enhancement in fluorescence.

5. DMSO is a permeation enhancer. In an effort to minimize the amount of DMSO used in the assay, we made our own CAM stock solution rather than using the commercial version.

6. Solutions of EH-1 in H_2O/DMSO can be stored frozen for at least 1 yr. CAM is subject to hydrolysis when exposed to moisture, and the CAM stock solution will take up some moisture from the air when it is made. It thus should be used within 30 d.

7. The cornea holder (which we made out of aluminum foil) was used because we found that immersion of the whole cornea in the EH-1/CAM solution resulted in extracellular CAM hydrolysis, probably due to leakage of enzyme from the edge of the tissue into the probe solution.

8. The sample holder is essentially a short piston in a cylinder, with a setscrew to hold the piston in place. The cylinder is made by using a lathe to bore a cylinder of diameter 0.5015 inch about 1.2 inches deep through the center of a short section of 1-inch lucite rod. The projecting end of the rod is then faced (made smooth) while it is still in the lathe, and then cut off about 1.1 inches from the faced end. The other end of the resulting cylinder is faced, and a hole for a nylon bolt (size 8 or larger) is drilled and tapped along a line that would intersect the center of the bore, about 0.5 inches from one face. This bolt acts as a setscrew for the piston, which is made by first turning lucite rod stock to an exact diameter of 0.5 inch in a lathe (rod stock is typically not truly round) and then cutting and facing an 0.75 inch section of the turned rod. When lubricated by tissue fluids or buffer solution, this piston moves easily with the 1.5 thousands clearance pro-

vided. It is important that the faces of the piston be smooth and unscratched. O-rings can be mounted on the sides of the piston if a water-tight seal is desirable.

9. The type of silicone or vacuum grease is not important, as long as it is somewhat sticky and not too liquid. Stiki-wax, which was used to apply adhesive to small notes before the advent of Post-Its, is easier to use than double-sided adhesive tape, and may still be available in some stationery stores.

10. This technique requires an inverted microscope to prevent loss of fluid from the cell. We used a Bio-Rad MRC 600 confocal mounted on a Zeiss IM-35, and a Zeiss Plan-Neo-Neofluar 25×/0.8 oil/water/glycerin objective set to oil. The FITC label was detected with the blue high sensitivity (BHS) filterblock. Ten scans were obtained and Kalman-averaged to obtain an image.

11. Sample viability is a concern, and should be verified in parallel experiments using an appropriate method.

12. We used porcine buccal mucosa dermatomed to 500 µm (which includes the epithelium and part of the connective tissue) and washed with cold Krebs pH 7.5 buffer before storage in the same buffer. The mucosal surface faced the donor chamber of the cell, and the serosal surface (the cut edge) faced the receptor side. The Krebs buffer was prepared fresh using twice-distilled water and equilibrated with carbogen for 15 min before use.

13. New razor blades come coated with a protective film, usually oil or silicone. This film should be removed before use, or it may form small globules on the sample which will interfere with imaging (and the protective material may also be fluorescent).

14. Be sure to use an isotonic buffer solution. Also, note that fluorescently labeled compounds used as flow probes may have unreacted label present when purchased, and can also lose their labels during the course of a long experiment. The presence of fluorescence in the tissue can then be an artifact. It is therefore advisable to test the purity and stability of the label, e.g., by HPLC.

15. The chosen concentrations depend very much on the fluorescent compounds used. FITC-labeled dextrans, e.g., are available from various sources with different labeling efficiency, and donor concentrations have to be adjusted accordingly.

16. We found Bouin's fixative to be superior to other formaldehyde-containing fixatives in immobilizing the intranasally administered dextrans and preserving the general morphology of the tissue well. However, Bouin's fixative is less appropriate for other uses, e.g., immunocytochemistry, as it contains high amounts of picric acid and methanol, both of which destroy cell organelles. When the loss of ultrastructural elements is unacceptable, Bouin's fixative can be replaced by other fixatives that contain smaller amounts of picric acid and replace formalin with paraformaldehyde. One such fixative is Zamboni's fixative, which consists of 2% formaldehyde and 0.2% picric acid *(12,13)*.

17. FITC-labeled aldehyde fixable dextrans with molecular weights of 3000 and 10,000 *(4)* were used as model compounds in our studies. These dextrans have covalently bound lysine residues that permit them to be covalently linked to biomolecules by aldehyde fixation.

18. The confocal microscope system used in our study was a Bio-Rad MRC600 equipped with a krypton/argon laser and mounted on a Nikon Optiphot microscope. Samples were simultaneously excited with the 488- and the 568-nm lines and imaged with Nikon Planapochromat air (10×/0.5) or oil immersion (40×/1.3 and 60×/1.4) objectives and a zoom of 1. Calcein and EH-1 fluorescence were detected using the Bio-Rad K1/K2 filterblock set.

19. The piston is lowered about 3/8" inches below the top edge of the cylinder, and kept in place either by the setscrew or by a finger inserted up the bore from the other side. The tissue sample is then placed on the head (top surface) of the piston. A thin layer of silicone grease is applied to the rim of the cylinder above the sample, and a small drop of liquid (e.g., buffer) is placed on top of the tissue. The coverslip is set carefully onto the grease and lightly pressed down on its edges to seal the top of the cylinder. The piston is then slowly advanced by pushing it up from the other side until complete contact between the top of the tissue surface and the bottom of the coverslip has been established (the drop of liquid ensures an air-free coupling between the tissue and the coverslip). Stiki-wax or double-sided tape is applied to the bottom rim of the cylinder, which can thus be firmly attached to a wide glass slide.

20. Calcein signal was found to bleach faster than EH-1 signal. On the MRC 600, full images can only be acquired consecutively; therefore the calcein image was collected first followed by the EH-1 image of the same field. For image acquisition, normal scanning speed was usually satisfactory (1 s/frame, 512 lines/frame; F2 on Bio-Rad COMOS) to obtain a good quality Kalman-averaged image, and slow speed scanning (3 s/frame, 512 lines/frame; F1) was not necessary.

21. Typical microscope settings for epithelium or endothelium imaging (Kalman averaging, n = 3–5) with the Bio-Rad MRC 600 using K1/K2 filterblocks, full horizontal box, zoom = 1, and scan speed F2, were: ND = 1–3%, aperture 3–5 units, gain 4–6, black level 4.5–5.0, enhancement off.

22. When flipping corneal button over, be careful to keep track of tissue orientation. The curvature of the button is minimal, and thus this is not a good indicator for which side is which.

23. NIH Image is a public domain program developed at the U.S. National Institutes of Health. NIH Image and the Bio-Rad confocal macros can both be downloaded from the NIH Image homepage on the Internet at http://rsb.info.nih.gov/nih-image/).

24. The object is to prepare a small rectangular block of tissue about 6 mm long, 1–2 mm deep, and anywhere from 0.5 to 2.0 mm thick for the surface against the coverslip. For epithelia, this latter dimension is usually the thickness of the particular epithelium.

25. If one side of the tissue is more difficult to cut than the other (e.g., contains connective tissue), place this side facing the cutting surface to avoid unnecessarily compressing the tissue.

26. One way to do this is to sprinkle a tiny amount of graphite powder on one side of the tissue. The powder grains are easy to see by eye and in the microscope, and are not chemically reactive. Scrapings from a pencil lead work well.

27. Be careful not to wrinkle the tissue slice, and avoid trapping a liquid film or air bubbles beneath it. Inspection with a magnifier is advisable.

28. Be sure to arrange the clay so that the left and right edges of the tissue strip are captured and a good seal is formed.

29. The coverslip needs to be loose to allow access to the donor and acceptor chambers in the clay.

30. Tissue autofluorescence varies between individual animals, and interior autofluorescence cannot be determined until a cross-section is made. If significant autofluorescence is present in the tissue at the gain, aperture, etc. settings to be used in the experiment, discard the preparation.

31. This can be accomplished in several ways, but the easiest is to epiilluminate the tissue with laser light, and increase detection sensitivity until its residual autofluorescence can be used to determine the position of the interface.

32. Repeat **step 11** at least once during the experiment, and at the end as well, to verify that no leakage along the interface has taken place.

33. Be sure to note all settings that are not automatically recorded by the imaging software, including the magnification of the lens used.

34. Check that the cannula is inserted into the nasal cavity and not the buccal cavity by holding a mirror in front of the animal's nostrils. Blow through the esophageal cannula, and water vapor should condense on the mirror.

35. After euthanasia and fixation the rat is decapitated. The skin and the mandibles are removed to expose the palate of the buccal cavity. Separate the front of the nose from the skull by cutting between the second palatal ridge and the first molars. Carefully excise the nasal septum by cutting the dorsal and ventral portions of the skull and removing these from both sides of the septum. Remove the ridge on the ventral part of the septum and divide the septum in three equal parts: anterior, medial, and posterior.

36. The tissue is post-fixed and stored in a 10% formaldehyde (pH 8.0) solution because formaldehyde fixation is reversible, and thus the tissue must be stored in formaldehyde to preserve the fixation. In addition, Bouin's fixative can quench fluorescein fluorescence to some extent, but when the tissue is stored in 10% formaldehyde (pH 8.0) overnight after Bouin fixation, the FITC fluorescence is bright enough for confocal microscopic imaging (fluorescein fluorescence intensity is maximal between pH 8.0–10.0).

37. Evans blue (FW 960.8, λ_{max} absorption = 611 nm) and DiIC18(5) (mol wt 960, λ_{max} absorption = 644 nm, λ_{max} emission = 665 nm) were used as counterstains because they fluoresce in the far red area, and thus avoid overlap with FITC fluorescence.

38. Confocal microscopic visualization. We used a Nikon Planapo 60×/1.40 oil immersion lens, 488 nm excitation for the FITC and 633 nm excitation for both of the red fluorescent probes, Evans blue and DiIC18(5). There is no crosstalk between FITC and Evans blue/DiIC18(5). The depth of imaging in the fixed tissue is generally not deeper than 15–20 μm. Depth resolution can be improved by the use of mounting media with a refractive index close to the refraction index of the immersion medium *(12,14)*.

Acknowledgments

We gratefully acknowledge support from the UCSF TSR&TP for D. I.; from the AACP New Investigator program and the UCSF Academic Senate for C. C.; and from NIDR-DE11275 for C. C. and D. I.

References

1. Doughty, M. J. (1995) Evaluation of the effects of saline versus bicarbonate-containing mixed salts solutions on rabbit corneal epithelium in vitro. *Ophthalmic Physiol. Opt.* **15(6)**, 585–599.
2. Poole, C. A., Brookes, N. H., and Clover, G. M. (1993) Keratocyte networks visualised in the living cornea using vital dyes. *J. Cell Sci.* **106(2)**, 685–691.
3. Merkus, F. W. H. M., Verhoef, J., Romeijn, S. G., and Schipper, N. G. M. (1991) Absorption enhancing effect of cyclodextrins on intranasally administered insulin in rats. *Pharm. Res.* **8**, 588–592.
4. Marttin, E., Verhoef, J. C., Cullander, C., Romeijn, S. G., Nagelkerke, J. F., and Merkus, F. W. H. M. (1997) Confocal laser scanning microscopic visualization of the transport of dextrans after nasal administration to rats: effects of absorption enhancers. *Pharm. Res.* **14**, 631–637.
5. Hoogstraate, A. J., Cullander, C., Nagelkerke, J. F., Spies, F., Verhoef, J., Schrijvers, A. H. G. J., Junginger, H. E., and Bodde, H. E. (1996) A novel in-situ model for continuous observation of transient drug concentration gradients across buccal epithelium at the microscopical level. *J. Control. Rel.* **39**, 71–78.
6. Bindra, B. and Heath, J. R. (1997) Adhesion of elastomeric impression materials to trays. *J. Oral Rehab.* **24**, 63–69.
7. Imbert, D. and Cullander, C. (1997) Assessment of cornea viability by confocal laser scanning microscopy and MTT assay. *Cornea* **16**, 666–674.
8. Rojanasakul, Y. and Robinson, J. R. (1989) Transport mechanisms of the cornea: characterization of barrier permselectivity. *Int. J. Pharmacol.* **55**, 237–246.
9. Huang, A. J., Tseng, S. C., and Kenyon, K. R. (1989) Paracellular permeability of corneal and conjunctival epithelia. *Invest. Ophthalmol. Vis. Sci.* **30(4)**, 684–689.
10. Imbert, D. and Cullander, C. (1997) Assessment of cornea viability by confocal laser scanning microscopy and MTT assay. *Fund. Appl. Toxicol.* **36(1)**, 44,45.
11. Westphal, M., Naush, H., and Zirkel, D. (1996) Cell culture of human brain tumors on extracellular matrices: methodology, and biological applications, in *Methods in Molecular Medicine Series: Human Cell Culture Protocols*, (Jones, G. E., ed.), Humana Press, Totawa, NJ, pp. 81–100.
12. Bacallao, R., Kiai, K., and Jesaitis, L. (1995) Guiding principles of specimen preservation for confocal fluorescence microscopy, in *Handbook of Biological Confocal Microscopy*, Plenum Press, New York, pp. 311–325.
13. Stefanini, M., De Martino, C., and Zamboni, L. (1967) Fixation of ejaculated spermatozoa for electron microscopy. *Nature* **216**, 173,174.
14. Hell, S. W. and Stelzer, E. H. K. (1995) Lens aberrations in confocal fluorescence microscopy, in *Handbook of Biological Confocal Microscopy*, Plenum Press, New York, pp. 347–354.

20

Measurement in the Confocal Microscope

Guy Cox

1. Prerequisites
1.1. Alignment

Accurate alignment of the microscope is vital for most measurements. Details of the adjustments will vary from system to system but the principles of alignment are universal. The methods given here should be appropriate for any spot-scanning confocal microscope but will not be directly applicable to Nipkow-disk, slit scanning, or spot/slit systems.

If the pinhole is not precisely in position the image will not be at its brightest when it is in true focus. This is guaranteed to destroy the accuracy of any measurement, particularly measurements of depth and intensity. If the pinhole size is adjustable, carry out the alignment with it fully closed then open it a fraction since no alignment can be perfect. As a check, if any doubt exists, take a vertical (X–Z) section of a test specimen with the pinhole fully closed—misalignment will show as double images of very thin structures (**Fig. 1**).

The axis of the scan must pivot around the back focal plane of the objective lens. Otherwise, vignetting will occur—the edges of the image will be dimmer than the center. This will affect some measurements more than others, but it will have most importance where structures are being segmented out by gray values. Some manufacturers provide a prism that fits in the objective position to check this. If not, provided that the microscope is "infinite tube length" (current designs from Zeiss, Leica, Nikon and Olympus all are, as are some older microscopes) just remove an objective and place a piece of card on the stage. *Do not use any reflective object such as a mirror or even a glass slide.* The circular patch of laser light should appear stationary. If it is not, the appropriate adjustment must be made. On a Bio-Rad system, e.g., the height of the transfer lens between microscope and confocal head is adjusted by a screw

From: *Methods in Molecular Biology, vol. 122: Confocal Microscopy Methods and Protocols*
Edited by: S. Paddock Humana Press Inc., Totowa, NJ

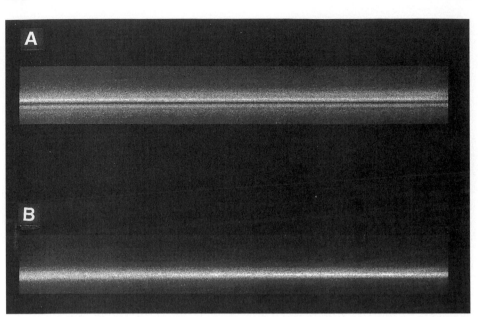

Fig. 1. Confocal reflection images of a mirrored coverslip. (**A**) XZ image with the pinhole misaligned. Note how the reflective surface appears artefactually as a double line. (**B**) With the pinhole aligned correctly the surface appears as a single line.

collar. After this adjustment, the confocal and conventional optical systems may need readjustment to make them parfocal, so that both the confocal image and that seen through the eyepiece are in focus. (This is worth doing purely as a matter of convenience, but it could also affect the magnification if it was severely out). Again taking the Bio-Rad system as an example, the whole confocal head is raised or lowered to make both sets of optics parfocal.

The field of view of the confocal optics should be centered on the optical view. It is inconvenient if it is not, and although measurements will not be affected, resolution might be, as all lenses perform at their best in the center of the field of view. The illumination should be even across the confocal image. Some microscopes are notably uneven at their lowest zoom settings—in this case it is better to use a higher zoom.

1.2. Contrast

The requirements for obtaining accurate measurements in the confocal microscope are not necessarily compatible with the best image quality as judged by eye. First, it is essential, when making any measurements of intensity or resolution, that the mean background intensity nowhere goes to zero. Although this is not so essential for measurements of position or depth, it is

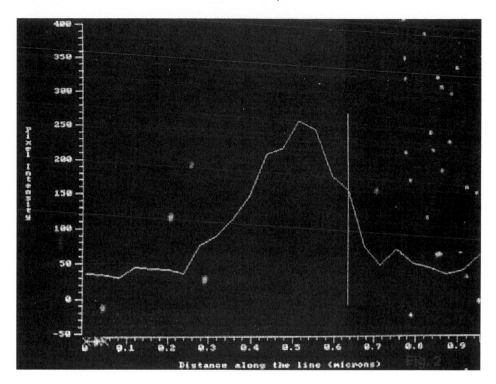

Fig. 2. Line profile through a 100-nm fluorescent bead, taken using a x63, NA 1.4 oil immersion lens. The width of the curve, taken halfway between the maximum and minimum values, gives the resolution of the system. To obtain a meaningful value the peak must not reach saturation intensity (255) and the background has to be above zero.

still advisable because otherwise the resolution, and hence the possible error, cannot be estimated (**Fig. 2**). In most cases the background level needs to be the dark current from the photomultiplier (PMT), as background fluorescence in the tissue may vary from sample to sample. Therefore the first step should be to remove the sample from the microscope and adjust the black level until a (just) nonzero value is present across the entire field of view. The resulting background will probably be higher than would be chosen for visual impact. On some systems automatic gain and/or black level controls are on by default— it is absolutely essential to turn these off.

It is even more important, for any sort of measurement, that the intensity at any point in the image never exceeds the maximum that the system can record. Some microscopes (e.g., Bio-Rad MRC 500 and 600, OptiscanF 900e) have the option of storing images in 16-bit form. In this case the maximum possible intensity is approximately 65,000 counts—a value that is unlikely to be exceeded with any real-life specimen. But in many cases microscopes only store 8-bit images, in which case the maximum recordable intensity is 255. This is all too easy to exceed, and it is very important to set the gain of the amplifier or the PMT voltage to a value at which even your brightest sample will not saturate the image. If the intensity values saturate at any point in an object then its true intensity cannot be measured, and neither can its position laterally or in depth.

1.3. Summary

Before you start, the microscope must be aligned as accurately aspossible, and the gain (or PMT voltage) and black level must be set to avoid any overflow or underflow.

2. Depth and Thickness Measurement

Confocal microscopes are commonly used for making measurements in depth—something that conventional microscopes, whether light or electron, cannot easily do. Their major rival in this field is the scanning probe (atomic force) microscope, but this can only measure a single surface. In the confocal microscope various types of measurement are possible, with varying degrees of accuracy. The relative heights of various parts of a single surface can be measured with extreme precision, whereas measurements of vertical depths and spacings within a three-dimensional solid or liquid are constrained to some extent by the resolution of the system, which is unlikely to exceed 0.5 μm in the axial direction. Because axial resolution is related to the *square* of the numerical aperture (NA), it is absolutely essential to use the highest NA lens that is capable of covering the desired area.

Common to all depth measurements is the need to take account of the refractive indexes of the specimen and of the medium in which it is mounted. Wherever possible these should be the same! If specimen, mountant, and the medium between sample and lens are all equivalent in refractive index no correction is needed—the depth measured by displacement of the slide or objective will be correct. This applies whatever the actual refractive index is. Examples are:

1. Sample mounted in permanent mountant, under a coverslip, viewed with an oil immersion lens [refractive index (n) = ~1.5 throughout]
2. Surface sample such as the ornamentation on a dry pollen grain viewed under reflected light in air ($n = 1$ throughout)
3. Living sample, viewed in water, using a water-immersion lens, with or without a coverslip ($n = \sim 1.3$).

In each of these cases, although the measurement will be correct, it is necessary that the lens is appropriate to the media used. In (1) it is obvious that an oil immersion lens is needed, whereas in (2) the objective lens must be corrected for use without a coverslip (the prefix epi- is often used). In **(3)** the lens must not only be designed for water immersion, but it will also specify whether or not a coverslip is required. The coverslip will not affect measurements because it will introduce an equal apparent displacement to all parts of the sample. But if the lens is not corrected for the medium the image will be poor, and resolution both laterally and axially will be compromised. This is a consequence of spherical aberration, which can only be corrected for a single working distance and optical medium.

Conventional "dry" (nonimmersion) objective lenses for biological use are intended for use with a coverslip, and with the specimen beneath mounted in a medium of refractive index ~1.5. The distance from the top of the coverslip to the sample should be 0.17 mm, and this distance will be found engraved on the lens. Provided that this distance is, overall, approximately correct, reasonably well-corrected images can be obtained over a distance of several microns, but relative depth measurements must be corrected for the difference between the refractive index of the medium around the lens and the medium containing the sample. **Figure 3** shows why this is so. Rays of light from the object are refracted away from the normal as they leave the coverslip, according to Snell's law of refraction. They therefore appear to emanate from the point indicated by the dotted lines. The correct measurement will be given by multiplying the apparent measurement by the refractive index of the mountant (1.5) over that of air (1)—in other words the correct figure will be 1.5 times the measured value.

Samples mounted in water should not, in theory, be measured with a dry lens but in the real world we often want to do this. We will get a reasonable image approx 0.2mm below the upper surface of the coverslip, and the measured depth should be corrected by multiplying by 1.3, the refractive index of water. Nevertheless a water immersion lens is strongly preferable—as well as not needing any correction, the image will be aberration free. Measuring depth or thickness in aqueous media under oil immersion should not be attempted—although a passable image can be obtained immediately below the coverslip the image quality deteriorates so rapidly with depth that measurements could not be trusted *(1)*.

2.1. Measuring Heights On A Surface

Variations of height in a single plane surface can be measured with considerable accuracy in the confocal microscope; this is very commonly done in the engineering sciences but also has applications in biology. A line profile can be taken by making a vertical (X–Z) section of the surface, or an overall view can be had by collecting a series of optical sections. The accuracy to which a sur-

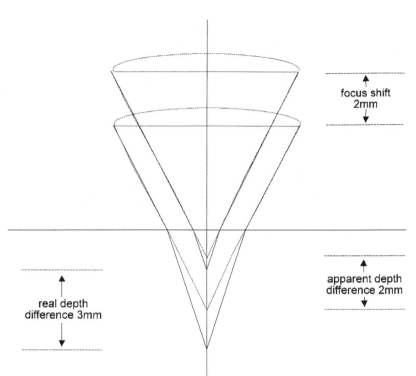

focus shift
2mm

apparent depth
difference 2mm

real depth
difference 3mm

Fig. 3. The relationship between real and measured depth when using a dry lens with a permanently mounted sample. (In real life the actual value could not be as large as 2 mm without causing other errors).

face can be located considerably exceeds the vertical resolution of the microscope (resolution being the vertical distance between two objects that can be distinguished from each other). The spacing between planes (or lines in an X–Z section) should therefore be as fine as is compatible with a reasonable size for the resultant file.

The resulting dataset, viewed as an X-Z section or a side view in a 3D reconstruction, will generally look very disappointing at first glance. The plane surface will be stretched or smeared in depth—possibly to the extent that measurement seems impossible (**Fig. 4A**). However a line intensity profile through the surface will show that there is in fact a strong maximum at the actual location of the surface. This can be used directly as the basis for measurement, or various image processing techniques can be used. Thinning or erosion algorithms *(5)* can produce a single-pixel wide line indicating the profile of the surface to the nearest depth step (**Fig. 4B**). Some systems (e.g., Zeiss LSM 410) include this feature as part of their standard software, but in other

Fig. 4. X–Z section through a sample consisting of a 3 μm layer of SiO$_2$ on a silicon substrate. Over the oxide is a thin layer of oil (not normally a recommended practice but deliberate in this case!). In the original image (**A**) although the actual position of the surface of each layer is actually the brightest part this is not apparent to the unaided eye, and accurate measurements of layer thickness would seem impossible. After only five iterations of an erosion (thinning) algorithm, which preserves the peak intensity (*5*), the actual position of each surface (to the nearest pixel) is clearly visible.

cases third-party software will be needed. Free software is available that will not only extract a surface in this way but also carry out hyperbolic interpolation to give subpixel accuracy (**Note 1**). Another alternative is to create a height-coded image, which is effectively a contour-colored map of the surface, plotting the depth of the brightest pixel as the intensity in the output image.

2.2. Thickness Measurements

When measuring the thickness of a cell or similar structure the simplest approach is again to take an XZ section, or multiple XZ sections. Where the boundary of the object is stained (e.g., where a cell is bounded by a labeled membrane or wall) the brightest pixel can again be taken as an indicator of the true position of the surface. The depth (with any necessary correction for refractive index) can then be measured from a line profile.

If the object to be measured is stained or labeled throughout its thickness the problem of determining the true edge of the structure is more complex. The best general approximation is to take the halfway point between the background level and the peak intensity. This will be reasonably valid provided that there is no refractive index difference between the object and the surrounding medium.

2.3. Summary

Refractive index effects must always be borne in mind when making this class of measurement. Measuring surface profiles and relative depths is straightforward and can be carried out to a higher accuracy than the depth resolution of the microscopes, even though the actual images may look poor. Measuring the thickness of objects that are labeled throughout is less accurate.

3. Length, Area, and Volume Measurement

3.1. Two-Dimensional Measurements

There is little here that is specific to the confocal microscope. Because the image is already in digital form these measurements are simple to do, and most microscopes include software permitting segmentation of single or multiple grayscale ranges for simple area measurement, and line cursors for measurements of straight-line lengths. More sophisticated measurements will require the files to be exported to specialist image analysis software. Because most confocal microscopes offer export to standard file formats, and many image analysis packages will directly read confocal microscope formats, this is no longer the problem that it was as recently as 3 or 4 years ago.

3.2. Surface Area and Volume

Volume measurement from a confocal dataset is straightforward, as a count of the voxels in the volume of interest (with appropriate corrections for scaling in the vertical dimension) is all that is required. Many packages, including the Lasersharp software provided with Bio-Rad confocal microscopes and 3D analysis packages such as Voxblast™ (**Note 2**) and VoxelView® (**Note 3**), provide a facility for seeding (identifying a voxel within the desired volume) and then setting intensity thresholds to segment out the required volume. This will give a numerical result very simply but in the case of complex structures it may still be preferable to segment out the structure of interest in the individual layers and measure each one. The considerations outlined above for determining the vertical extent of the object will still apply.

Surface area is much less simple—a count of voxels making up the surface is *not* a measure of the actual area. Some specialist 3D packages will fit a surface to a 3D dataset and measure it but at the time of writing commercial systems that can do this remain rare. Alternatives are (1) approximating the shape to a regular solid (cuboid, ellipsoid) and estimating the area from that or (2) adopting a stereological aproach (below).

3.2.1. Stereological Approaches for Cases in Which the Volume of Interest is Not Fully Contained Within the Confocal Dataset

Measurements made on a random sample of sections can provide very accurate estimates of surface area and volume and this is often a more appropriate approach to determining these parameters in many biological systems. Textbooks on stereology will offer a bewildering array of solutions that may well be necessary in difficult cases but two very simple formulae will be usable in a wide ranges of circumstances (2).

These are:

$$V_V = A_A$$

In other words, the volume of a structure or compartment, per unit volume (V_V), is the same as the measured area of that component in a section, per unit area (A_A). Many confocal microscopes will enable the area of interest to be segmented out, either by gray levels or by drawing, and its area (and that of the full image) calculated automatically. This will (after measuring a suitable number of images) give an accurate estimate of the volume of the structure of interest.

$$S_V = (4/\pi)L_A$$

Thus by measuring the length of intercept of a membrane (or other surface) per unit area of section (L_A), and multiplying it by $4/\pi$, or ~1.27, we can estimate the surface area of our structure in a unit volume (S_V).

Both these measurements depend on sections being taken at random. If your structure has a particular orientation in the specimen you must cut (real or optical) sections in a range of directions, and you must sample the section so as to include all parts equally. Picking views that contain the structure you are measuring will **not** give the correct result! It is probably best, to avoid any unconscious bias, to sample your specimen in a grid pattern, taking images at a series of predetermined coordinate patterns.

3.2.2. Summary

Length and 2D area measurements are common image analysis problems and easily carried out with image analysis software. Volume measurements are conceptually equally simple but require manual techniques or 3D analysis software. 3D surface area measurements require specialist software but are probably better carried out with stereological techniques.

4. Fluorescence Intensity Measurements

Making relative intensity measurements is relatively simple, while making absolute measurements is absolutely impossible. Within its limitations the confocal microscope is probably the most accurate tool available for measuring intensity within a defined voxel space, although fluorescence photometry may be preferable for measuring the integrated intensity from a specific organelle, at least if that organelle can be isolated from the cell. In either case bleaching of the fluorochrome is likely to be at least as significant a factor as instrumental parameters *(3)*.

4.1. Linearity Calibration

Before making any actual measurements the linearity of the response should be checked using neutral density filters. If the transmitted light detector uses the same PMT as the fluorescence detector, an initial check can be made using the neutral density filters which control the incoming laser power. On a Bio-Rad MRC 500/600, for example, these provide 1%, 3%, and 10% of full power, which should enable a calibration graph of adequate accuracy to be constructed. Ideally this should be done with each available laser line although one could hope that the detector would be equally linear at all wavelengths even if it is not equally sensitive.

If the transmission detector is a separate system this calibration must be done using a fluorescent sample. In this case great care must be taken, if using the neutral density filters in the beam path from the laser, to ensure that the fluorescence does not saturate (all available molecules in the excited state) at the higher intensities. Otherwise saturation of the fluorochrome could be mistaken for nonlinearity of the detector. To avoid bleaching a solution of dye is best (e.g., DiI in immersion oil) and by using several dilutions one can check whether the response is linear in relation to concentration. Alternatively, fluorescent plastic sheet (available from any sign-maker) is reasonably resistant to fading.

It is preferable, however, to calibrate the system by placing a series of neutral density filters in front of the detector. On a Bio-Rad MRC500/600 this can be done simply enough, provided a single-channel filter block is used, by placing neutral density filters (which do not need to be any particular size) inside the "tunnel" for the second filter block so that they cover the hole leading down to the detector. (The cover plate must be put back in position each time, of course.) Many other microscopes will offer a similar spot at which a neutral density filter can be placed in the path of the fluorescent light. Kodak sells gelatine neutral density filters that can easily be cut to any desired shape or size.

One way or the other, this exercise will lead to a calibration curve relating detected light intensity to PMT output. Ideally this should be linear, and in practice this will probably be true provided that one keeps the brightest pixels at least 10% below the maximum of the system.

4.2. Measurement

The black level should still be at the "no sample" setting established in **Subheading 4.1.** Place the brightest sample you have in the microscope and (quickly, so that bleaching is avoided) set the gain to keep the maximum intensity comfortably below saturation. From now on the gain, PMT voltage, and black-level controls must not be touched. If they are calibrated, write down the values, and if they are lockable lock them at these values. (It is a good idea to unlock them once the experiment is over because the next user may well do some damage by forcibly turning a locked knob).

There are three measurements you will normally need to make. First, measure an unlabeled sample, to get a value for background fluorescence with no label present. Second, measure your labeled control sample. Then measure your various experimental treatments. The relative differences in the amounts by which the fluorescence in the control and experimental samples exceed the background fluorescence of the unlabeled sample, corrected if necessary for nonlinearity in the detection, will be a reasonable measure of the differences in fluorescence resulting from the experimental treatments.

4.3. Ratio Imaging

Many fluorochromes can be used to measure pH or other ion concentrations, as the wavelength of their emission or excitation peak (or both) will change with the concentration of the ion in question. Examples of curves for both emission-ratioing and excitation-ratioing dyes are shown in **Fig. 5**. It will be clear from these that because all curves are distinct, *in principle* measurements of intensity at any two wavelengths could be used to determine which particular curve we are on, and therefore the ionic concentration present. In practice this is less simple because the signal-to-noise ratio of the image will not be good enough for any reasonable accuracy unless we choose points where large changes can be expected. In general, for a given integration time, a confocal microscope will always give a worse signal-to-noise ratio than a wide-field CCD camera. During a 1-s exposure acquiring a 768×512 pixel image, each point in the sample will be sampled for 1 s in the wide-field camera, but only for 2.5 µs in the confocal microscope! This disadvantage is offset by the ability to sample in a particular focal plane, so that the measurement is not degraded by interference from cells or free dye in higher or lower planes.

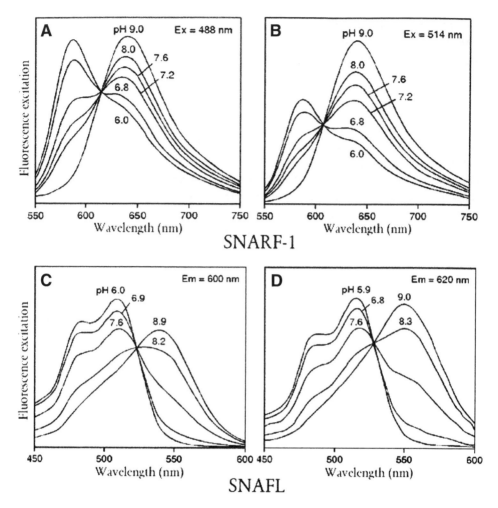

Fig. 5. pH-sensitive ratiometric dyes suitable for confocal microscopy. (A) Emission curves of carboxy-SNARF-1 from pH 6.0–9.0 with excitation at 488nm. (B) Corresponding curves using 514 nm excitation. (C) Excitation curves of carboxy SNAFL-1 from pH 6.0–8.9 measuring emission at 600 nm. (D) Excitation cuves of carboxy SNAFL-2 measuring emission at 620 nm. Both would be suitable for ratiometric measurement in a confocal system using 488 nm and 568 nm excitation. (Data reproduced from Haugland 1996, by kind permission of Molecular Probes, Inc.)

The great advantage of ratiometric measurement is that it is the *ratio* of the intensities at the two wavelengths that defines the ion concentration—the absolute intensity is unimportant. It is therefore unaffected by dye concentration, uneven loading, partitioning among cell components and bleaching. It *will*, however, be affected by background fluorescence so care needs to be taken to minimize autofluorescence. Dynamic range can also be a problem especially with confocal microscopes that store only 8-bit images. It is important to be sure that even high dye concentrations do not take the intensity at either wavelength to saturation.

In wide-field ratiometric imaging excitation ratioing is often preferred. For calcium measurement the dye of choice is fura-2. Images are taken using excitation wavelengths of 340 and 380 nm with a 510 nm emission filter, and the ratio is taken between the two images. In the confocal microscope suitable wavelengths are not available, even from ultraviolet lasers. In fact, because confocal microscopes only have very limited excitation wavelengths the possibilities are much more restricted than with a mercury arc source, so emission ratio measurements are often more practical. They offer the added advantage that, because almost all confocal microscopes have two or more photomultipliers, both images can be captured simultaneously.

Sometimes mixtures of dyes are used for emission ratiometric imaging, e.g., fura red and fluo-3 to measure calcium concentration. Using 488 nm (argon) excitation, ratio imaging can be accomplished by detecting images at 520 and 650 nm [in practice fluorescein isothiocyanate (FITC) and rhodamine filter sets would be used]. The problem with this approach is that is difficult to ensure equal loading of the two dyes in all cells of a population. Single dyes are preferable, and more are now being developed with confocal microscopy in mind. However, the discussion of individual dyes is beyond the scope of this chapter.

The fundamental protocol for making an emission ratiometric measurement is to record images at each wavelength (which can normally be taken simultaneously), subtracting the background level (determined previously), then displaying the ratio of the two images. Confocal microscope manufacturers normally provide (at extra cost) software for doing this automatically. In the case of the Bio-Rad confocal microscopes, e.g., this software will do these operations as the scan proceeds, so that the ratio image, as well as the two original images, are displayed live. It will also display graphs of ion concentration against time for multiple points over the image. However, if speed is less of a consideration it is quite possible to make ratiometric measurements without additional software.

The detection wavelengths will need to be chosen with some care. As **Fig. 5** shows, SNARF-1 will work very effectively with the two-channel filters designed for 514 nm excitation in an argon-ion system, such as Bio-Rad's D1/ D2 or A1/A2 combination. It cannot, however, be measured efficiently with filters designed for two-channel detection with an argon/krypton laser, as the fluorescence at 550 nm is very low and the low pH peak at ~580 nm will be blocked as it lies close to the 568 nm excitation wavelength. However, the barrier and dichroic filter sets designed for argon-ion lasers (A1/A2) will work quite effectively on an argon/krypton system using 488 nm excitation only and may prove a much less costly alternative than purchasing a custom filter set.

Excitation ratioing will require sequential collection of the two images. On systems in which the emission line selection is under software control, and there is a macro language available to automate the process, the time delay can be made quite small. On older or simpler systems, however, filters will need to be changed by hand and time-course measurements will be difficult. However, if a suitable dye is available it may be worth taking the trouble. The precision of the laser lines (relative to the rather wide-band filters usually used for emission ratioing) can give a very accurate measurement if they match the characteristics of the indicator. **Figure 5** shows an example—carboxy SNAFL-1 will give very effective ratio images from the 488 nm and 568 nm lines of an argon/krypton laser.

In any ratio imaging, accurate calibration is essential if actual numeric output is required. Because both dissociation constants and fluorescent properties may be influenced by the environment inside a cell, this calibration cannot be done in vitro but must be done on the cell system being investigated. The principle is to clamp or buffer the concentration of the ion in question to a known value, or series of values. EGTA can be used to produce defined calcium ion concentrations, and nigericin is used to calibrate pH. In general the measured intensity ratios obtained from these known concentrations can be entered into the microscope's ratiomentric software so that actual concentrations are shown on the live display.

4.4. Summary

First check that your system is linear in its response—and draw up a calibration curve if it is not. Set gain and black levels to be well clear of overflow and underflow. Then measure an unlabeled sample as well as a labeled control and experimental samples. For ratiometric measurements filters and/or laser lines should be chosen to optimise the response and calibration should be done in conditions as close as possible to the experimental ones.

Notes

1. 3-D View, by Iain Huxley. A modified version of NIH Image, the freeware image analysis system developed at the National Institute of Health, USA. Available from http://www.physics.usyd.edu.au/physopt
2. Voxblast. Vaytek Inc, 305 West Lowe St., Fairfield, IA.
3. Voxel View. Vital Images Inc., PO Box 551, Fairfield IA.

References

1. Sheppard, C. J. R. and Török, P. (1997) Effects of specimen refractive index on confocal imaging. *J. Microscop.* **185,** 366–374.

2. Underwood, E. E. (1970) *Quantitative Stereology.* Addison-Wesley, Reading, MA.
3. Pawley, J. (1995) Fundamental limits in confocal microscopy, in *Handbook of Biological Confocal Microscopy*, (Pawley, J. B., ed.), Plenum Press, New York, pp. 19–38.
4. Haugland, R. P. (1996) *Handbook of Fluorescent Probes and Research Chemicals.* Molecular Probes, Eugene, OR.
5. Cox, G. C. and Sheppard, C. J. R. (1998) Appropriate image processing for confocal microscopy, in *Focus on Modern Microscopy*. World Scientific Publishers, Singapore, in press.

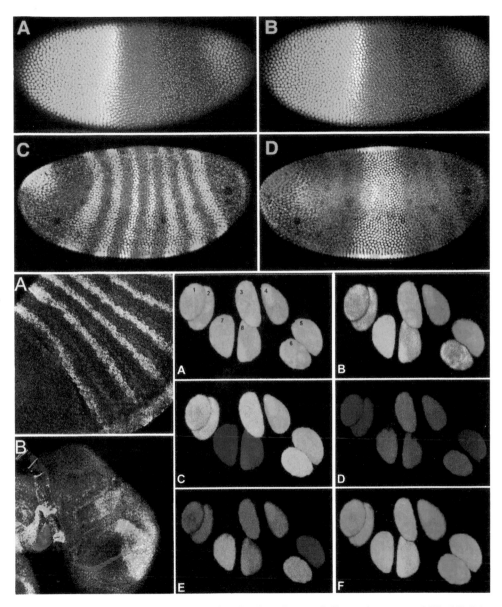

Plate I (*Top Sequence:* **[A-B]** Fig. 21-2 A-B, *see* full caption on p. 379; **[C-D]** Fig. 21-1 B, D, *see* full caption on p. 377 and discussion in Chapter 21. Lower Left Sequence: **[A-B]** Fig. 5-1 A-B, *see* full caption on p. 99 and discussion in Chapter 5. Lower Right Sequence: **[A-F]** Fig. 18-13 A-F, *see* full caption on p. 336 and discussion in Chapter 18).

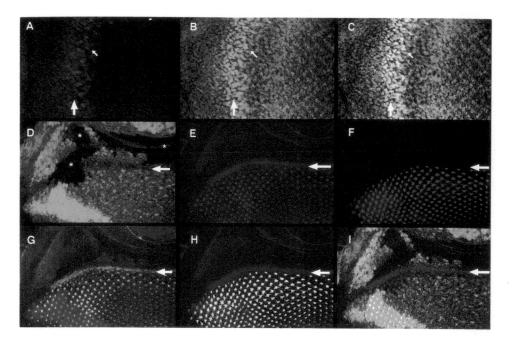

Plate II (Fig. 4-1 A-I, *see* full caption on p. 81 and discussion in Chapter 4).

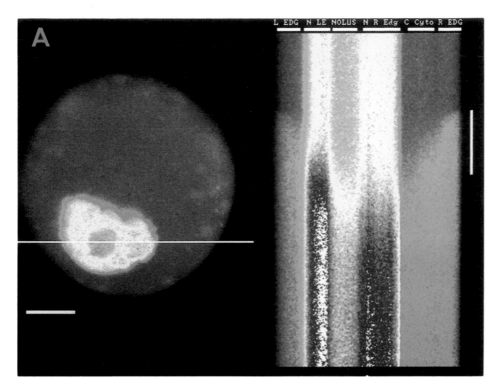

Plate III (Fig. 16-8 A, *see* full caption on p. 285 and discussion in Chapter 16).

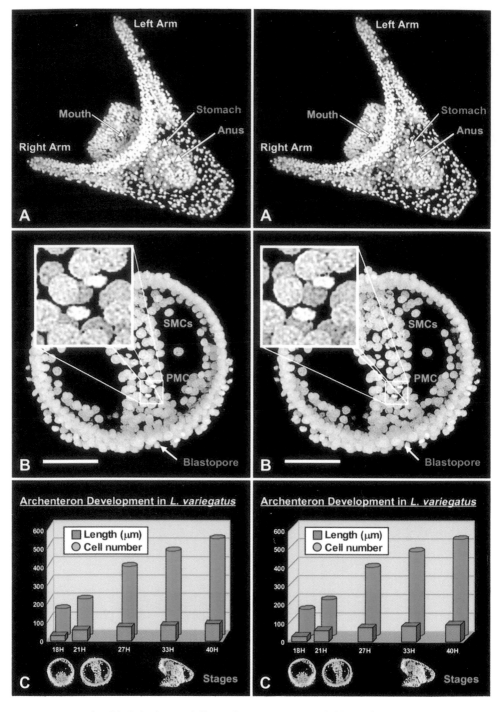

Plate IV (Fig. 22-8 A-C, *see* full caption on p. 397 and discussion in Chapter 22).

21

Presentation of Confocal Images

Georg Halder and Stephen W. Paddock

1. Introduction

Confocal microscopy is routinely used to produce high-resolution images of single, double-, and triple labeled fluorescent samples. The images are collected as single optical sections (2D imaging), as Z-series (3D imaging), as time-lapse series (2D over time), or as Z-series over time (3D over time or 4D imaging). Because the images are in a digital format, they can be further manipulated using a range of software.

This chapter covers methods of presenting confocal images for publication purposes, and is focused mainly on manipulating single optical sections or Z-series projections using relatively simple and commonly available software such as PhotoShop™ (Adobe Systems, Mountain View, CA). Many programs are available for constructing a 3D representation of a Z-series of confocal images either using proprietary software that is supplied with the confocal microscope or secondary software on a separate computer. 3D and 4D methods have been reviewed elsewhere *(1–3),* and *see* Chapter 18.

Most of the information contained in a confocal image of a biological specimen is related to the spatial distribution of various macromolecules. Images of different macromolecules are collected at different wavelengths. At the present time images collected at three or four different excitations are relatively routine using the laser scanning confocal microscope (LSCM) but more images at different wavelengths are theoretically possible given enough lasers and filter combinations. However, many such multiparameter images rapidly become complex and difficult to interpret when more than three of them are colorized and merged unless the images contain many regions of nonoverlapping structures, e.g., chromosomes painted with fluorescently labeled DNA probes *(4).* A convenient way of displaying two or three images is to use the red, green,

From: Methods in Molecular Biology, vol. 122: Confocal Microscopy Methods and Protocols
Edited by: S. Paddock Humana Press Inc., Totowa, NJ

and blue channels of an RGB color image where any overlap (colocalization of fluorescent probes) is viewed as a different additive color when the images are colorized and merged into a single three-color image (5,6).

A relatively simple method for displaying three color confocal images using Adobe Photoshop (Adobe Systems Inc., Mountain View, CA) and a Macintosh computer is described. Photoshop is available not only for Macintosh but also for PC and UNIX machines. It is now in its fifth version, and it has evolved over the years into an extremely powerful yet affordable image manipulation program, and is used extensively in the graphic arts and publishing industries.

Several applications of the three-color merging protocol in addition to displaying confocal images are outlined below including mapping color to depth in Z-series, mapping color to time in a time-lapse series, the production of red/ green or red/blue stereo anaglyphs from Z-series, and merging confocal and transmitted light images.

These digital methods are not confined to images produced using the confocal microscope, and can be applied to any digital images imported into Photoshop collected with many different kinds of imaging devices including both light and electron microscopes, and images that have been scanned from other sources. Photoshop is currently the program of choice for compiling these images from different sources, including the confocal microscope into figures for publication. It is used by most of the journals, and therefore it is relatively easy to transfer images directly to them with a reduced risk of errors at the final publication stage, and at a resolution that is preserved from the microscope itself.

Color hard copies of the images can be produced directly from Photoshop using a 35-mm slide maker (e.g., Lasergraphics), dye sublimation printer (e.g., Tektronix Phaser IISDX), or color laser printer (Tektronix Phaser 350 or 560). These devices are controlled directly from Photoshop as Plug-ins, which avoids transferring the images to yet other programs (7).

Images of fluorescently labeled specimens were collected using a Bio-Rad MRC600 LSCM or more recently an MRC1024 LSCM (Bio-Rad Laboratories, Life Sciences Division, Hercules, CA) using previously described methods although all of these methods are compatible with digital images collected with other confocal systems. Selected images were transferred from the host IBM microcomputer of the confocal microscope to a Power Macintosh 8500/150 (with 80 MB of RAM) microcomputer via Ethernet using Fetch 3.0.1. (usually as binary files).

Images were imported directly into Adobe Photoshop (version 3.0.5, and more recently version 4.0.1), either as RAW or as TIFF files. In the case of RAW files, the actual dimensions of the Bio-Rad confocal image must be entered, e.g., a typical image size is 768×512 pixels with a 76-byte header.

The header is specific to the Bio-Rad proprietary "pic" file format, and may vary with imaging systems from different companies. This information on the image files should be freely available from each confocal company. TIFF files are opened directly into Photoshop. For example, Bio-Rad collect their images in their own *.PIC format, which can be converted into TIFF or PICT files relatively easily. The *.PIC is occasionally confused with the Macintosh PICT files by some software.

If necessary, confocal image files may be converted to TIFF files using any one of a variety of programs such as Confocal Assistant (written by Todd Brelje, Dept. of Cell Biology and Neuroanatomy, University of Minnesota Medical School and available from the BioRad web page; www.microscopy.bio-rad.com), NIH Image 1.57 (Wayne Rashband at the NIH and available at no cost from http://rsb.info.nih.gov/nih-image/), or Graphic Converter available from LemkeSoft on the web at www.lemkesoft.de. Harvey Karten has published a most valuable web page on manipulating confocal images using NIH Image: http://rsb.info.nih.gov/nih-image/more-docs/confocals.html. This page also contains many useful details of the image formats from many different confocal imaging systems.

2. Materials

2.1. Computer Hardware

1. Powermac or PC with at least 32 MB RAM (the more RAM the better)
2. Backup devices: Optical drive, Zip drive, CD writer
3. Hard copy devices: Printers, Slide Writers

2.2. Computer Software

1. Adobe Photoshop (www.adobe.com)
2. Fetch (copyright Trustees of Dartmouth College)
3. NIH Image
4. Graphic Converter
5. Confocal Assistant
6. Database (*see* Chapter 23)

3. Methods

3.1. Importing Images Using Fetch

There are several options for transferring files from the confocal computer into PhotoShop. If the files are directly compatible with PhotoShop then it is a relatively easy process of transferring the files directly to the second computer usually via Ethernet or using a floppy, optical, or CD.

1. Turn on the confocal workstation and the Macintosh computer.
2. Open Fetch 3.0.1

Usually it is best to save a shortcut to the confocal computer in Fetch, which enables rapid entry directly into the confocal images directory so that long computer addresses and passwords are not typed in each time. For example the shortcut to our MRC1024 microscope looks something like this:

Host:Internet address of the computer
User ID:Computer internet name
Password:*******
Directory:c:\ls_user\paddock\images

3. Images are best transferred as binary files directly into a folder on the Mac hard drive. Check "binary" in the Fetch menu before highlighting the files to be transferred and then check "Get File," and place the files into a specific confocal images directory on the Macintosh computer (*see* **Note 1**).

3.2. Opening Confocal Files in Photoshop

Many commonly used file formats, e.g., TIFF files, are opened directly into Photoshop. We usually transfer the *.pic files from the BioRad confocal microscope as RAW files and convert them to TIFF files using the batch conversion feature of Graphic Converter. Alternatively *.pic files can be opened as RAW files in Photoshop. Often the proprietary software that comes with the microscope has a facility for converting files or a second program can be used. We are currently using Graphic Converter because it converts files from the confocal microscope in a batch mode using the Macintosh computer. To open RAW files in *.pic format:

1. Open the file in Photoshop by opening Photoshop and selecting File and Open. Select the confocal images folder. Note that the RAW image files may not be shown if "Show All Files" is not checked.
2. Select an image to open and fill out the image dimensions, e.g.:

Width	768 pixels
Height	512 pixels
Channel	1 for a single grayscale image
	3 for a three color image made up of three grayscale images.
Check	"Not-interleaved" for both single and triple images
Depth	8 bit
Header	76 bytes for BioRad pic file

3. Select OK and a message will say "specified image is smaller than file: open anyway?" This is good and ignore it — it pertains to the header.

3.3. Merging Grayscale Images

Images of multi labeled samples collected at different wavelengths can be merged by copying and pasting between different channels within Photoshop

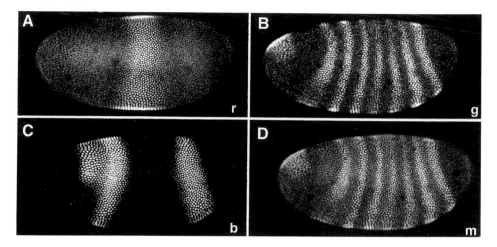

Fig. 1. A three-color image is made up of different images in the **(A)** red (r), the **(B)** green (g), and the **(C)** blue (b) channels of an RGB image. Different color combinations can be made simply by rearranging and copying the images to different channels. The merged image **(D)** is shown here in black and white and also in Color Plate I.

(Fig. 1). This method can also be used to rearrange the colors in a two or three -color image (*see* **Subheading 3.3.1.**).

Three-color images are produced in Photoshop by pasting each of the grayscale images from the confocal microscope into the red, the green, and the blue channels of an RGB Color image (*see* **Notes 2–4**). In the following description of the process, the location of various commands within the Photoshop program is included in (*italics*) after each operation.

1. Open three grayscale images in Photoshop.
2. Construct a blank RGB image (*File, New*) or copy from one of the grayscale images (*Select All, Copy, New, Paste*). The image must be the same size and pixel resolution as the three original grayscale images from the confocal microscope — in our case 768 × 512 pixels, and 72 pixels/inch. The size and resolution of an image can be determined and adjusted (*Image, Image Size*). This new RGB image should be black because the background in fluorescence images is usually very close to, or at black (black = 0, white = 255). Another color can be chosen for the background for cosmetic purposes, but a black mask for the actual region delineated by the image should be constructed, so that the new background color does not interfere with the information in the actual image.
3. Open the Channels palette (*Window, Palettes, Show Channels*).
4. The three grayscale images are now ready to be pasted into the newly-created RGB image. Select the first of the grayscale images (*Select, All*), and copy it into the required red, blue or green channel of the new RGB image (*Edit, Copy*), by clicking on the newly created RGB image with the mouse, and then on the required

channel in the Channels palette column. For example, for the red channel click on the red window in the column. Finally the grayscale image is pasted into the RGB image (*Edit, Paste*). The image will now appear in the Channels palette column in both the red and the RGB channel.

5. Select the second and third grayscale images, and copy and paste them into the green and blue channels of the RGB image using the same routine. The result of these manipulations is a single three color merged image, which is displayed by clicking on the RGB image line in the Channels palette column. There is no loss of bit depth information from the three original source images because three 8-bit images are merged into a single 24-bit image.

3.3.1. Rearranging Colors in an Image

1. Open the double- or triple-label image.
2. Select Window and click on Show Channels. Click on the arrow in top right of the Channels box. Select Split Channels which will split the image into three separate grayscale images comprising the red, green, and blue channels.
3. Click on the arrow in the top right of the Channels box. Select Merge Channels; select RGB Color and the Specify Channels, and then rearrange the images within the menu.
4. An alternative method is is to copy and paste images between the channels. This is especially easy for two-color images.

3.3.2. Double-Label Images

1. A double label image is simply an RGB image that is composed of grayscale image in each of the red and the green channels with a blank black image in the blue channel (**Fig. 2**). This method can be used to construct red green or red blue stereo pairs (*see* **Note 5**).

Extra colors can be included in double-label images by placing two versions of the same image into two of the three channels, with the second image of the merge in the third channel. For example, a purple and green image is produced by pasting the same image into the red and the blue channels to give purple, and the second image is placed into the green channel (**Fig. 2**). Additional color combinations are red and light blue where light blue comprises the blue and green channels or blue and yellow where yellow comprises the red and the green channels. Here overlaps of expression invariably appear white in the image because all three channels now contribute to the overlapping signals. These additional color combinations can be useful when making a multipanel figure of several double-label images where the expression patterns of more than three proteins are displayed in separate panels so that different colors are assigned to individual proteins.

3.3.3. Adding Color to a Single Grayscale Image

A grayscale image can be colorized by copying it into one of the three color channels of an RGB image and adjusting the colors using Levels. However, the

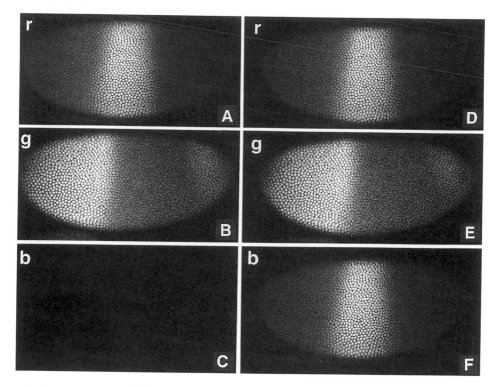

Fig. 2. Structure of double-label images. A red green image (**A–C**) is made up of two different grayscale confocal images in the red (A) and the green (B) channels with a blank black image in the blue (C) channel. A green and purple image is made by copying the same image into the red and the blue channels (**D, F**) with a different image in the green channel (**E**). (*See* Color Plate I, following p. 372.)

image then appears rather flat (*see* **Note 6**). If only one color is to be used then the appearance of the image can be much improved by displaying it using a dynamic color table (or look up table, LUT). This method works well for images with many different gray values. To produce a color table ranging from black over green and yellow to white:

1. Open the grayscale image.
2. Select Indexed Color (*Image, Mode, Indexed Color*).
3. Select Color Table (*Image, Mode, Color Table*).
4. Highlight a portion of the table by dragging from the top left square inwards.
5. Slide to color and select first color (usually black) — click OK.
6. Select second color, e.g., dark green — click OK.
7. Select a second portion of the color table by dragging from the last square of the former selection.

8. The first color selected is identical to the color in the first selected square. Click OK if you want produce a continuous color table.
9. Select the second color, e.g., yellow — click OK.
10. Repeat **steps 7–9** by highlighting the last portion of the color table and assigning color values from yellow to white.
11. The color table will be displayed — click OK and view its effect on your image.
12. Specific color tables can be saved for later.

3.3.4. Transmitted Light / Confocal Image Merging

Many confocal imaging systems have the facility to collect a nonconfocal transmitted light image, which usually comes into the imaging system on one of the confocal channels as a grayscale image. For example, in the Bio-Rad system the blue channel is used. An improved merged image of transmitted light and confocal images can be produced using the following method (Harvey Karten, *personal communication*).

The following description assumes that the transmitted light image (Nomarski, phase contrast, dark field etc.) is in the blue channel with the confocal images in the red and green channels or a blank image in the red or green channels for single confocal images merged with transmitted light images.

1. Make sure that the Mode = RGB.
2. Go to the blue plane (containing the transmitted light image). Copy the blue plane image into the Copy Buffer.
3. Add an additional layer. Call it "Transmitted." Set the opacity to 40%.
4. Paste the blue plane image into the new layer, and to make it gray, paste it into separate R, G, and B channels.
5. Return to the original image and erase the blue layer. Make sure that the blue plane is replaced with a black background.
6. Play with the opacity function of the Nomarski layer to vary the ratio of the confocal image and of the background layer.

3.4. Processing Images in Photoshop

The amount of processing required to "clean up" an image is highly subjective, and is best described as an art. This phase of the process is largely up to the individual investigator and also to the information and noise in the image. Of course an ideal image will not require any processing in Photoshop. It should be remembered that most of the actions performed in Photoshop can be done in a color darkroom except much more quickly and easily. It is often tempting to over process the images, and so it is advisable to always save a basic image, and then save subsequent files as versions of the basic image. A useful feature of Photoshop 5 is the ability to undoe more than one operation.

It is always faster and more reliable to work directly from images copied to the hard disk rather than working directly from a storage device or server, provided that the hard disk is not close to being full. It is also essential to keep track of the available RAM, and not to have too many programs open at the same time as running Photoshop. In general one should multiply the image size by at least three to estimate the amount of RAM needed to process images with Photoshop — this is because the image itself takes up memory and then Photoshop makes copies of the image each time it performs any operation on it. In addition it is a good idea to occasionaly restart the computer during a long session in order to defragment the memory.

Here we describe some of the basic operations that we routinely use on confocal images. We advise you to practice and to play with the program, and make a few test prints before deciding on any one method of manipulating the images.

1. Change the color levels in the resulting merged image using Levels (*Image, Adjust, Levels*) to adjust the red, green and the blue values in different channels of the image. The "Levels" option is extremely useful for adjusting the black and white points of the image. This is the first operation that is usually applied to an image. Here the channels are selected independently and the effects of changing levels can be viewed directly on the screen. Fine tune the brightness and contrast (*Image, Adjust, Brightness/Contrast*) of the image for presentation purposes.

Variations (*Image, Adjust, Variations*) is another useful feature in order to adjust brightness and contrast of an image. This feature displays several different settings together on the screen which enables the user to easily compare between the different effects, and to more easily make a choice.

2. Confocal images often benefit from a sharpening routine. In general "Unsharp Mask" (*Filter, Sharpen, Unsharp Mask*) works better than sharpening of the images (*Filter, Sharpen*), and is usually one of the last operations to be performed on an image (*see* **Note 10**). There are three variables, Amount, Radius and Threshold, to adjust when using the Unsharp Mask filter, and these can be adjusted interactively. Users should experiment with these on their own images. A good place to start with confocal images is a radius value of 2 and a threshold value of 1.
3. Especially useful features allow addition of graphics and compilation of the images into multipanel figures (*see* **Note 11**). Graphics are subsequently more easily edited if they are pasted into a separate layer in the image rather than permanently replacing pixels in the actual image itself. Graphics often appear much sharper when added on a different layer using a second program such as Adobe Illustrator. It is advisable to add the graphics within PhotoShop on a different layer so that changes, reformats for different journals or deletions can be made. In earlier versions of Photoshop this feature was not available so that text was added by actually replacing the pixels in the image. This meant that an unlabelled image had to be stored in addition to each of the labeled ones (*see* **Notes 12** and **13**).
4. Save the images to the hard disc of the computer in Macintosh format by selecting Save As (*File, Save As*). Images are usually saved in the TIFF format,

and LZW-compressed, which uses a lossless compression scheme. We have found that other compression methods such as JPEG compression can introduce artifacts which may be compounded each time a file is resaved. Images are eventually archived to optical disk or CD-ROM (*see* **Note 14**).

4. Notes

1. Files will not transfer if the Fetch preferences are not set correctly. The main symptom is that the file names are replaced by numbers, and the files cannot be opened. Always check the files after transfer before deleting them from the confocal computer or at least before making some form of backup!

2. Using PhotoShop it is a relatively simple task to experiment with various color combinations by rearranging the images into different channels. For example, in the triple-labeled *Drosophila* embryo, the green *hairy* stripes appear light blue in the blue *Kruppel* domain, and in a different color combination using the same images, the red *hairy* stripes appear yellow in the green *Kruppel* domain. This is achieved by rearranging the component grayscale images of the three color RGB image into different channels using Split Channels (*Window, Palettes, Show Channels, Split Channels*) and recombining the images using Merge Channels (*Mode, RGB Color*) or by copying and pasting the images between channels. Using the Specify Channels option the component grayscale images are assigned to different channels of the RGB image. The colors in the final image do not therefore always correspond to the actual colors of the specimen.

   The combination of colors within a three-color merged image is important for clearly conveying the biological information collected by the microscope. The true emission colors of two of the most commonly used fluorophores, rhodamine and fluorescein are, conveniently, red and green respectively, and overlapping domains of expression are yellow. Also some of the commonly used nuclear dyes that are excited in the near ultraviolet (UV), such as Hoechst 33342, emit in the blue. These are the colors observed by eye in a conventional epifluorescence microscope equipped with the appropriate filter sets for simultaneous double-label imaging — now available from most of the microscope manufacturers. However, the third channel in a triple-label sample prepared for confocal analysis usually emits in the far red, e.g., Cyanine 5, which is conveniently shown as blue in digital images whereas the real Cyanine 5 emission is often extremely difficult to visualize by eye and not so easily depicted in a digital image. By rearranging the grayscale images, the best combination of colors that conveys the maximum amount of information, and best color balance can be achieved.

3. In addition to displaying the relative distribution of up to three different macromolecules within cells, this method of combining the three images can be used as an alternative to 3D reconstruction for displaying depth information within a specimen. Using the LSCM, a series of images from different focal planes within the specimen is collected into a single file or Z-series. These images

maintain the *X, Y,* and *Z* registration from the specimen, and are the same size and pixel resolution. The simplest method is to extract the three images from such a Z-series, export them as single image files into Photoshop, and then merge them as before, so that the colors red, green, and blue are assigned to structures at different depths within the specimen. Alternatively, to gain a more accurate representation of the Z-series it is advisable to split the images into three groups and merge each group using the confocal software or NIH image, and subsequently combine them into a three color image in Photoshop.

4. In a similar way, single images of three different time points can be extracted from a time-lapse movie sequence file collected using the confocal microscope and three color merged. These files are identical in format to Z-series files except that time has now replaced the Z-dimension. Here, color differences are used to summarize changes in the positions of structures over time in a single image.

5. Two-color stereo anaglyphs can be constructed in Photoshop by placing one Z-series projection in the red channel and the second offset image in the green channel for a red/green stereo pair or in the blue channel for a red/blue stereo pair.

6. Single-color images often lack contrast when printed on a dye sublimation printer. The problem can be overcome, somewhat, by producing a single RGB image, as before, and adjusting the relative levels of the three channels so that there is information in all channels for printing (*Image, Adjust, Levels*) or using a color table applied to the grayscale images.

7. Photoshop can also provide a bridge to further manipulating images because the files are compatible with many other programs. For example sequences of confocal images of development have been manipulated using Photoshop, and subsequently transferred to a commercially available morphing program such as Elastic Reality (Avid Technology, Tewksbury, MA), and processed into short animated sequences of development *(8)*. These sequences can be further edited and compiled using Adobe Premiere or Adobe After Effects (Adobe Systems, Mountain View, CA), and viewed as a digital movie using QuickTime software directly on the computer or exported to VHS video tape for presentation purposes. For more detailed information on the production of digital movies the reader is referred to Chapter 24.

8. The presentation of time-lapse series for publication presents special problems of representing 3D and 4D data on a journal page, although with the popularity of digital publishing and the World Wide Web such video sequences can be accessed more easily.

9. In Photoshop 4 and now 5 the "Actions Palette" can be used to perform the same manipulations on a series of images.

10. Some confocal images may be contaminated with parallel lines from vibrations or a weak laser line in laser scanning systems or from a wobbling disk in the disk scanning systems. These images can be improved by using the "Blur" filter followed by Unsharp Mask.

11. In some instances a Z-series profile or *X–Z* section may appear flattened because the incorrect Z-step was selected. A relatively easy way of stretching such an image to the correct proportions can be performed in Photoshop using the image size feature (Image, Image Size). Here the image can be resized and stretched in

the Z direction by clicking on the "do not constrain proportions" box and entering the correct Z value while leaving the width the same.

12. The reorganization of panels in a composite image requires the selection of individual panels. An easy way to achieve this is by activating a channel with no image/colour information and to use the magic wand tool with a tolerance of zero. Alternatively, select the white spaces between the panels with the magic wand tool (*tolerance 0*), select Inverse (*Select, Inverse*) and then deselect unwanted panels by holding down the option key while selecting unwanted panels.

13. To align text labeling of panels in a composite image the text is first placed in the top left hand corner of each panel; it is then selected and the text is moved to the appropriate positions simultaneously.

14. It is not usually advisable to store image files on the computer hard disk or on a server because space can be limited on a multiuser confocal instrument and also hard disks are notorious for unpredicted crashes. It is therefore prudent to archive image files as quickly as possible after acquiring them. There are several options for archiving image files, including zip drives, optical drives and CD writers. Ideally copies of the most valued files should be stored in at least two different locations, and preferably with at least one copy in a fireproof safe.

Acknowledgments

The work of G. H. was supported by an EMBO fellowship. We thank Harvey Karten for advice on merging transmitted light with confocal images.

References

1. White N. S. (1995) Visualization systems for multidimensional CLSM images, in *Handbook of Biological Confocal Microscopy*, Plenum Press, New York.

2. Thomas, C., DeVries, P., Hardin, J., and White, J. (1996) Four-dimensional imaging: computer visualization of 3D movements in living specimens. *Science* **273,** 603–607.

3. Mohler, W. A., and White, J. G. (1998) Stereo-4-D reconstruction and animation from living fluorescent specimens. *BioTechniques* **24,** 1006–1012.

4. Schrock, E., du Manoir, S., Veldman, T., Schoell, B., Wienberg, J., Ferguson-Smith, M. A., Ning, Y., Ledbetter, D. H., Bar-Am, I., Soenksen, D., Garini, Y., Ried, T. (1996) Multicolor spectral karyotyping of human chromosomes. *Science* **273,** 494–497.

5. Waggoner, A. S., DeBiasio, R., Conrad, P., Bright, G. R., Ernst, L. A., Ryan, K., Nederlof, M., Taylor, D. L. (1989) Multiple spectral parameter imaging. *Methods in Cell Biol.* **30,** 449–478.

6. Paddock, S. W., Langeland, J. A., DeVries, P. J., and Carroll, S. B. (1993) Three-color immunofluorescence imaging of *Drosophila* embryos by laser scanning confocal microscopy. *BioTechniques* **14,** 42–48.

7. Kiehart, D. P., Montague, R. A., Rickoll, W. L., Thomas, G. H., and Foard, D. (1994) High-resolution microscopic methods for the analysis of cellular movements in *Drosophila* embryos. *Methods Cell Biol.* **44,** 507–532.

8. Paddock, S. W., DeVries, P. J., Buth, E., and Carroll, S. B. (1994) Morphing: a new graphics tool for animating confocal images. *BioTechniques* **16,** 448–452.

22

The Preparation of Stereoscopic 3D Illustrations of Confocal Data Sets for Publications and Slides

Gabriel G. Martins, Alan T. Stonebraker, and Robert G. Summers

1. Introduction

In this chapter we describe a simple method to prepare 3D illustrations from stereo-pairs of images from confocal sections. Methods for preparing both stereo slides and stereo images for publication using Microsoft's PowerPoint® 97 and CorelDraw® 8 are described. The parallel and cross-eyed viewing methods are also described and used to prepare the 3D illustrations. The procedure assumes previous preparation of stereo-pairs from the confocal data sets. Therefore, stereoscopy theory and methods of preparation of stereo-images are not described in this chapter but can be found in the operating manuals for the confocal microscopes and in more detail in *(1–4)*.

1.1. Considerations Before Starting

Images prepared for a book or for slide projection must be prepared with different values of pixel shift or rotation. In other words, if you try to project slides of stereo-pairs prepared to print in a textbook, they will appear too "deep" and distorted for the audience. This is due to the phenomenon of "distance deformation" that leads to Z-stretched images when viewed from long distances *(2)*. For slide projection, try to use half the pixel shift (or half the rotation angle) that you would use for publication stereos. Note that this is just a "rule of thumb" and does not necessarily apply to all images and compositions. Often, you will have to try different pixel shifts or rotations, and choose the best. If desired, the formulas used to calculate the rotation or pixel-shift necessary for correct 3D viewing can be found in **ref. 4**.

Make sure both images in a stereo-pair have exactly the same brightness/contrast and sharpness settings, as well as the same magnification.

From: *Methods in Molecular Biology, vol. 122: Confocal Microscopy Methods and Protocols*
Edited by: S. Paddock © Humana Press Inc., Totowa, NJ

1.2. Simple Methods for Viewing Stereoscopic Images

There are basically two approaches for viewing stereo-paired images in 3D. Some methods rely on the use of different colors for both images (e.g., anaglyphs of red-green/blue, producing monochrome 3D images), whereas others rely on the side-by-side display of both images. The latter is a preferable method because it provides a genuine separation of the paired images, allows the display of full-color 3D images, and avoids red/green vision problems.

One way of viewing these 3D images is to use a stereo viewer that causes the eyes to merge the paired stereo-images; unfortunately, these viewers are not always available. Another way is to merge the images yourself using the cross/parallel-eyed methods. It is well worth the time spent learning these viewing methods.

In this chapter, we introduce a simple method for mounting stereoscopic 3D images using Microsoft PowerPoint® 97 for preparation of slides and CorelDraw® 8 for preparation of plates for publication. These methods require the ability to either cross or diverge the eyes to merge the stereo-paired images on screen while working on them. If you are not familiar with these viewing methods read the instructions in the Methods section. Most confocal images can be viewed with either of the methods.

2. Materials

1. A stereo-pair originated from a confocal data-set (or other method)
2. Computer system with:
    Microsoft Powerpoint® 97 and/or Corel Draw® 8 (or similar drawing/presentation software)
    Slide printer and/or paper printer
3. 3D viewer for side-by-side prints (stereoscope)
4. Stereo slide projection system *(3)* with:
    Two slide-projectors mounted on an adjustable rig
    Silver screen
    Two polarizing filters for projectors
    Polarized 3D glasses
    3D viewers for "twin" slides
5. Accessories for mounting slides:
    Stereoscope (3D viewer for side-by-side prints)
    Small fluorescent light box
    35-mm Glass slide mounts (e.g., Gepe mounts)
    Alignment set for mounting 3D film "chips," including alignment grid
    Ruler

Note: Some of the materials can be acquired through the companies mentioned in the Websites section (page 400).

- Simulated stereo-view -
Location of stereo-image

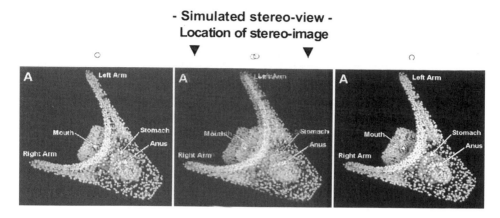

Fig. 1. Representation of the stereo-image resulting from merging the stereo-pair in **Fig. 7A** with the the cross-eyed method. The small circles on top of the image should be used to achieve correct stereo viewing. The goal is to fuse the two circles in the center (location of the stereo-image). Vertical displacement of the circles is corrected by slightly tilting the head and horizontal displacement by changing the degree of convergence of the eyes.

3. Methods

3.1. Viewing Stereo-Images

With paired stereo-images, the trick is to "persuade" each eye to see only its respective image. Once the mind accepts it, the viewer perceives a 3D view. In theory it sounds simple, but accomplishing a comfortable 3D view is difficult and takes practice and patience at first. There are basically two methods (use Fig. 7A as an example).

3.1.1. The Cross-Eyed Method

The idea is to cause the eyes to focus onto a close point (like the tip of the nose!), while looking at a stereo-pair. This way the right eye will focus on the image on the left side, and the left eye vice versa. When both images are focused, they fuse into a central image (*see* **Fig. 1**) that the brain interprets as a 3D image. To practice, try placing your index finger onto Fig. 7A and slowly bring your finger toward your nose. Keep focusing on the tip of your finger but also try to see what happens to the image in the background. It should start to look something like **Fig. 1**. You will perceive three images on the page, the one at the center being somewhat visually confusing.

There is a point where both the finger and the image on the center become less confusing (when your finger is at approx 4 inches from your nose). This middle image is the fusion of the stereo-pair and is gradually perceived as a 3D image. You may have to make small adjustments to your finger, by moving it toward and away from you very slowly.

Also, tilt your head slowly to either side to visually position the two middle circles at the same vertical level. These circles should merge into a single one when correct stereo is achieved. Do not let your eyes uncross, but allow them to focus on the center image (on the location of stereo-image; *see* **Fig. 1**), and remember…RELAX YOUR EYES, WAIT, BE PATIENT AND PERSISTENT! It may take a while, but with practice, your brain will get used to this unusual view and finally perceive the image three-dimensionally.

All the stereo images in this chapter are prepared for cross-eyed viewing, as we find this to be the easiest and most useful method to learn. However, be aware that the conventional display of stereo-pairs for publication is the parallel-eyed method, since it is the method used by most 3D viewers and stereoscopes. We also suggest that everyone obtain stereo-glasses (available from EM and 3D suppliers mentioned in the References section) to view published stereo-images in greater detail.

3.1.2. The Parallel-Eyed Method

Instead of crossing your eyes, the idea is to diverge them as when focusing on a distant object. This method is normally used to view the recently popular random dot stereograms, and is the most common in publications, although it is more difficult and restrictive then cross-eyed viewing. Hold Color Plate IV in front of your eyes and try to focus on a distant point ahead (such as the wall across the room). Then, keeping the eyes diverged, focus on the stereo images on the page. As with the cross-eyed method, you will see three images, the one at the center being somewhat confusing. Again…BE PATIENT…! This method is more difficult to master (see also the "Showing 3D Images" section).

Note: Viewing Fig. 7 (made for cross-eyed viewing) with the parallel-eyed method will produce an image that is inverted with respect to the viewer. This is effective only on a transparent specimen (e.g., a confocal image); otherwise, an unnatural pseudoscopic image is obtained.

3.2. Creating the 3D Illustrations

The procedure described can be easily adapted for any other program with the same capabilities. Be sure to read the section "Techniques and Tricks" (*see* **Notes 1** and **2**), even if you are an experienced computer user.

To prepare slides, start by formatting the page to twice the width of a normal slide (Width-height aspect ratio of 3:1), to accommodate both images of the stereo-pair, side by side. If you are preparing confocal images, the background should be set to black to conform to the unstained background of your confocal image. You should also use a *threshold* function after acquiring the confocal slices (or set the contrast) to create a pure black background on your image. This

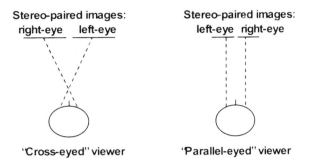

Fig. 2. Diagrammatic representation of the two standard stereo viewing conventions, as seen from above the viewer.

will prevent the appearance of visually confusing spots and blurs on your 3D image. Remember to perform exactly the same transformations on both images of the stereo-pair. To prepare plates for publication using CorelDraw®, create a graphic page of the desired size, according to the instructions for authors. Then, create a black box 5 inches wide to be used as background for the illustration.

To view the stereo image on the computer screen, the cross-eyed method should be used since the parallel-eye method does not work on stereo-images wider than 5 inches (2× the average interpupillary distance—approx 13 cm), while the width of computer screens may vary from 11 inches to 16 inches. With the cross-eyed method the right-eye image is placed on the left side of the stereo-pair and vice versa for the left-eye image (for the parallel-eyed method it is the opposite; *see* **Fig. 2**). Accordingly, insert the right-eye image on the doubled slide (pixel shift to the left or volume rotation to the right) and position it on the left half. Then repeat with the left-eye image and place it in right half. It should look somewhat like **Fig. 3**. After positioning, the images should be cropped to eliminate any areas that do not contain important objects or points of interest. When preparing stereo-pairs for publication, usable area is precious because the image size is constrained to 5 inches, because of the requirements for parallel-eyed configuration. If cropping is done, the images should be resized identically to fill the printable area, and repositioned.

To create a 3D image comfortable for viewing, it is essential to have perfect horizontal alignment of both images of the stereo-pair. To do this, first activate the *guides*. Then move one of the images (using the <ALT> key + mouse) to make homologous points of both images coincident with the horizontal guide (*see* Fig. 5). Notice that, in PowerPoint, up to eight vertical and 8 horizontal guides can be created. Repeat this process for several points from top to bottom on the stereo-pair (*see* Fig. 5). If the images are exactly at the same zoom (magnification) then all points should be simultaneously adjustable at the same

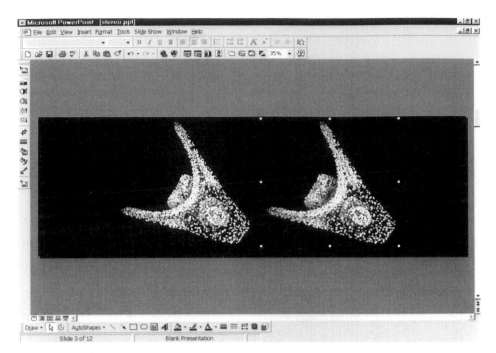

Fig. 3. Appearance of PowerPoint after inserting the stereo-images, and during align-ment of the stereo-pair. Note the use of a page with the double width of a normal page formatted for 35-mm slide. The left eye image is placed on the right side and vice versa.

horizontal level; otherwise, one of the images will have to be resized until all the homologous points coincide with the guides.

Now that the stereo-pair is created and aligned, the 3D image should be moved inside the "stereo-window," in order to make it appear more natural and comfortable to the viewer (**Fig. 4**). The "window" refers to the plane in which the image is printed; in this case, the computer's screen. Start by creating ver-tical guides on the center, on the sides and exactly in the middle of each frame, as shown in **Fig. 5**. This will produce a grid that can be seen in stereo and that is <u>exactly</u> in the plane of the "window." The nearest point in the object (in this case, the left arm of the sea urchin pluteus) should lie directly behind the grid. To relocate the object, move the frames of the stereo-pair closer (towards the center), causing the 3D image to move toward the back of the "window" (inside the screen). Moving them apart causes the 3D image to move toward the viewer, in front of the "window" (outside the screen).

Understanding this concept is not easy and also takes practice, but this method should make it easier. After preparing and aligning the stereo-pair, you can add labels and other objects to the 3D illustration.

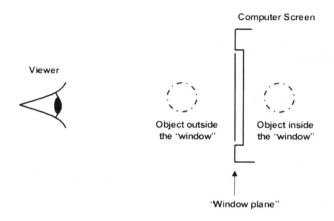

Fig. 4. Diagrammatic representation of the stereo "window." For example, in **Fig. 7B** the embryo is inside the "window" and the inset is outside.

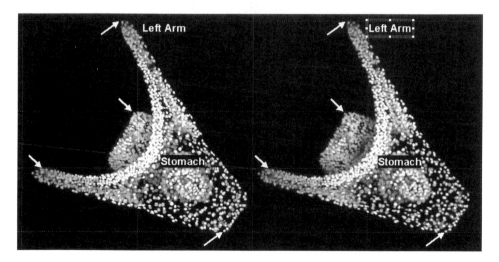

Fig. 5. Alignment of the stereo-pair and placing of labels in the 3D illustration. Note the horizontal guides (*dashed lines*) that are being used to align both images to coincide on the homologous points indicated by *arrows*. Vertical guides are placed exactly in the middle of each frame, creating a 3D grid at the level of the "window" plane. The 3D image of the embryo is placed behind this "window."

3.3. Labeling of Stereoscopic 3D Illustrations

Start by creating a textbox and placing it close to a point of interest on the left image. Then copy that textbox to the right image and place it close to the homologous point. Be sure that the textboxes are in perfect horizontal align-

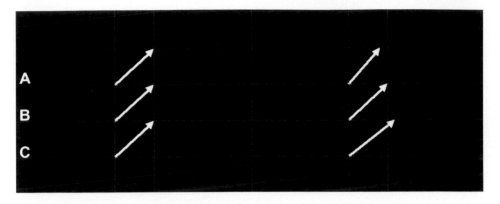

Fig. 6. 3D arrows. These arrows were created in PowerPoint as paired images. Then, the right arrows were resized to create a 3D stereoscopic effect. (**A**) Arrow pointing inside the "window." (**B**) "Flat" arrow. (**C**) Arrow pointing outside the "window." The guides (*dashed lines*) are used to show the relative sizes and positions of the stereo-paired arrows, and the "stereo-window" plane.

ment: if you are not sure, select both and use the *align* function, before viewing in 3D.

To position the label in a precise location in space, select one of the textboxes. While viewing the image in 3D (crossing your eyes), move the selected text-box to the left or right using the cursor keys (i.e., closer or farther apart, resulting in a label that moves towards or away from the viewer, respectively). It is best to start by placing the labels close to their homologous structures. However, there are restrictions to placing labels. For example, the label "stomach" in **Fig. 5** cuts into the image, and we cannot embed the label within the original bitmap (under the embryo's skin). In such a situation, it is best to use an arrow (preferably a 3D arrow) pointing to the structure of interest (*see* Fig. 7A).

To create arrows, start by drawing an arrow on the left frame connecting the label to the structure. Then, copy both the arrow and the label to the right image and position them on the appropriate homologous point. This will produce a flat arrow and label (**Fig. 6B**). To make it 3D, the right arrow will have to be resized to simulate the left eye perspective of an arrow in space. Once again, horizontal alignment should be maintained. Reducing the length of the right arrow will produce a 3D arrow pointing inside the window (**Fig. 6A**), while enlarging it will produce an arrow pointing outside the window (**Fig. 6C**). Care should be taken to minimize this resizing; otherwise, the 3D arrow will be confusing and difficult to perceive comfortably (being too "deep"). Creating 3D images that are too "deep" is a common mistake that leads to uncomfortable viewing.

After creating the right arrow, reposition it to match the location in space of both the arrow and structure. Colors can also be used to make the image more interesting visually (*see* Color Plate IV). Using the same principle, it is possible to create other objects in 3D such as insets, magnification bars, geometric figures, or circles/dots to help viewers achieve correct stereo, like in **Fig. 7A**. Any program that allows drawing 3D objects (such as 3D graphs in PowerPoint®) can be used to create the stereo-paired views of objects (by rotation or horizontal shift) that can then be used in the illustration. It is possible, e.g., to create a stereoscopic 3D graph (*see* Color Plate IV or **Fig. 8**), using these principles.

3.4. Preparing for Printing or Publication

3.4.1. Printing the Stereo-Slides for Projection

After finishing the stereo illustration in PowerPoint®, all the objects should be grouped on two separate frames (left and right) to prevent any accidental displacement of labels or arrows. To do this, select all objects of the left frame and group them, and then repeat the same operation for the objects in the right frame.

To prepare the slides for this projection method it will be necessary to create a new presentation containing one frame per slide (stereo-projection systems use two projectors simultaneously, for the left and right frames). While keeping the presentation with the illustration opened, create a new blank presentation and format the slides for 35-mm film (in Page Setup). Next, define a black background. Go back to the illustration, select the left grouped frame and copy it to the clipboard (Edit > Copy). Then go to the new blank presentation and "paste" that object into one slide. It is also a good idea to add a very small (almost unnoticeable) text label in the corner of the slide, briefly describing the slide and whether it is the left or right frame of the stereo-pair. Then create a new slide and repeat the same operation for the right frame.

Because the two frames are now separated into different slides, the image will have to be repositioned in the "stereo-window" using a slightly different method. Start by identifying the point that will be nearest to the viewer (in the case of **Figs. 5** and **7A** it seems to be the tip of the left arm). Activate the guides and position the left frame on the slide. Place one vertical guide exactly at that "nearest point" on the left frame. Then, change to the right frame slide and position the whole frame so that the same homologous point coincides with that guide, and be sure to maintain the horizontal alignment.

Finally, run a PowerPoint slide-show and by using <PageUp> and <PageDown> repeatedly you should see the objects in the illustration undergoing a slight rotation, and not stepping horizontally or vertically. The nearest point should look like a fixed point. It is also possible to use the slide sorter

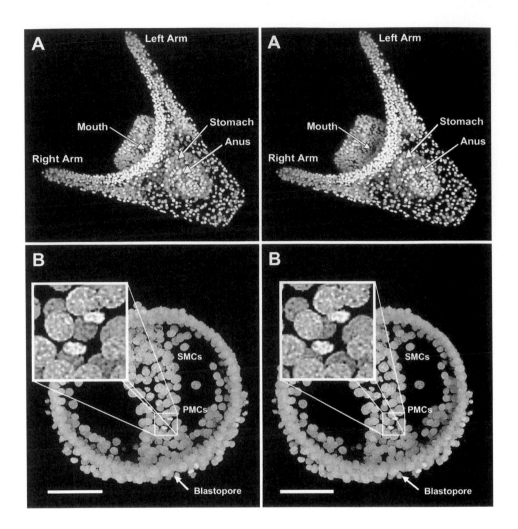

Fig. 7. Examples of 3D illustrations prepared in PowerPoint, from confocal slice-series of sea urchin embryos. The two circles on top of the figure can be used to obtain correct stereo viewing with the cross-eyed method. (**A**) Several anatomical features of the embryo are labeled with 3D arrows and labels. (**B**) The inset is outside the stereo "window," and the embryo is inside. Note the use of magnification bars in 3D. For a color version of these illustrations prepared for the parallel-eyed method, *see* Color Plate IV. Viewing these images with the parallel-eyed method will produce a pseudoscopic effect. Also, try to cross your eyes to see Color Plate IV for the same effect. Note that the pseudoscopic effect of the graph is very confusing.

view at 100% zoom to view both slides side by side, and confirm the stereo quality by crossing your eyes. Remember that the right-eye slide should be on the left side and vice versa.

This operation will save you time when trying to align the "film-chips" later, if the company that processes the film is careful enough to mount all the slides accurately (which very rarely occurs). We suggest the manual mounting of the "film-chips" using glass slide mounts (e.g., Gepe mounts available from vendors mentioned in the References section). During manual mounting it will be necessary to go through the process of realigning the chips to maintain the 3D image in the stereo "window" and at perfect horizontal alignment. When aligning the slides on a fluorescent box, keep a separator of 0.5 inch between both mounts. A transparent aligning grid and instructions to mount slides are available through Reel 3D Enterprises (*see* References section).

3.4.2. Printing Plates with Stereo-Pairs for Publication

The same principles apply for the preparation of stereo-plates for publication. However, in this case, stereo-images are printed side-by-side on the plate. Since convention dictates the publication of stereo-pairs for the parallel-eyed viewing method, you must switch the frames that were prepared with the cross-eyed method (*see* **Fig. 2**). Keep in mind that in this case, the stereo-pair width is restricted to 5 inches. Also, a white line separating the two frames should be added. *See* Color Plate IV (or **Fig. 8**) for a version of some of the stereo-images presented on this chapter, as prepared for publication.

To keep the stereo-image inside the "window" use vertical guides exactly in the middle of each frame (–2.5 inches and +2.5 inches from the center). Then move the frames apart to place the image inside the "window" or closer to bring it outside the "window." In this case, when you use the cross-eyed method to view a parallel-eyed stereo-pair, you will see an inverted 3D image—a pseudoscopic image (with confocal images, the pseudoscopic image is also easily perceptible, unlike stereo-images from SEM or the 3D graph in **Fig. 8C** or Color Plate IV C). The object will be inside the "window" if it looks "outside" and vice versa (try viewing **Fig. 5** with both methods)! You may often find situations where there is a compromise between using printable area and keeping the image inside the "window." Reducing the size of the frames may be necessary, for placing the image inside the "window." If detail is important, be aware that images outside the "window" can also be perceived (*see* **Figs. 7B**, **8B**, and Color Plate IV C).

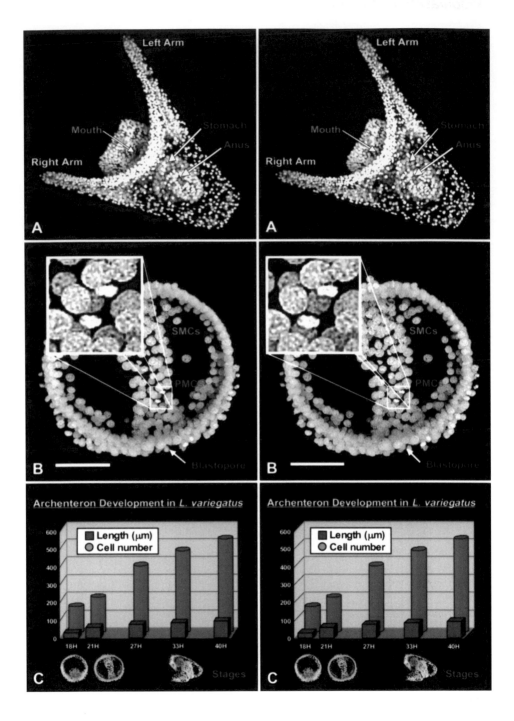

3.5. Showing 3D Images

3.5.1. Projection of Stereo-Slides

More detailed information about how to set up a stereo-slide projection system can be found elsewhere *(2,3)*. In brief, two slide projectors are necessary to simultaneously project the left and right frames on a silver screen. These two projected images have to be in perfect register (both horizontally and vertically superimposed). The left and right images are then separated with the use of polarized filters in front of the projectors' lenses, set at a 90° angle to one another (45° and 135°). Viewers also have to wear polarized glasses to allow each eye to see only the frame for that eye. These glasses and filters can be ordered through several companies, and information about some of these can be found on "Websites" at the end of this chapter.

3.5.2. Viewing Stereo-Images with Glasses and Stereoscopes

Other stereoscopes for slides include twin 35-mm 3D slide viewers, which can also be purchased in the same way. These are particularly useful for presentations of posters at scientific meetings, where no projection capabilities are readily available. They also constitute an easy way of assessing the final quality of your printed slides before projection, but keep in mind that deformation due to smaller viewing distance will be noticeable (the 3D image will look somewhat flat, i.e., "squeezed," when compared with the 3D image resultant from proper slide projection).

For viewing published stereo-images, unless the viewer is familiar with the cross/parallel-eyed viewing techniques, stereo-glasses have to be used. These

---

Fig. 8. *(previous page)* Example of 3D illustrations prepared for the parallel-eyed viewing method (standard method for publication). This figure is reproduced in Color Plate IV, following page 372. (**A**) 40-h *L. variegatus* sea urchin pluteus larva. The nuclei of cells are visible and several anatomical features are labeled in 3D. (**B**) 21-h *L. variegatus* sea urchin gastrula. The inset represents a magnification of a mitotic figure within the archenteron wall. Some mesenchymal cell types are labeled (SMC = secondary mesenchyme cells; PMC = primary mesenchyme cells). (**C**) Stereoscopic 3D graph representing the continuous increase in cell numbers that accompanies archenteron length increase during development in the sea urchin *L. variegatus*. The stereographs were created in Excel97® by rotation (8° apart) and by adjusting sizes for perfect horizontal and vertical match. For this type of illustration it is preferable to create only the graphs in Excel and then add the legends, titles, and other objects in PowerPoint® or CorelDraw®, using the methods described for preparation of slides and stereo-images for publication. Titles and legends were repositioned to create the 3D stereoscopic effect and stereo-images of embryos were added to depict some of the stages.

can be acquired in the same way as the polarized glasses. Their prices vary with quality and generally, plastic glasses are of poor quality. There are also tricks to "force" people to see stereo-images. A common method is to use a piece of card (letter size) and place it vertically between the two images. Then, the viewer can look down from the top edge of the card (with the nose touching the edge of the card) and allow the eyes to focus only on the respective image. It is a simple method for beginners to learn how to diverge the eyes, for the parallel-eyed method, and it creates a less confusing 3D image. There are also similar card-mounts with lenses available through Reel 3D Enterprises, Inc. and other suppliers.

4. Notes

4.1. PowerPoint® Techniques and Tricks

When pasting or inserting images, use the function "*Paste Special...Picture.*" Microsoft® applications handle bitmaps in the *"Picture"* format more effectively.

Use the <CTRL> key to copy objects: Holding the <CTRL> key while moving an object with the mouse will create a copy of that object and place it in the new location. This is most useful for creating the labels and arrows for the second image of the stereo-pair.

Use the <ALT> key for fine adjustments: Holding the <ALT> key while moving an object with the mouse will move the object freely and not within the predefined grid. This function is essential for the fine adjustment necessary for horizontal alignment and comfortable stereo viewing.

Use the <SHIFT> key for multiple selections: It is possible to select two or more objects without having to select an area. Pressing the <SHIFT> key and then clicking on several objects will enable multiple selection. This function is useful for moving paired labels of the stereo-pair at the same time, and ensures that they remain aligned.

Use the <SHIFT> key to move objects in one direction: Objects can be moved horizontally or vertically pressing the key <SHIFT> while moving them with the mouse. This is useful to maintain horizontal alignment when copying labels.

Group objects that are to be handled together: Use the command Draw > Group to group a multiple selection of objects into a single object.

Use *Guides* to align objects: Activate guides with <CTRL> + G, or menu View > Guides. They are most useful to verify on screen, the alignment of objects, and are not printed with the illustration (*see* **Fig. 5**).

Move objects with the arrow keys: After selecting an object (as when clicking on the border of a textbox) that object can be moved with the arrow keys on the keyboard. However, this method does not allow for finer adjustments in PowerPoint.

To align objects:

1. Select the objects you would like to align.
2. Go to the Align menu via DRAW > Align or Distribute, and then align on *top* or *bottom*.

Turn off the automatic spell-checker: The red lines that recent versions of spell-checkers use to underline misspelled words are confusing and often prevent correct stereo viewing.

When resizing objects, be sure to preserve the original aspect ratio by using only the corner handles of an object + <SHIFT> key, and never the side handles.

4.2. CorelDraw® Techniques and Tricks

Use the <CTRL> key to constrain object movement in one direction: Hold down <CTRL> while selecting and moving an object.

Use the standard COPY and PASTE functions to create multiple instances of objects such as labels or arrows. Be sure to constrain movement of the copied object in the horizontal direction if it is to be paired with the original object.

Use the arrow keys to make fine adjustments to an object's position. The default "nudge" setting is too coarse in CorelDraw; to change it, go to the menu TOOLS > Options > Workspace > Edit. Set Nudge to 0.02 inches, or to a comfortable setting. This function is essential for the fine adjustment necessary for horizontal alignment.

Group objects that are to be handled together: Use <Ctrl> + G to group selected objects.

Create guides to check alignment of homologous points: Click and drag a horizontal guide from the horizontal ruler onto the illustration.

To align objects:

1. Select the objects to be aligned.
2. Go to the Align menu via ARRANGE > Align & Distribute, or the "Align…" button located on the upper right portion of the toolbar. <CTRL> + A is the shortcut for this option.
3. Select the appropriate alignment button and press "OK." Align TOP, CENTER, or BOTTOM are the settings used to align stereo elements in the horizontal direction.

Use the <SHIFT> for multiple selections: It is possible to select two or more objects without having to select an area. Pressing the <SHIFT> key and then clicking on several objects will enable multiple selection. This function is useful for moving paired labels of the stereo-pair at the same time, and ensures that they remain horizontally aligned.

When resizing objects, be sure to preserve the original aspect ratio by using only the corner handles of an object and never the side handles.

Note: In CorelDraw®, be aware that text placed on a stereo-pair sometimes appears "jumpy" in stereo, i.e., the characters in a word do not all appear to be within the same plane. This is only a display anomaly, and should not appear in the printed piece.

Acknowledgments

Figures 1, 3, 5, 7A, 8A were adapted from Martins, G. G., Summers, R. G., and Morrill, J. B. (1998) *Dev. Biol.* **198,** 330–342, with permission of Academic Press Inc.

References

1. Chen, H., Swedlow, J., Grote, M., Sedat, J., and Agard, D. (1995) The collection, processing and display of digital three-dimensional images of biological specimens, in *Handbook of Biological Confocal Microscopy*, (Pawley, J., ed.), Plenum Press, New York, pp. 197–209.
2. Ferwerda, J. G. (1987) *The world of 3-D, A Practical Guide to Stereo Photography*, 3D Book Productions, The Netherlands, 300 pp.
3. Wergin, W. P. and Pawley, J. B. (1980) *Recording and Projection of Scanning Electron Micrographs.* SEM/I, SEM Inc., AMF O'Hare, IL 60666, pp. 239-250.
4. White, N. (1995) Visualization systems for multidimensional CLSM images, in *Handbook of Biological Confocal Microscopy*, (Pawley, J., ed.), Plenum Press, New York, pp. 211–254.

Websites

Stereoscopy.com
http://www.stereoscopy.com
A good starting point for general 3D resources. Numerous links.

Reel 3-D Enterprises, Inc.
http://www.stereoscopy.com/reel3d/index.html
Purveyor of 3D instruction books, images, and equipment by mail order.

How to make 3D pictures by computer
http://www.stereoscopy.com/3d-info/index.html
Basic theory about stereoscopy and principles for preparing stereo-images from photographic pictures.

Stereoscopy and Illusions
http://www.lhup.edu/~dsimanek/3d/3dpage.htm
Various links to related sites, documents on 3D, and optical illusions.

3-D Scanning Laser Confocal Microscopy
http://www.cs.ubc.ca/spider/ladic/confocal.html
An extensive site for confocal microscopy resources, from specimen preparation to volume visualization. Includes postings of upcoming courses and meetings related to confocal microscopy and 3D. A must for the confocal microscopist.

Confocal Assistant
ftp://ftp.genetics.bio-rad.com

An anonymous FTP site where Confocal Assistant® V4.02 can be downloaded. This program can be used to create the stereo-pairs from confocal datasets, using both the pixel-shift and rotation methods.

Alan Gordon Enterprises, Inc.

http://www.A-G-E.com

Supplier of a large selection of stereoscopes. The catalog can be requested at this site or at: 1430 Cahuenga Blvd., Hollywood, CA 90029, USA.

23

Information Management of Confocal Microscopy Images

*Traditional Text-Based Databases
and Image Gallery Databases*

Harvey J. Karten

1. Introduction

Imaging methods, whether they use film, videotape, or digital capture methods, are only of use if you are able to readily categorize and retrieve the information and images that are produced. Modern imaging techniques are capable of generating large numbers of images of various dimensions and content. Major obstacles confronting the confocal microscopist include how to keep records of (1) how the various images were collected, (2) where they are stored, (3) what they look like, (4) the important features of the images, and (5) how to keep track of modifications of these images. The digital nature of confocal images lends them well to the use of computer-based databases for information storage, classification and retrieval. This chapter deals with these problems and the application of simple databases for maintaining and tracking images and information of confocal image datafiles.

This chapter will outline a few useful and inexpensive database tools that prove of value in the specific context of confocal imaging. A brief introduction to the use of image databases for keeping track of confocal images was discussed in my chapter on Use of NIH-Image in Fluorescence and Confocal Microscopy. This chapter can be downloaded from my website at http://www-cajal.ucsd.edu. Confocal microscopists may find that chapter useful in dealing with transferring files between different computer systems, and for post-acquisition processing of confocal images using NIH-Image.

From: *Methods in Molecular Biology, vol. 122: Confocal Microscopy Methods and Protocols*
Edited by: S. Paddock © Humana Press Inc., Totowa, NJ

1.1. What is a Database?

Databases are collections of information stored in an orderly manner, designed to sample and facilitate the retrieval of information based on logically structured queries. Databases are widely used in government, hospitals, industry, and laboratories. Every scientist relies heavily on databases such as that provided by the National Library of Medicine (Medline), the phone company directory, and the table of contents and index of a book. However, surprisingly few microscopists take advantage of modern computer based database tools for keeping track of their images.

1.2. How is a Database Organized?

A database File contains a grouping of Records. Each Record contains the information about a single image or series of images. A Record contains Fields. The various types of Fields may include designated categories of information, such as the particular microscope used, the magnification, laser lines used, dyes, location of file storage, animals, plane of section, a copy of the image itself, etc.

1.3. What Do I Want to Store in My Database?

The types of information of importance include information about the nature of the picture, the manner of its collection, and a gallery of images to allow rapid review and selection of particular images. An ideal database should also be integrated with the software used for image acquisition on the confocal microscope. The database should then be available for use on other computers, independently of the confocal microscope. Although some confocal microscope systems do provide this type of information, such as the new Zeiss LSM 510, it is not a standalone program, and cannot operate separately of the program specifically associated with the CLSM.

2. Two Types of Databases

I will deal with only two simple types of databases: (1) traditional text oriented databases with limited ability to display images and (2) Image Gallery databases.

Traditional text-based databases, can be used to store information about the nature of the images collected, the manner of collection, names of the files, date, comments and similar data, and a thumbnail image of the microscope image. There are a number of inexpensive programs that can be used for this purpose, including FileMaker Pro® (FileMaker Inc.) and Access® (Microsoft® Corp.). These types of programs are mainly used to store textual information, with limited ability to also store images. These programs have very limited ability to simultaneously display large numbers of images. This type of database is particularly useful in providing tools for searching for specific textual information. They are described in further detail elsewhere in this chapter. These programs are most effective when used at the time of initial collection of confocal microscope images. Additional comments can subsequently be added to each record.

A sample Record of an image file is shown in **Fig. 1**. This contains information in a FileMaker Pro Template. The format of the Template is easily constructed and can be adapted to the preferences of individual users.

Individual Fields may include a group of predetermined values, such as various planes of section, as shown in the "Pull-down" menu shown in **Fig. 2**. This facilitates data entry, and establishes a rigorous protocol to ensure that all desired information is recorded by the user.

One of the most useful features of a program such as FileMaker Pro is the ease with which you can display the data in alternate formats. You can easily modify the layout, and instantaneously switch from one layout to another, in order to concentrate on a particular aspect of your data. You can make a layout that will provide a tabular summary that mainly emphasizes the Title, the date of collection, the plane of section, the species, the dye used, or any other feature. Several alternate columnar layouts are shown in **Figs. 3–5**. The lengths of the columns are only limited by the size of your computer monitor.

This type of database is mainly based on textual content. However, Fields for images can also be inserted into the Record. However, the nature of programs such as FileMaker Pro limits their suitability for gathering and storing large numbers of images. To limit the size of such files with many images, the original image is modified to produce a smaller scale "thumbnail" image. The process of generating thumbnails and then storing them in the correct location can be very tedious and time consuming.

Image Gallery databases are designed to present large numbers of "thumbnail" images on the computer to enable the viewer quickly find a desired image based on visual content, rather than on descriptive phrases. These programs automatically scan a disk for files containing specified types of images, such as Bio-Rad *.PIC files, then generate and store thumbnail images of files, the size of the file, the location and the format of the file, and various parameters pertaining to the original mode of collection of the images. The user can then view a "gallery" of thumbnail images from many files (*see* **Fig. 6**), zoom in on a selected file and examine it in detail, search for selected files based on keywords, date, file names, etc. This type of program typically is used hours or days after the original confocal image is collected. Most of these types of programs were developed in relationship to graphics arts as well as satellite imaging. However, Image Gallery databases provide only limited ability to also

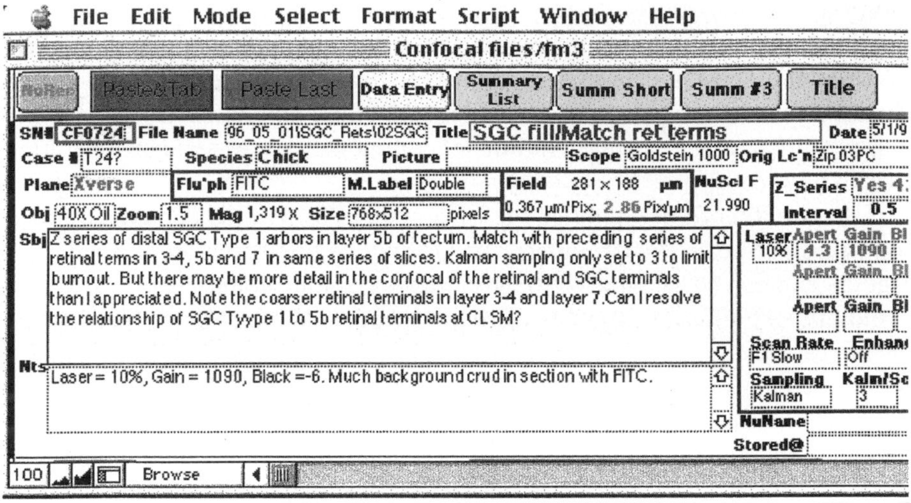

Fig. 1

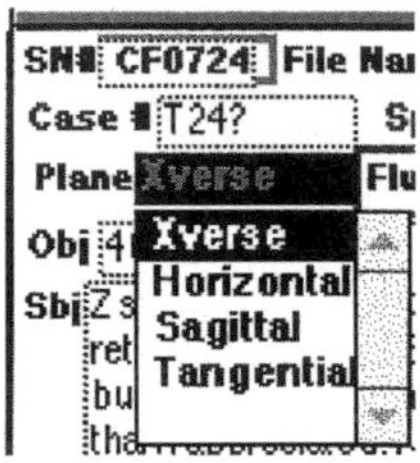

Fig. 2

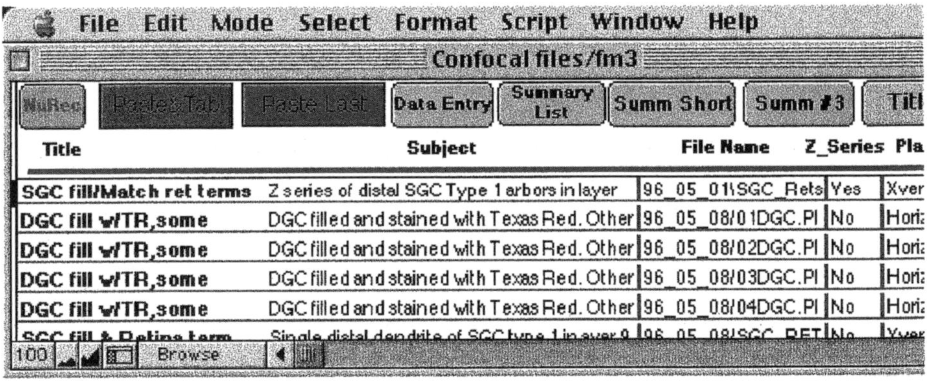

Fig. 3

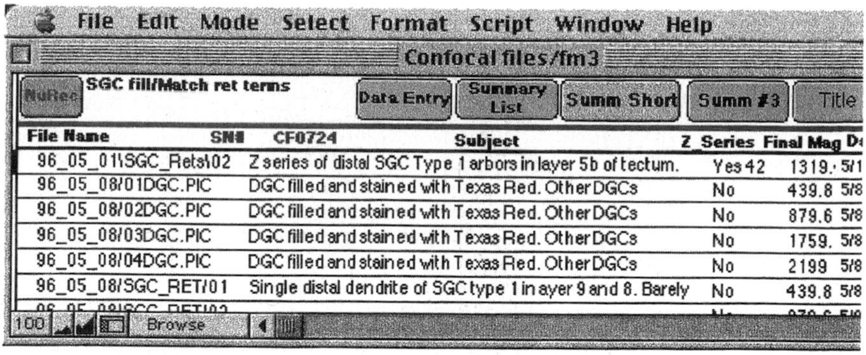

Fig. 4

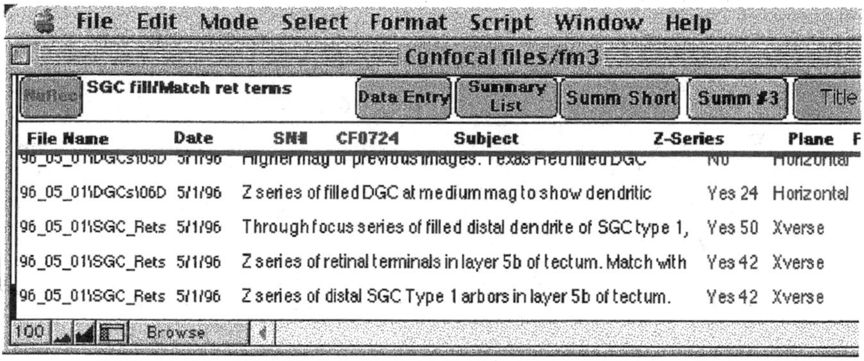

Fig. 5

store detailed information regarding method of data collection, detailed descriptions of content, relationship to other images, interpretation and analysis of the image content, etc. All such information is usually stored in a single field of very limited size. As in the first type of database, searching the database depends upon the use of text based descriptive features. Some representative programs of this type include Cumulus 4.0 (Canto Software), Portfolio™ 3.0 (Extensis Corp.), and ThumbsPlus® 3.1 (Cerious® Software, Inc.). Although all three of these Image Gallery programs perform similar operations, they differ is some important ways. The relative advantages and disadvantages of each of these programs are summarized later in this chapter.

Various lower cost programs have been developed for use with Photo CD storage. These programs do not seem to have the capability required for our purposes, but may be helpful when you are first familiarizing yourself with these types of programs.

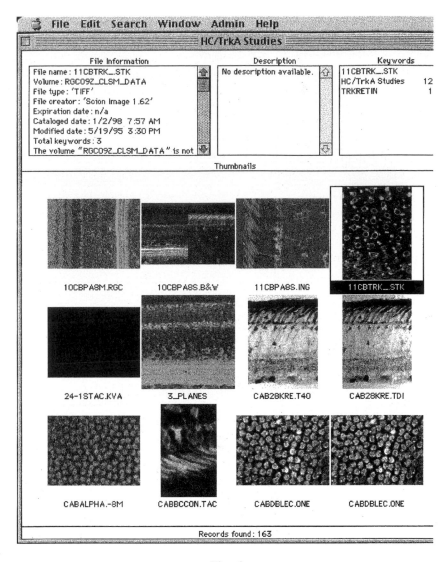

Fig. 6

3. Where Should Data About the Image be Stored: *With the Image, in a Separate Database, or in Both Places?*

Some of the software provided by confocal microscope manufacturers for collection and storage of images also includes the capability of storing a limited amount of information about the method of data collection and for comments about the image. This information is stored either in the same file as that containing the image itself, or in an associated file.

Bio-Rad includes some basic information about each image in the file itself. Leica stored similar information in a separate text field with a similar file name, but with a distinct three-letter extension indicating that it contained information about the collection parameters. When examining the file with Leica software, the file containing the text field is also displayed. However, if the image is viewed with other software, such as Adobe® Photoshop®, this information is not displayed. Olympus (FluoView) stores this information in the image data file as a specific TIFF field, although it is only available when the full data file is opened. Although the collection parameters were automatically stored with the file, most users do not include comments within the same file about the nature of the material, interpretation of the results, related experiments, etc. Perhaps they assume that they will write this down later, or that it will be self evident, though most commonly, I forget to enter such information.

The data in the original files may include information about the magnification, gain, offset, pinhole, resolution in micrometers/pixel, filters, look-up table (LUT), mode of accumulation (Kalman, accumulate to peak, etc.) and some limited textual information about items of note in the image. However, this information is generally only available as each file is individually opened. There is no simple means of examining all this information about all your images at a single time. Thus, you could not easily and rapidly search for all your images stained with a particular antibody, or of a unique preparation.

Various filters (a "filter" is software that translates a file from, e.g., Bio-Rad format to Photoshop format) have been developed that allow Bio-Rad, Leica, Zeiss, and Olympus files to be transferred to more general platforms (Mac, PC, and SGI) and more common file formats (Photoshop, NIH-Image). However, this transfer invariably resulted in loss of the collection parameters. Photoshop and NIH-Image do not allow hidden textual information to be stored in their files. Thus, if you transfer your file to either of these programs, the information is no longer associated with the file. Although you may be able to return to the original files to extract that information (see below), that is an extremely inefficient way of storing and comparing information.

Unfortunately, the various confocal microscope manufacturers have been unable or unwilling to develop a single uniform standardized file format for dealing with these problems. Different confocal microscope manufacturers have chosen to use different data file formats.

These formats vary depending upon several factors:

Are the collection parameters and users notes stored in the same file as the image data itself?

1. Do the image files also contain data about collection parameters, your comments, scale bars as separate from the actual image (versus a scale bar that has been written directly on the image with resultant loss of the underlying data)?

2. The nature of the image (single, double or triple label, Z-series of single label, Z-series of double label, Z-series of triple label, etc.)

In view of this deficiency, I strongly recommend that information about each image also be stored in a separate database file containing information about all your images.

4. External Databases, Both Image Gallery Type and Traditional Records/Fields Types to Store all Collection Parameters

The second way of handling this information is to record the same information in a traditional database, consisting of multiple Records with multiple Fields. Each single record is used to store information about each image set. Each Field contains distinct categories of information pertaining to each record, e.g., an alternate title that more suitably describes the content of the file, the lens used, the filter, plane of section, stain, and so forth. Users can easily add additional Fields to their records, reflecting their personal needs and preferences. Modern database software also provides means of storing one or more images in each Record, several formats for displaying the same data, sorting the data, searching for specific information, etc.

A well-designed database template will also impose some discipline on your record keeping, and prompt you to insert all required information.

5. Text-Based Database

Using either FileMaker Pro or Microsoft Access, you should make a template reflecting the type of information you find important for your research. A sample record is shown in **Fig. 1**.

5.1 What File Name Shall I Call this Image or Set of Images?

It is often easier to take a good picture than to think of a file name for it. This is particularly bothersome when you have to use a DOS based file naming convention (8 + 3 = filename.ext), but also taxes my imagination even when using a Mac- or Windows NT®-based system with the option of long file names.

If you rely heavily on the file name to indicate the content there are potentially serious disadvantages. If you capture a triple labeled immunofluorescent stained section of the peripheral retina from animal NR12345, at 725× magnification, and only have a limited number of files, you can make up some name on the spur of the moment, such as "triplRet.tif," and hope that you remember what filename you used, what it contains, and where it is stored. As the number of images increases, you may find your creativity in designing file names is severely taxed. You can have a nonspecific descriptor, but a file name such as MX4bixi.tif is not much better. If you take one image a month, you can live with either approach. If confronted with a confusing group of names on your computer, you will likely consider shifting to hand drawn representations of your slide. Each person must develop their own strategy. My current solution is heavily oriented toward computerized databases. It solves some problems, but not all of them.

6. Using a Database to Designate a File Name

In addition to a Field that automatically assigns a serial number to each image, a database program can generate a formal file name (e.g., BR805689.TIF), or you can choose a less anonymous title, such as CalRet01.tif to indicate that this is an image containing a stain for calretinin, and is the first image in the series.

Image Title: This is a longer and generally far more comprehensible title that conveys more information about the file content, e.g., "Monkey retina stained for calretinin and calbindin: Z-series in transverse plane."

Several additional fields may consist of Pull-down lists that include commonly used values, as shown in **Fig. 2**. These may include: plane of section (transverse, horizontal, sagittal); objective lens used (10X, 20X Plan Apo, 40X oil, etc.); fluorescent dyes used; the number of sections in a Z-series and their spacing, scanning speed, PMT, gain, offset, special filters, pinhole diameter, laser power, etc. Some of this information may be redundant of information also stored in the original image file, but is far more readily retrieved from this database. In addition, you can also insert a thumbnail image of the original confocal image to help identify the content of the file.

The most valuable Field in the Record, however, is that containing a detailed description or commentary on the image, its relevance to your work, comments about unusual aspects of the manner of preparation, why this particular image was saved, what you learn from it, etc. You can also add technical comments regarding errors in data collection, irregularities in specimen quality, etc. This capability is particularly important, and requires that the Database program support variable length fields. All Fields in FileMaker Pro are variable length fields, and are fully indexed, so you can continue to add comments into a field without constraints on their length. Once indexed, you can then search for particular phrases or words. Many database programs, including Microsoft Access 97, require that the user define the length of each field, frequently limiting it to 255 bytes. Only a single variable length field is permitted in each record, and the contents are often not indexed. Thus searching for particular words in this field cannot be accomplished. This is a severe constraint when you want to enter lengthy comments about an image.

However, many users may find that both FileMaker Pro 4.0 and Microsoft Access 97 are effective for maintaining records of data collection. While both also provide Relational Database tools, I suggest that initially you keep things simple and use a flat-file, nonrelational organization. FileMaker Pro has the advantage that the files are fully cross-platform compatible; i.e., the same file can be examined without difficulty on both a PC and a Mac PowerPC. Microsoft Access is not presently available in a PowerPC version. If you only work on a Windows NT platform, either program should prove satisfactory.

I suggest that you keep a copy of the Database Program open as you collect your original images, and update your file as you collect each image or set of images.

7. Image Gallery Databases

As mentioned previously, there are several image database programs specifically designed to store, display, and retrieve images. They all have limited text display capability, but provide a useful "Gallery" display of thumbnail images. FileMaker Pro does not provide this type of utility. However, you can generate Thumbnail images that can then be pasted into FileMaker. One of the most useful qualities shared by all the Image Gallery Programs, including ThumbsPlus, Cumulus, Portfolio, and Multi-Ad Search® (Multi-Ad Services), is their ability to automatically scan your hard disk and generate thumbnail images of each file. Each of these programs also allows you to store images contained on removable media, such Jaz, Zip, and Mag-Optical disks.

In addition, Canto Cumulus permits insertion of the thumbnail image into a FileMaker Pro database. The new editions of Extensis Portfolio and Thumbs-Plus promise to provide similar functions in the PowerPC version. All of these programs also allow users to view a full screen image of the thumbnail, and to directly open the original file in the program that produced the original image. These programs also provide information about the size of the files, its current storage location, file format, image dimension, and keywords. None of these programs will generate a simple Z-projection of an extended Z-series. The thumbnail of a Z-series is invariably the first image in a Z-series. Because the first image is usually at the margin of the most interesting data, the thumbnail is of little value. If your set of images have been transferred to an image processing program, such as NIH-Image, I suggest that you make a Z-projection of the stack of images in a Z-series, and place that as the first image in the stack. Label the image so that you not confuse it with the main part of the stack when you perform further operations on the Z-series. The resulting thumbnail image will be that of the first image in the stack, containing the Z-projection.

In addition to the Image Gallery display of large numbers of thumbnail images, these programs also permit the user to view a List of images, and the thumbnail image of a single file at a time, as shown in **Fig. 7**.

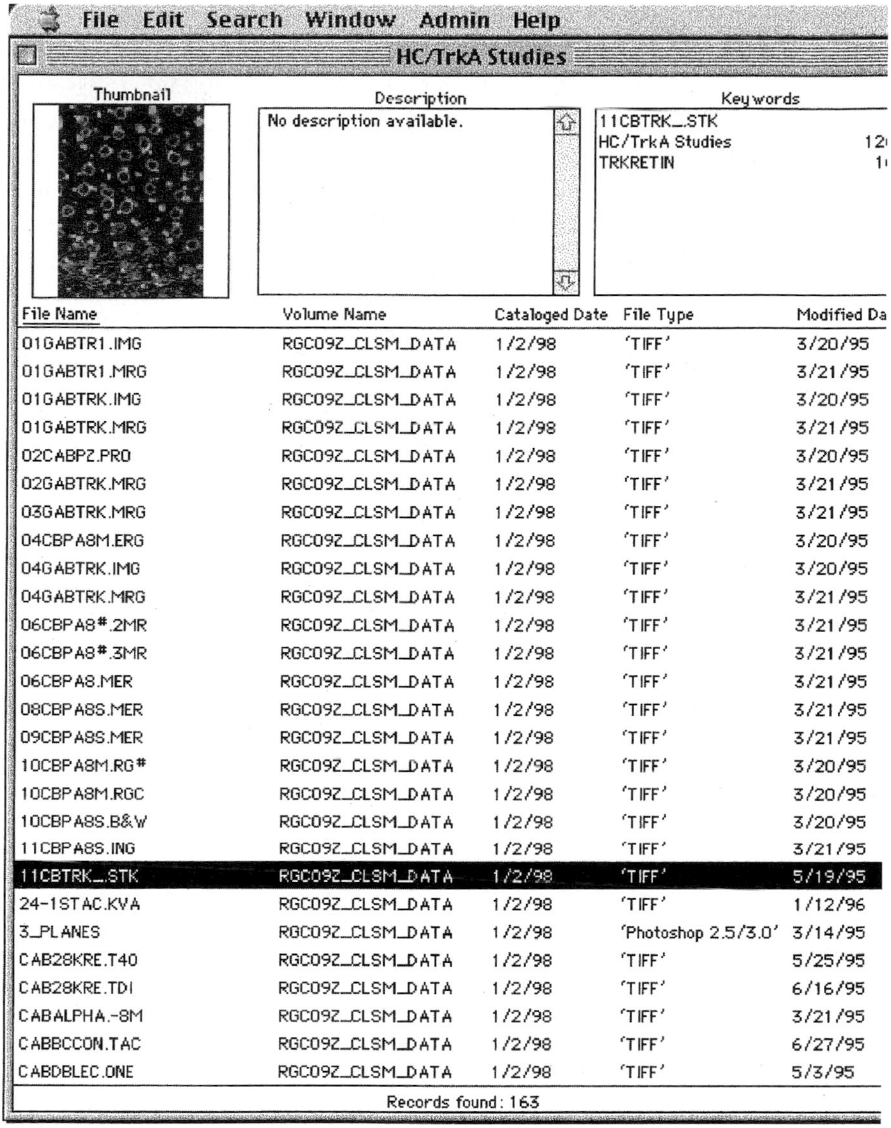

File   Edit   Search   Window   Admin   Help

HC/TrkA Studies

Thumbnail	Description	Keywords
	No description available.	11CBTRK_.STK HC/TrkA Studies 12 TRKRETIN 1

File Name	Volume Name	Cataloged Date	File Type	Modified Da
01GABTR1.IMG	RGCO9Z_CLSM_DATA	1/2/98	'TIFF'	3/20/95
01GABTR1.MRG	RGCO9Z_CLSM_DATA	1/2/98	'TIFF'	3/21/95
01GABTRK.IMG	RGCO9Z_CLSM_DATA	1/2/98	'TIFF'	3/20/95
01GABTRK.MRG	RGCO9Z_CLSM_DATA	1/2/98	'TIFF'	3/21/95
02CABPZ.PRO	RGCO9Z_CLSM_DATA	1/2/98	'TIFF'	3/20/95
02GABTRK.MRG	RGCO9Z_CLSM_DATA	1/2/98	'TIFF'	3/21/95
03GABTRK.MRG	RGCO9Z_CLSM_DATA	1/2/98	'TIFF'	3/21/95
04CBPA8M.ERG	RGCO9Z_CLSM_DATA	1/2/98	'TIFF'	3/20/95
04GABTRK.IMG	RGCO9Z_CLSM_DATA	1/2/98	'TIFF'	3/20/95
04GABTRK.MRG	RGCO9Z_CLSM_DATA	1/2/98	'TIFF'	3/21/95
06CBPA8#.2MR	RGCO9Z_CLSM_DATA	1/2/98	'TIFF'	3/21/95
06CBPA8#.3MR	RGCO9Z_CLSM_DATA	1/2/98	'TIFF'	3/21/95
06CBPA8.MER	RGCO9Z_CLSM_DATA	1/2/98	'TIFF'	3/21/95
08CBPA8S.MER	RGCO9Z_CLSM_DATA	1/2/98	'TIFF'	3/21/95
09CBPA8S.MER	RGCO9Z_CLSM_DATA	1/2/98	'TIFF'	3/21/95
10CBPA8M.RG#	RGCO9Z_CLSM_DATA	1/2/98	'TIFF'	3/20/95
10CBPA8M.RGC	RGCO9Z_CLSM_DATA	1/2/98	'TIFF'	3/20/95
10CBPA8S.B&W	RGCO9Z_CLSM_DATA	1/2/98	'TIFF'	3/20/95
11CBPA8S.ING	RGCO9Z_CLSM_DATA	1/2/98	'TIFF'	3/21/95
11CBTRK_.STK	RGCO9Z_CLSM_DATA	1/2/98	'TIFF'	5/19/95
24-1STAC.KVA	RGCO9Z_CLSM_DATA	1/2/98	'TIFF'	1/12/96
3_PLANES	RGCO9Z_CLSM_DATA	1/2/98	'Photoshop 2.5/3.0'	3/14/95
CAB28KRE.T40	RGCO9Z_CLSM_DATA	1/2/98	'TIFF'	5/25/95
CAB28KRE.TDI	RGCO9Z_CLSM_DATA	1/2/98	'TIFF'	6/16/95
CABALPHA.-8M	RGCO9Z_CLSM_DATA	1/2/98	'TIFF'	3/21/95
CABBCCON.TAC	RGCO9Z_CLSM_DATA	1/2/98	'TIFF'	6/27/95
CABDBLEC.ONE	RGCO9Z_CLSM_DATA	1/2/98	'TIFF'	5/3/95

Records found: 163

Fig. 7

8. Where Shall I Store My Images?

Most confocal microscopes are purchased for multiuser environments. The images generated in a single day can occupy several gigabytes or more. This would quickly overload even the largest hard disk, and most confocal facilities provide a means of storing data on removable media or, via a network, on remote computers located in the individual user's lab. Moving the new data off the original collection site should be a standard practice on all computers, in order to free up space for other users. DO NOT ERASE THE ORIGINAL FILES UNTIL YOU ARE CERTAIN THAT YOU HAVE A VALID COPY ON ANOTHER COMPUTER. We now routinely transfer copies of the original files to a CD-ROM for long term stable storage. Your database files should indicate where the file is currently stored, and alternate backup storage sites, such as CD-ROM disks.

9. Lack of Standard File Format: Comments on Image Software

All the Image Gallery software described in this chapter will provide useful image galleries and thumbnails of common file formats, such as TIFF/tif, JPEG, PICT, and any file that includes it's own "Preview" image, such as Illustrator®, Photoshop, Canvas™ (Deneba Software), and others.

However, the nature of the file format generated by various confocal microscopes differs, depending upon the software written for each microscope. Thus, Bio-Rad uses a unique file format, Leica a simple TIFF format, the older Zeiss format was also TIFF, whereas the newer formats of Zeiss vary depending upon the manner in which the data was collected. The recently released Zeiss LSM 510 provides optional choice of 8-bit or 12-bit collection. But it is not yet clear if the 12-bit images are immediately converted to 8 bits, or can be stored as 12 bits. Preliminary information from Zeiss indicated that they have not yet released their file specifications. They presently are apparently using a proprietary format, but are reportedly considering shifting to TIFF 6.0 standards, with Multi-TIFF format. The new Olympus FluoView collects images in 12-bit/channel, but permits saving the files in either 12-bit/channel Multi-TIFF, in 8-bit TIFF or in 24- (RGB) bit (8-bits/channel). The NORAN confocal is based on the SGI and stores files in 8-bit format. To add further to the confusion, a Z-series of sections may be stored in a single file or as a series of individual files. A double or triple labeled Z-series can be stored in a single large file, or with each labeled Z-series as a separate file, or each section of each Z-series as separate files.

The lack of a uniform data file format is a continuing source of confusion, and hopefully, with the maturation of the confocal microscope industry, will eventually lead to a more widely used format, as is now happening in the field of radiology (DICOM) and satellite imaging (HDF).

This variation in file structure imposes special problems on both the user and on the manufacturer of both Image Processing and Image Gallery Database software. The confocal microscope community represents only a small number of users, thus most authors of Image Gallery software designed for a mass consumer market, have not found it worth their effort to write software to meet the specialized needs of Bio-Rad, Olympus, and others.

Recently, however, some of these software companies are indicating interest in attracting confocal microscopists as customers. The most notable amongst them, to date, is ThumbsPlus (Cerious Software), which now supports 12/16 bits/channel and can directly read the Olympus Multi-TIFF files, and Bio-Rad *.PIC files. Extensis Portfolio and Canto Cumulus have indicated that they may support the Bio-Rad format in future editions, but are unwilling to state their plans at the present time.

Part of the problem derives both from the often nonstandardized format (as in Bio-Rad) or the limited number of likely customers interested in 16-bit Multi-TIFF files. The use of specialized file formats, such as those containing only two channels, or a red and green channel with a gray scale nomarski image, poses problems that Extensis and Canto refuse to address. Fortunately, Cerious Software, Inc (ThumbsPlus) has been extremely helpful in this regard.

10. Three Important Strategies When Collecting Images

10.1. Use Folders (Directories) and SubFolders (Subdirectories) to Cluster Files

Give the folder a name indicating the experiments contained within, e.g., calbindin_retina. Subfolders might include: Monkey Calbindin, Pigeon Calbindin, etc. Within the Monkey Calbindin folder, I would have additional folders representing each experiment.

10.2. Use a Database Program as You Collect Images

Using a simple, low cost database program, such as FileMaker Pro 4.0, information about each image is entered as a separate Record in the FileMaker Pro database. The program automatically assigns a serial number to each record. The Record of that image contains a Field where you insert the name of the Folder/subfolder containing the original data. A Z-series is treated as a single record (This is OK for Bio-Rad, when a Z-series is stored as a single file. May have to work out a different strategy for Leica or those optional Bio-Rad images, where a Z-series is stored as a series of single files). The Automatic Serial Number allocation is now used as the file name for storing the original images. Just to be safe use serial numbers with no more than five characters in the file name if you frequently collect extended Z-series. Bio-Rad, Leica, Zeiss, and Olympus may all overwrite the characters 8, 7, and 6 of the filename, when storing a lengthy Z-series. Thus, you might have thoughtfully called your file "myfilenw.tif." But if you have multi-label image or a Z-series, it will be renamed as "myfile01.tif, myfile02.tif, myfile03.tif...myfilexx.tif."

10.3. Use an Image Gallery Database on Completion of a Work Session

Once you have completed a worksession collecting images on the confocal microscope, and transferred files to their final storage media or location, you should immediately generate thumbnail images with a program such as ThumbsPlus. It will also facilitate your being able to recall critical features of those images, and insert suitable keywords into the correct thumbnail.

11. Comparison of Various Image Gallery Software

A number of Image Gallery software programs are suitable for use in conjunction with confocal microscopy. This is only a partial listing of such programs. The selection reflects only my personal experience, and is not intended to be comprehensive. The major criteria determining the choice of programs included: (1) cross-platform compatibility, as many microscopists work on both PC and Mac computers; (2) ability to work across a network of computers; (3) ability to work with large number of file types, including Bio-Rad, Olympus, Zeiss, and Leica; (4) current or potential ability to read 12/16 bits/channel images; and (5) ability to read data collection parameters stored in the original files.

The only program currently able to directly read Olympus FluoView 12-bit files and Bio-Rad native files is ThumbsPlus. Extensis and Canto have been less emphatic about such plans for the future. In view of the widespread use of Bio-Rad confocals, and their unique file format, this is a serious limitation on the use of Portfolio 3.0 and Cumulus 4.0. However, if you routinely convert your Bio-Rad files to NIH-Image, the resultant 8-bit TIFF file can be cataloged by all of these programs. ThumbsPlus has some limited ability to display all the images in a Z-series, although with somewhat limited performance. Cumulus, Portfolio, and ThumbsPlus indicate that they may soon be able to read and display a movie in PICS format.

One of the valuable operations provided by ThumbsPlus is that which allows the user to crop and change the brightness, contrast and color balance of the thumbnail image, without having to modify the original data file.

Based on these several considerations, there is no single program that fully meets all current needs of the confocal microscopist (*see* **Table 1**). The one that comes closest is ThumbsPlus, as it is presently the only one able to read Bio-Rad 8 bit files, Olympus 12/16 bits/channel Multi-TIFF images, can read the collection parameters of both confocal data files, and works with Windows® 3.11, as well as MacOS, Windows NT, and Windows 95®. The ability to work with Windows 3.11 is of value for users of Bio-Rad's Lasersharp 1024 which operates under IBM's OS/2. OS/2 supports simultaneous processes under Windows 3.1, but not NT or 95. The Olympus FluoView operates under Windows

Table 1
Comparisons of Image Databases

	Portfolio 3.0	Search 3.1	Canto Cum 3.0	ThumbsPlus 3.1	Confocal system
Mac	Y	Y	Y	Y	FV, BR (OS/2)
Windows 3.11	N	N	N	Y	
Windows 95	Y	N	Y	Y	FV, Z, L
Windows NT	Y	N	Y	Y	
Networkable	Y	N	Y*	Y	
8-bit TIFF	Y	Y	Y	Y	BR, Z, L, N, MD
12/16-bit TIFF (FV)	N?	N	N?	Y (Excellent)	FV, (BR?), Z
Bio-RAD *.PIC	N	N	N	Y (Mac version)	BR
Make thumbnails	Y (Fixed sizes)	Y	Y (Fixed sizes)	Y (Variable size)	
Modify thumbnails	N	N	?	Y (Excellent)	
Editing software	N	N	N	Y (Good)	
Read CLSM parameters	?	N	?	Y (Excellent)	FV, BR
Image gallery	Y (Fair)	Y (Good)	Y (Good)	Y (Good)	
List® view	Y (Good)	Y (Good)	Y (Good)	Y (Fair)	
Linkage to full database	N	N	Y (FM Pro)	ODBC?	
Full path of file	Y	Y	Y	Y	
Web enable	?	N	Y	Y	
Drag & drop thumbnail	Y/N	Y/N	Y	Pending?	
AppleScript	?	N	Y	N	
OLE	?	N	Y?	Y*	
Price	$100	$100	Basic = $100 Apple Script = $600 Network = $1,900	Single user = $70 Multi-User = $250	

Abbreviations used to indicate manufacturer of different confocal microscope systems: FV = Olympus FluoView; BR = Bio-Rad; Z = Zeiss; L = Leica; N = Nikon; MD = Molecular Dynamics.

3.11, although there is some expectation that a Windows NT version will be available in the near future. The only shortcoming of ThumbsPlus is the need for further development of the List View interface and the need for support of Drag and Drop on the Mac. It is fully cross platform compatible.

ThumbsPlus, Portfolio, and Cumulus are able to work across a network, although with varying reliability and great variations in cost. ThumbsPlus and Portfolio include this capability for approximately $100 (U.S.). Canto Cumulus charges an additional $1800 for this capability. Although Cumulus is an excellent program, you will have to decide if the robust industrial level performance of Cumulus justifies the enormous added expense. However, Cumulus 4.0 has a number of excellent functions, including very smooth interactions with File-Maker Pro, ability to handle vast numbers of images at very high speed, and the most sophisticated search engine of this group of programs. Portfolio 3.0 is a program that has been on the market for a long time, under different names. Initially marketed by Multi-Ad as Search 1.0, then sold to Adobe and marketed as Fetch 2.0 and now marketed by Extensis as Portfolio 3.0. The major change in Portfolio is the introduction of cross platform compatibility, increased numbers of file filters for reading a broader range of file types, and ability to work over a network. The program is very easy to learn, generates thumbnails very quickly, and is supported by a major corporation. Hopefully, Extensis plans to maintain an interest in this excellent, but frequently orphaned, product.

12. How Many Versions of the Image Should I Save?

There is no perfect image of a specimen. This is all the more true in the case of a confocal image. The number of variables can be overwhelming, and include selection of the field, orientation of the image, noise in the images, fading, depth of focus of the picture, change in contrast, brightness, color balance, etc. As a result, you may find that you have several different images of the same preparation, each showing slightly different features. Each such sample adds additional burden to your storage and retrieval capability.

Once you have collected an image, however, you invariably find it desirable to perform various procedures on the dataset that are collectively referred to as "Image Processing."

Image processing may consist of a range of operations varying from simple change in brightness or contrast, to cropping an image, generating stereo views, animating a sequence, reslicing the stack of images in alternate planes, changing colors, selectively editing parts of the image to heighten detail, and then saving the new set of images in a location associated with the original file. This requires massive amounts of storage capacity. This is manifest at three different points in the use of CLSM: (1) data collection, (2) image processing and analysis, and (3) preparation of final image for publication.

Inventing file names for such images will again tax your imagination. I suggest that they should be only slight modifications of the name of the original file, or else you may not be able to keep track of their heritage, in the event that you want to go back to the original file to modify or resample the image that provided the interesting variant image.

Each new file can either be associated with a new Record in your master database, or you can dedicate a Field in that record to indicate what Image Processing operations you performed on the original dataset, and where it is stored. My own preference is to keep this information in the same Record as a history of this image set. This strategy will help track related images as you prepare for final publication.

13. Conclusions

13.1. Why do You Need a Database?

Most scientists are told of the importance of taking careful notes from their first visit to a laboratory. Some of them even do so! Some people are naturally inclined to planning out every detail of their lives in advance, including how they will store their collections of rocks, stamps and confocal images even before they start saving them.

But very few scientists seem to have planned out a global strategy for their data storage. I start collecting rocks, mix in some stamps in the same box; and add in photos, books, and miscellaneous other lifetime accumulations. When the box becomes too full, I worry about separating the items into more sensibly organized groupings. How many new boxes do I need? How shall I label them? Do I have to take everything out before I reclassify them?

In the case of complex digital images, it is important to have some means of self imposed discipline to be sure that you record all relevant information at the time of image collection. It is equally important to decide in advance where you will store the files and how they should be grouped.

To deal with the large numbers of big files associated with confocal microscopy, and to be able to find things when you need them, I strongly urge the reader to use the widely available and inexpensive database programs now available for the personal computer.

13.2. Using a Database to Help Name, Sort, Identify, and Retrieve Images

Several confocal microscopy software packages already provide a means of storing specific details about each image, such as collection parameters, file size, magnification, resolution of image, and extensive comments about the image and its scientific significance. This information, however, does not readily allow you to organize the information in a manner commensurate with your personal preference for grouping information, or for comparing the contents of large numbers of files. In addition, most software provided with confocal microscopes has very limited capacity for searching for unique or related files, e.g., "show all the calbindin containing images of the monkey and mouse retinae in transverse section." Specialized Image Database Programs, such as Portfolio, ThumbsPlus, and Cumulus provide excellent means of automatically scanning a disk, generating and displaying many images in thumbnail format, with some limited information about each image, mainly limited to some key words, location of the file, etc. Image Database programs have advantages for displaying large numbers of thumbnail sketches. However, they have limited ability to categorize multiple types of information, and generally only are useful for generating thumbnail images and comments well after you have finished collecting a large number of images. They are not useful as a means of assigning filenames, storing extensive comments, setting designated fields of the nature of the material, species of animal, fluorescent dyes used, special filters, laser power, confocal apertures, etc.

Traditional database programs, such as FileMaker Pro 4.0 and Microsoft Access provide a range of such essential functions, can be used simultaneously while collecting confocal images, but are far less effective in dealing with large numbers of images. Individual thumbnails can be displayed in these programs, but they cannot automatically scan a disk and generate thumbnails, nor can they display them readily in gallery format.

I anticipate that in the near future, these two formats will converge in a single database design.

24

Morphing Confocal Images and Digital Movie Production

Eric Hazen

1. Introduction

Visualizing change over time is essential to understanding biological events such as embryogenesis and development. More and more specimens can be visualized in the living state due to improvements in confocal microscopy, and the development of novel probes such as the green fluorescent protein. However, it is still necessary to dissect and fix many tissues before staining with antibody probes, and viewing them with the confocal microscope. In this case, still images from different samples at different stages in development can be imaged. If enough of these images are collected, an animation could be created, simulating the actual process, although this would require a very large number of images, and this is not usually a practical option because of the number of animals required. Instead, images are collected from a few representative points along the developmental timeline and these static images, when arranged in order are used to build up a visual narrative of the actual process.

Computer technology provides an alternative method for simulating change over time: a process called morphing, which was originally developed for the movie and advertising industries to create visual effects. Morphing is essentially a controlled fade of one image into another *(1–3)*. The computer is given two images, a starting image and an ending image, and it calculates all of the intermediary images necessary to produce the desired effect. The appearance of motion is achieved by specifying motion paths from shapes in the first image to shapes in the second image. While the starting image fades into the ending image, the shapes appear to be stretched into one another. Because images from only a few selected time points are used, a morph can only be an approximation or hypothesis of the true motion. Because of these assumptions,

From: *Methods in Molecular Biology, vol. 122: Confocal Microscopy Methods and Protocols*
Edited by: S. Paddock Humana Press Inc., Totowa, NJ

morphing should be used as an instructional tool rather than a method for data analysis or quantification.

Using confocal images, morphing allows a digital movie file to be created that simulates motion or change over time from one still image to another still image; the computer models the intermediary stages of development *(4)*. This digital movie file can then be played directly from a computer, it can be recorded to videotape, put on a CD-ROM, viewed from the World Wide Web, or compiled into a longer digital movie sequence for presentation purposes in class or at meetings. Very few images need to be collected because the computer creates all of the intermediary frames. Moreover, the process is relatively quick and easy and does not require expensive equipment, although some familiarity with computer image processing and graphic and animation formats is necessary.

The tools that are required for morphing are available for the Macintosh, PC, and SGI environments as well as most other computer platforms. However, for the purposes of this article, morphing using an Apple Macintosh will be described. These computers are inexpensive, easy to use, and common in research environments. The techniques described here work on any platform, although some details may differ depending on the software selected. For information on software available for other computer platforms, *see* **Subheading 2.2.3.** Movie clips relating to to this chapter can be found on the author's website, http://www.academicis.org/is/multimedia

2. Materials

2.1. Hardware

2.1.1. The Computer

Because morphing is a fairly graphic and processor intensive application, a relatively high-end system is preferred. A Power Macintosh is strongly suggested. Its superior processor power enables it to calculate the objects' motion paths in a process called "rendering" much more quickly.

The system used to create most of the morphs shown in this chapter was a Powermac 8500/120 with 80 MB of RAM, a 2 GB hard drive, 4 MB of VRAM and a 17-inch monitor capable of displaying 24-bit color at 1152×870 resolution. A 230 MB magneto-optical drive was used for temporary data backup and final data files were written to CD-ROM for archiving and distribution.

2.1.2. Random Access Memory (RAM)

RAM requirements depend on the size of the images and the complexity of the shapes defined in the morph. 32 MB of RAM is needed, but 64 MB of RAM or more is recommended. Virtual memory can be used to compensate for inadequate RAM, but it will slow down the machine considerably.

2.1.3. Hard Drive

The size of digital movie files is dependent upon their intended use. For example, files destined to be printed to videotape are very large, while those intended for viewing on the World Wide Web are small. Enough hard drive space must be available to store several copies of the movie, because several versions may need to be created. More detail on hardware requirements are covered in **Notes 4.1–4.5**.

2.1.4. Display Monitor

A large monitor that is capable of high resolutions is strongly recommended. Shapes must be drawn precisely and it is extremely advantageous to be able to zoom in on the image while still viewing a large portion of the image. At least 4MB of VRAM is recommended in the display card in order to display 24-bit color on the monitor at high resolution.

2.1.5. Archiving Data

Data files must be backed up for safety, and eventually, they must be removed to make room for the next project. There are many options for archiving data. Writable CD-ROM drives have become much more affordable. The media is inexpensive and durable and CD-ROM drives ship as standard equipment on most computers. Magneto-optical and removable magnetic media such as Sy-Quest and Iomega drives work well, although because of its low cost of media, portability and data stability, the CD-ROM is often the preferred method of image archiving.

The hardware requirements described so far are sufficient to create morph movie files for the web. But playing directly from the computer for presentation purposes or recording to video may require additional hardware. Most Macs ship with multiple video out ports. Some have an additional VGA monitor port, whereas others have RCA or S-video out ports. With the right adaptor, these video out ports can be used to drive a standard video monitor or a video projector.

2.1.6. Digital Video Hardware

For recording digital movies to videotape, a digital video card is necessary. The AV Macs ship with a digital video card installed. Fairly good quality video can be produced with a standard AV mac. But for professional quality video, a third-party video card that specializes in video output is required, such as a TARGA or Radius card. Digital movie files destined for videotape have high resolutions (at least 640×480 pixels) and fast frame rates (approx 30 frames/s). The computer must be able to transport a large amount of digital video data in a short amount of time. These movie files are too large to be loaded into RAM; consequently, they must be spooled off of the hard drive. The standard Mac SCSI bus is incapable of transferring data fast enough to produce professional quality video. If professional quality output is desired, a SCSI accelerator card and an AV hard drive or hard drive array are also necessary. However, if professional quality is not absolutely necessary, a standard AV mac will do. Newer AV Powermacs such as the 8500 and 8600 are particularly good choices. They ship with better quality video circuitry than previous AV macs, their internal SCSI buses are twice as fast (10 MB/s) as the standard mac SCSI bus (5 MB/s), and if demands exceed their stock capabilities, they can be very easily upgraded.

Some examples of video cards include, for the Macintosh

> VideoVision, Radius Inc., Sunnyvale, CA
> TARGA 2000, Truevision Inc., Santa Clara, CA
> miroVIDEO, Miro Computer Products AG, Braunschweig, Germany

and for Windows

> TARGA 2000, Truevision Inc., Santa Clara, CA
> DPS Spark, Digital Processing Systems, Markham, Ontario, Canada
> miroVIDEO, Miro Computer Products AG, Braunschweig, Germany
> Rainbow Runner Studio, Matrox, Dorval, Quebec, Canada

2.2. Software

There are several stages in the process of creating a digital movie file, of which morphing is only one. The first stage is image processing. Raw images that have been collected on a microscope or other image collecting device are resized, reformatted, and cleaned up to prepare them for morphing. For image editing functions such as resizing; rotating; cleaning up the background, and adjusting brightness, contrast, and color balance, Adobe Photoshop is the industry standard. However, until version 4, Photoshop did not support batch processing—the automation of image processing functions. There are several applications that specialize in batch processing. Debabelizer, from Equilibrium Software, a commercial utility that is available for Macintosh and Windows, is a powerful utility that contains a host of image processing and conversions functions. Although it does not contain as many features as Debabelizer, the shareware utility Graphic Converter will perform batch conversions between virtually every graphic format ever used and will also perform many simple functions such as color table conversion, resizing, and sharpening as batch processes (*5*).

There are several morphing packages available for each of the major computer platforms. However, for the purposes of this chapter, we will be discussing Avid's Elastic Reality. Elastic Reality is professional caliber software that is capable of producing high-quality effects and it is available for most major computer platforms. *See* **Subheading 2.2.3.** for a list of additional morphing software packages.

The morph file created by the morphing software will be in form of a digital movie file. The most common format is QuickTime. The QuickTime digital video format is available on several platforms including Mac, Windows, and UNIX and it has been approved as an ISO standard. The movie file can be played back as it is, it can be modified by adding text or graphics, and it can be combined with other movie files to create a larger movie file. For the movie file to be played back, movie player software is needed. Apple's QuickTime MoviePlayer application works well. There is a shareware QuickTime player called Peter's Player that handles very large movie files slightly better than MoviePlayer does. For combining multiple files and adding titles, effects and sound, video editing software is required. Adobe Premiere is the industry standard. Avid's Videoshop is not as powerful, but less expensive, and provides all the basic functions necessary for compiling digital movies.

2.2.1. Software Companies

Adobe Photoshop, Adobe Systems, San Jose, CA
Adobe Premier, Adobe Systems, San Jose, CA
Debabelizer, Equilibrium Software, Sausalito, CA
Elastic Reality, Avid Technology, Inc., Tewksbury, MA
Director, Macromedia, Inc., San Francisco, CA
Media Cleaner Pro, Terran Interactive, Inc., San Jose, CA
Norton Utilities, Symantec Corporation, Cupertino, CA

2.2.2. Alternative Video Editing Software

Macintosh

Media Cleaner Pro, Terran Interactive, Inc., San Jose, CA
Videoshop, Strata, Inc., St. George, UT

Windows

MGI VideoWave, MGI Software Corp., Richmond Hill, Ontario L4B 1H8, Canada

2.2.3. Alternative Morphing Software

Macintosh

Gryphon Morph, Gryphon Software Corporation, http://www.gryphonsw.com/
Kai's Power Goo, Metatools, Inc., http://www.metatools.com/goo/goo.html
MovieFlo', MetaFlo', The Valis Group, http://www.valisgroup.com/

Windows

> Gryphon Morph, Gryphon Software Corporation, http://www.gryphonsw.com/
> Kai's Power Goo, Metatools, Inc., http://www.metatools.com/goo/goo.html
> MovieFlo', MetaFlo', The Valis Group, http://www.valisgroup.com/
> WinImages, Black Belt Systems, http://www.intermarket.net/blackbelt/ sw_wi_dl.html
> Plastic Morph, Algobit Software (Shareware), http://www.algobit.com/morph/ morph.html

3. Methods

3.1. Step 1: Planning the Morph Project Based on Available Hardware Resources and End-Product Goals

Before any raw images are modified, it is essential to plan out the project and decide on frame size (the size of each image in the animation measured in pixels, also called resolution) and frame rate [the rate at which images flash on the screen, measured in frames /s (fps)]. Although images and animations can be scaled down in resolution and in frame rate, it is not advisable to scale up. Image data is lost in scaling down and motion data is lost in decreasing frame rate (frames are dropped). These lost data are irrecoverable. It is better to make a movie which is large in frame size and frame rate and scale down later than to make it too small. You can't regain resolution once it is lost. Being aware of the requirements for the project from the beginning can save a significant amount of time. Image resolution and frame rate values are dependent upon the projected use of the morph and the hardware resources available. It is beyond the scope of this chapter to discuss possibilities in depth. Only general guidelines can be given. *See* **Notes 1–5**.

3.2. Step 2: Preprocessing the Images in Photoshop

For a brief discussion of image formats *see* **Note 6**, and for more information on image processing, refer to these sources:

> The MIT 15.566 Digital Imaging Resource Homepage: http://web.mit.edu/ beblack/www/digim.htm
> The Kodak Digital Learning Center: http://www.kodak.com/daiHome/DLC/
> Digital Image Center of the University of Virginia Library: http:// www.lib.virginia.edu/dic/info/webinfo.html

3.2.1. Rotating the Images

The first process applied to images to prepare them for morphing is often rotating. For the morph to appear realistic, the two images must be oriented properly in relation to one another. There are a few ways to rotate an image. If possible, it is best to either flip or rotate by 90° units. Both functions simply remap each pixel to a different location on the image; there is no loss of data in this operation. Finer degrees of rotation involve recalculating each pixel in the image which results in data loss. Photoshop has excellent algorithms for calculating rotations, but nonetheless, for optimal image quality the number of fine rotations should be kept to a minimum, preferably one. Photoshop versions 3 and above offer an excellent way to easily perform image rotation and accurately using layers. Select one of the images and copy it. Then go to the other image and, in the layers palette, choose "New Layer." Copy the first image into the new layer. The opacity of the second layer can be adjusted using the opacity slider, allowing both images to be visible at the same time. The images can then be independently rotated and compared to one another immediately. When the images are in register, the second layer is copied and pasted into a new window and saved as a new image.

3.2.2. Resizing the Images

After rotating, the images may need to be resized. Subjectively, the morph will lack realism if the images are not the proper size relative to each other. From a technical standpoint, the resolution of the two images being morphed must be equal in order for the morph software to process them. If they are not, the software will usually resize them for you to make their resolutions equivalent (and distort the images in the process). Because it is preferable to have control over resizing, it is recommended to do it first before importing them into the morph software. The resolution of the final movie must be decided upon here so that the images can be sized to match that resolution.

There are two ways to resize images. The borders can be changed leaving the center of the image unchanged, or the image can be scaled. The first method involves removing pixels from around the edges if the image is being made smaller, or adding pixels around the edges if the image is being made larger. The image data in the center remain unchanged. It is accomplished by either selecting a section from the image and "cropping" it, or, to make the image bigger, the entire image is selected, copied, and pasted into another image of the desired resolution and background color. The second method involves recalculating all the pixels in the image and recreating a new image of a different resolution. This operation results in a modification of image data across the entire image. It is accomplished by selecting "Image Size" from the "Image" menu and specifying a new size for the image.

The reason care is taken to explain these two different resizing methods separately is because care should be taken in using them. In most cases, the first involves the least amount of image data loss and should be used whenever possible. The second method should be used sparingly because every time the image is scaled more data will be lost and eventually, the image will be unusable. It is best to scale an image only once. If an image is scaled improperly, it is best to undo the scale and redo it to avoid scaling more than once. It is better to maintain the image's original aspect ratio when scaling it. Photoshop has excellent algorithms for recalculating an image at a different aspect ratio, but the image will look best if its aspect ratio remains unchanged.

3.2.3. Cleaning Up the Image

The next step in preparing the images is often referred to as "cleaning up." In this step, the background of the images are made identical, anything contained in the images that is unwanted is removed, and anything that is desired is added. There are two reasons for cleaning up the background of images. Because morphing is used as an instructional tool and not a data analysis method, anything that is not directly involved with the morph should be removed because it will move and stretch, distracting attention from the focus of the morph. Second, the movie file will compress more efficiently if the background is a solid color. There are many ways of cleaning up the background. One simple and effective method is to use the paintbrush tool to paint over unwanted features in the images.

Once the images are rotated, resized, and cleaned up, the finishing touches can be put on them. The brightness, contrast, sharpness, and color are adjusted if necessary *(6)*.

3.3. Step 3: Setting up the Morph Project

Avid's Elastic Reality runs on most common hardware platforms and produces excellent results, but regardless of the software used, the principles are the same. *See* **Subheading 2.2.3.** for information about other available morphing software. An example of a morph used during a transition in a video on evolution demonstrates the key concepts in creating any morph. The morph took about 6 h to create. Rendering time on a Powermac 8500 was only about a minute.

3.3.1. Preprocessing the Images

First, the images were brought into Photoshop and resized so that the organism in each image were approximately the same length. Image B was cleaned up and cropped so that each image had the same size background. They were then copied and pasted into a 640 × 480 pixel black background in order to create images of the desired resolution while keeping the aspect ratio of the original image. If the images were simply scaled to 640 × 480 pixels, they would appear distorted. Image A was originally blue, but it was changed to brown by adjusting color levels in Photoshop to make a visually smoother transition.

3.3.2. Importing Images Into Elastic Reality

The next step was to import the images into Elastic Reality and begin defining shapes (**Figs. 1** and **2**). Elastic Reality uses Beziere curves to specify shapes. The resolution of the correspondence lines can be manipulated to increase the accuracy of the shape mapping. In order to define the shapes accurately and avoid overlap (Elastic Reality does not allow overlapping shapes), the image must be magnified many times and the resolution of the correspondence lines must be very high.

When the shapes in Image A move to become the Image B shapes, they take everything in their vicinity with them, almost as if the entire image is made of rubber and it is being stretched and squished. In most cases this is a desired effect because it makes the transition appear more natural, but sometimes it is unwanted. This effect can be prevented by specifying shapes to act as boundaries. It is often impossible to set up all the shapes perfectly before rendering the morph although this does not usually present problems. Rendering times for a typical morph on a Powermac are only a few minutes. The morph can be rendered and viewed and shapes and correspondence lines can be added and adjusted until the desired look is achieved (**Figs. 1** and **2**).

3.3.3. Rendering the Morph

When all the shapes have been defined and joined and the correspondence lines have been adjusted, a movie file can be created. The process is called rendering. It involves calculating the movement from Image A to Image B and creating all the images that make up the movie. Parameters such as frame rate, total number of frames, image size and compression format must be specified. Rendering usually takes a few minutes after which the movie file can be played back. **Figures 1D** and **2D** show montages of rendered morphs.

3.3.4. Post-Processing the Movie Files

Post-processing is often as important and time-consuming as preprocessing. It is not absolutely necessary—the morph movie will play on its own, but in order to give it a professional, polished look or to prepare it for a special application, it must be processed further. Post-processing can involve many steps including adding titles and transitions to the movie file, incorporating it into a larger movie, recompressing it at a smaller resolution, and/or outputting it to videotape, CD-ROM or onto the web.

Titles screens and credit screens can be added to a movie file without expensive software. One crude method for adding a title screen at the beginning or credits at the end of the movie file is as follows: Render the morph as a series of numbered PICT files (or some other file format, but PICT works best on

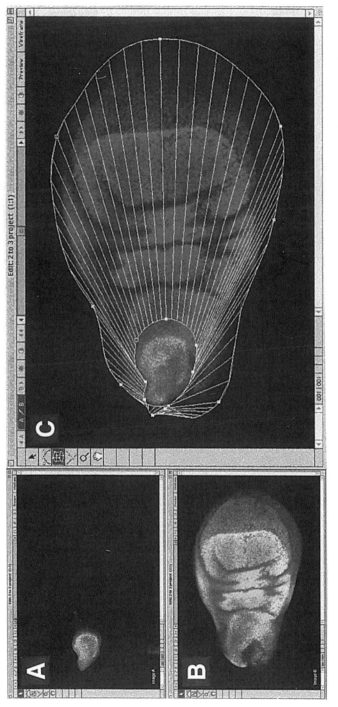

Fig. 1. Morphing between two confocal images. *Drosophila* second instar wing imaginal disc (**A**) morphed into a third instar wing imaginal disc (**B**). The motion paths from A to B are drawn in (**C**), and the resulting morphed sequence in (**D**). The sequence demonstrates the growth of the wing disk and the expression of the apterous gene (images courtesy of Jim Williams and Steve Paddock).

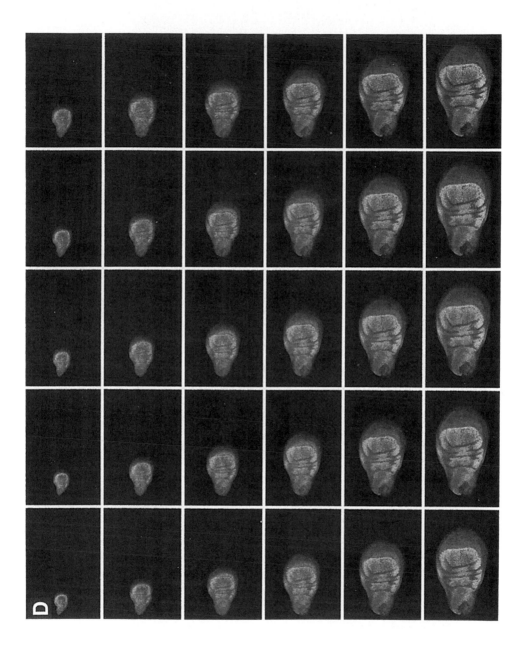

431

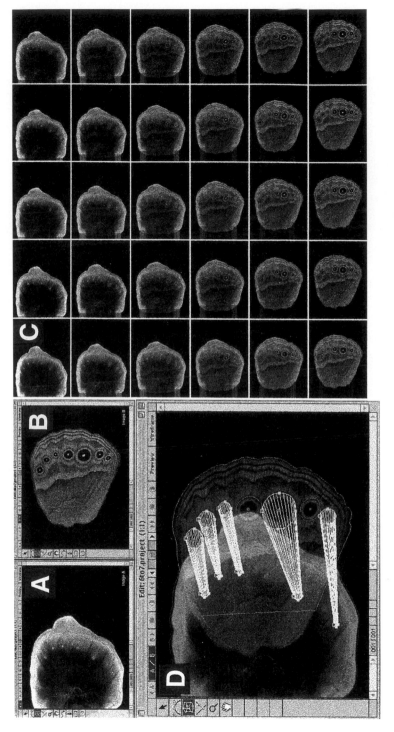

Fig. 2. Morphing between a confocal image and an image from a flat bed scanner. A butterfly fifth instar hind wing imaginal disc (**A**) is morphed into the adult hind wing of the butterfly *Bicyclus* sp. (**B**) with the motion paths shown in (**C**), and the resulting morphed sequence in (**D**). The morph demonstrates the development of the wing and the relationship between distal-less expression in the developing wing and the position of the eyespots on the mature wing. (Images courtesy of Julie Gates.)

Macintoshes). It is important not to start the sequence at 1 if you want to add a title in the beginning because the title screen will occupy the first several images. The number of copies of the title screen that will be needed is determined by multiplying the number of frames per second of the movie by how many seconds the title is to appear. For example, if the movie is 10 fps and the title is to be visible for 3 s, then 30 copies of the title screen are needed. The images will all be named something like "Morph0001.PICT" with the number corresponding to their sequence in the morph. Next, create a title screen in any drawing or painting program such as Photoshop, making sure that it is the same resolution and color depth as the movie. Name the files the same as the movie files with the same number of digits. If they are intended to go before the morph, make the numbers smaller, if they are intended to go after the morph, make the numbers larger. Make as many copies of the image as necessary. Finally, use a software utility to compile all of the numbered images into a movie making sure to correctly specify the frame rate (or frame duration), compression algorithm and anything else the utility requires. On the Macintosh, examples of such utilities are Apple's "QuickTime Converter" (Freeware) and "Movie Conversion" (Shareware). Adobe Premiere can also compile PICT files into a movie, but it is easier with one of the movie converter programs.

The previous technique for adding a title and/or credit screen to a movie is given for two reasons: to illustrate the basic process involved in editing digital movie files, and to show that it can be done without expensive software. There are software products available, however, that make digital movie editing faster and easier. An excellent example is Adobe Premiere, and its more sophisticated cousin Adobe After Effects, which are available for the Macintosh and the Windows platforms, and have become the industry standard in digital movie compilation (**Fig. 3**). There are other software products available that vary in terms of cost and number of features. Refer to **Subheading 2.2.2.** for a short list of alternative software products available for Windows and Macintosh.

3.3.5. Creating the Movie File

After the movie has been finalized, and all of the various title screens, credits screens, and everything else that will be included in the movie is ready, the final movie file can be rendered. Characteristics of the movie such as frame rate, compression type, color depth, and frame size have been discussed in the section on planning the morph project based on available hardware resources and end-product goals. These characteristics can either be preserved here or modified to use the movie for another purpose. For example, suppose a movie file has been created that is intended to be recorded onto videotape. The morph was created at 640×480 resolution and 30 frames/s. The final movie should be compiled at these same settings using a compressor that yields high

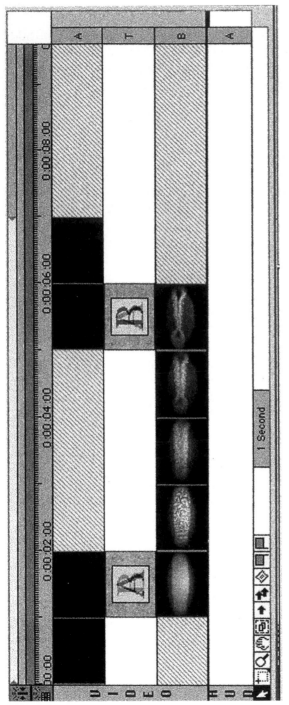

Fig. 3. Compiling a movie of *Drosophila* gastrulation using Adobe Premiere. It is a simple example of a morph movie with a fade to black at the beginning and at the end. The movie project consists of four files: a plain black image (called "Black 768 × 512," shown twice because it is used twice in the movie), the first image of the movie (called "first frame"), the last image of the movie (called "last frame"), and the morph movie itself (called SNA1.morph). The files that make up the movie (referred to by Premiere as "clips") are shown in the "Project" window. The "Construction Window" is a sort of storyboard or timeline that shows how the movie goes together. Clips can be placed into the Construction Window in any order desired. Transitions can be placed between clips to make them interact with each other in interesting ways. This example uses a Cross Dissolve to create a simple fade. Clips can also be modified in many other interesting ways. Transparency can be manipulated to combine multiple clips, filters can be applied, motion can be specified, frame rate and aspect ratio can be adjusted, and many other operations can be performed. For the ambitious artist, Premiere contains an abundance of powerful features, but for compiling a series of clips and adding some simple transitions, other, less expensive programs are more than adequate. (Images courtesy of Scott Weatherbee.)

image quality such as JPEG or Animation. But if, perhaps, the movie is also intended to be included on a web site, the movie can be recompiled at a lower resolution and a lower frame rate using a compressor that produces a small file size such as Cinepak. If the original morph file is created at a high resolution and frame rate, there will be room to experiment with different settings to find out what will work best for the intended purpose. Also *see* **Notes 6–9**.

4. Notes

1. As far as digital video is concerned, bigger is better. Data rates should be as high as available hardware permits. A higher data rate generally translates into better quality video. The size of a movie file is determined by five factors: image resolution, frame rate, color depth, length of the movie (or total number of frames), and compression. Typical resolutions are 640×480 pixels for NTSC video (recommended for videotape), 720×486 for NTSC 601 and 512×768 for PAL. Frame rates are typically 30 fps for NTSC and 25 fps for PAL. To calculate the size of a segment of video, multiply the number of pixels in a frame by the number of bits per pixel by the number of frames per second by the length of the segment in seconds. For example, a segment of NTSC video at 640×480 pixels in 24-bit color at 30 fps would take up 27 MB of hard drive space per second of video. Not only would this require gigabytes upon gigabytes of hard drive space for a few minutes of video, but it would also require the computer to be capable of streaming 27 MB of data from the hard drive to the video card every second. Very few computers are capable of these things. Therefore, shortcuts must be taken. One of the most useful shortcuts is compression. Compression involves crunching down the image data to make the file smaller.

   Unfortunately, there is a trade-off for decreased file size and that is image quality. Some compression formats (or codecs) require more processor power to compress and/or decompress than others. JPEG is a very popular codec because of its excellent combination of image quality and compression ratio; however, it requires an unusually large amount of processor power to compress and decompress it. Usually, a special video board is required that handles the decompression, freeing the computer's CPU to perform other tasks. High-end video boards usually have one proprietary compression format that they are very good at decompressing and playing (often it is a form of JPEG). The best quality video is obtained with these specialized boards. QuickTime version 3.0 offers better compression algorithms that allow higher quality video at higher compression rates. This may change the current preference for JPEG and allow much better quality video to be played on computers without expensive, high end video boards. Compression is not the only short cut that can be used to decrease file size. Any of the five previously mentioned factors that determine movie file size can be adjusted. The only advice that can be given here is to try different settings and see what looks the best and will play smoothly on your system.

It is beyond the scope of this text to discuss all data rate issues adequately enough to determine whether or not any given system is capable of performing a given video application. The actual data rate that is achieved on a system is dependent on many factors. However, the two most important components that determine data rate on a machine are the I/O bus (on Macs it is usually SCSI, on PC's it is usually IDE) and the video card.

The I/O bus is what the hard drive is connected to. The hard drive (or hard drives) is extremely important when doing digital video. Because data rate is defined in terms of megabytes per second, obviously, movie files with high data rates will be very large. A large amount of hard drive space is absolutely necessary. It is difficult to do any serious work with a hard drive any smaller than 2 GB. If the system has only one hard drive, it may be best to partition the drive with a large space that will be devoted to digital video work. This allows for isolation and containment of data so that drive errors can be fixed and the drive can be defragmented without disturbing the other files on the drive. The speed (or data transfer rate) of the I/O bus and hard drive combination is equally important. The faster the drive is, the better. Unless there is enough RAM to load the entire movie file into memory (which is rarely the case), the movie will have to be spooled off of the hard drive. The best solution is a fast/wide SCSI bus and an array of fast/wide hard drives. With this setup, data throughputs of around 15 MB/s can be achieved. But it is no longer necessary to combine multiple hard drives into an array to obtain adequate data throughput. Hard drives keep getting faster and faster. Some single drives can achieve sustained data throughput of 10 MB/s or more. If 10 MB/s or less is enough, a single drive is adequate and much cheaper. If the additional speed of a SCSI accelerator card is desired, it is recommended to buy a "bus mastering" card. A bus mastering card is capable of exchanging data directly with the video card, bypassing the computer's CPU.

To record digital video to tape, a special video card that is able to convert the computer's video signal to an NTSC or PAL signal and send it out to a video input device is needed. AV Macintoshes (including the Powermac 8500) come standard with video in and out capabilities. VHS-quality (non-broadcast) video can be achieved with a standard AV mac. If the quality achievable by the standard AV Powermac is not enough, a video card that provides hardware acceleration is necessary. Essentially, the video card contains a processor that does all the calculations necessary to compress and decompress the video data, thus freeing the computer's CPU to attend to other tasks. The best quality video is achieved with these cards. Refer to **Subheading 2.1.6.** for a list of video cards available for each of the major computer platforms.

To put all of this information into perspective, here are some examples of digital video systems that the author has experience in working with:

Macintosh Digital Video Workstation 1:

Type: Apple Power Macintosh 8500
Processor: PowerPC 604, 120 MHz
RAM: 80 MB

VRAM: 4 MB
I/O buses: Apple fast internal 10 MB/s, Apple external 5 MB/s
Hard drives: 2 GB Internal SCSI HD
Other drives: 4X Internal CD-ROM, 230 MB External Magneto Optical
Video card: Apple Built-In
Monitors: AppleVision 17 inch, Panasonic TV monitor

The Powermac 8500 has good quality video in and out circuitry and a fast (10 MB/s peak) internal SCSI bus. The PowerPC 604 processor is excellent for rendering morphs. The system has enough RAM to render large models and work with very large images. The Magneto Optical drive makes backing up data easy. The 4 MB of VRAM allows two monitors to be connected. The main weaknesses of this system is the video card, the lack of hard drive space, and the speed of the I/O bus. The Apple display card does not provide hardware acceleration for movie playing. The best quality video is achieved with cards that do provide hardware acceleration. There is not enough hard drive space to store all of the data files necessary to create a movie of a few minutes in length. Files had to be copied to Magneto-Optical and deleted from the hard drive often to free up space. The internal SCSI bus is faster than a standard SCSI bus, but it is not fast enough for professional quality digital video.

Macintosh Digital Video Workstation 2:
Type: Apple Power Macintosh 8100
Processor: PowerPC 601, 110 MHz
RAM: 64 MB
VRAM: 2 MB/4 MB*
I/O buses: Apple 5 MB/s, ATTO Silicon Express fast/wide 20 MB/s
Hard drives: 2GB Internal SCSI, 4.3GB Quantum Atlas External Fast/
                Wide SCSI
Other drives: 2X Internal CD-ROM, External Iomega Zip Drive
Video card: Apple Built-In, TARGA 2000 Pro NuBus
Monitors: Apple Multi-Scan 17 inch, Sony TV monitor
*The Apple Built-in graphics card had 2 MB of VRAM installed and the TARGA card had 4 MB of VRAM.

The PowerPC 601 processor, although not as powerful as the 604, is good for rendering morphs. The amount of RAM is adequate for manipulating large graphics files and rendering large morphs, although more RAM would be helpful. The main strengths of this system is the TARGA 2000 Pro video card and the ATTO Silicon Express SCSI accelerator card. The TARGA board provides hardware acceleration for digitizing and playback. The ATTO SCSI accelerator is a Bus Mastering card which means that it will transfer data directly to and from the TARGA card, bypassing the computer's CPU. The hard drive connected to the ATTO SCSI accelerator card is a Quantum Atlas 4.3 GB AV hard drive. This system can sustain video data throughput of 10 MB/s. If an array of multiple AV hard drives was connected, 15 MB/s data rates could be achieved. But a rate of 10 MB/s allows very large video files to be played smoothly resulting in high quality video.

2. Generally, the same issues involved in recording to videotape also apply to presenting a movie directly from the computer. In both cases the goal is to get the best looking movie to play on a given machine as possible. The only difference is, rather than outputting to a video recording device, the signal is outputted to a video display source such as a large monitor or an overhead projector.

3. If the movie file is destined for CD-ROM, data rate is the main concern. Generally, CD-ROM's have a slower transfer rate than hard drives do. A 2x CD-ROM drive, no longer the standard but still in common use, will transfer data at a maximum rate of 300 KB/s (kps). A 4× CD-ROM will transfer data at 600 kps, a 6× at 900 kps, and so on. By comparison, most SCSI hard drives will transfer data at a rate of at least 1000 kps or 1 MB/s. Movies played directly from a CD-ROM are usually small in frame size (240×180 or 320×240), slow in frame rate (10 or 12 fps) and compressed in a format that will result in extremely small data rates even at the expense of image quality. An excellent compressor for this purpose is QuickTime's Cinepak compression. Using JPEG compression set to very high compression (very low quality) also works well, but may not play back as smoothly. Another good way to get data rates down is to convert the images to 8-bit color or greyscale, if possible. This will result in a file that is approximately one third the size of the 24-bit color version of the file.

4. Issues involved in preparing a movie for use on the web are similar to those of CD-ROM. Files should be as small as possible. No one will want to wait for 3 h while a 10 MB movie file downloads. The main difference between the web and CD-ROM is that for CD-ROM, data rate is more important than total file size, whereas for the web the main concern is total file size and not data rate. There is no established maximum size that files on the web should be. It depends on many factors such as speed of the browsers' connection and how interested in downloading the movie they will be. Movie files on the web usually range from a couple hundred kilobytes to several megabytes. The best thing to do is create the movie file larger (and better quality) than necessary and scale down to the desired file size, recompiling the movie with either a smaller image size or higher compression. *QuickTime version 3.0 offers streaming video and better compression algorithms that allow higher quality video at higher compression rates. This will be especially useful for serving animation files on the web.*

5. If a third party, or non-Apple, digital video card is used, it is recommended to read the instructions that came with it very carefully. They should contain detailed instructions on how to configure the system, what hardware works with it and what doesn't, and how to connect the digital video card to the the computer, the hard drive array (if present) and all the external video devices. Special digital video cards often require strange little tweaks that have to be done to the system in order to obtain optimal performance. The best quality will be achieved only if these strange little tweaks are performed. Usually, video cards come with special programs for playing movie files. These special programs are optimized for their corresponding video card and will not work well (if at all) with other cards. When recording the movie to videotape, it is best to use the special optimized movie

player. Each program works slightly different. Read the instructions that came with the video card to find out exactly how to use the special movie player software.

A V Macintoshes, unfortunately, do not ship with detailed instructions on how to obtain the best quality video. Here are some suggestions: The manual that came with the computer should explain how to connect the computer to the VCR. There are several types of video cable. All A V macs have composite video ports and some have S-Video. S-Video produces a better quality signal because the luminance (greyscale information) and chrominance (color information) are carried on separate wires. If the computer and the VCR both support S-video, it is best to use it. Connect the video-out on the computer to the video-in on the VCR. There are several tweaks that can be made to a system in order to maximize the level of quality it is capable of producing. Defragmenting the hard drive that contains the movie file increases the rate at which it can be read by eliminating the need to search all over the drive to find each part of the file. There are several defragmenting utilities available. One example is Norton SpeedDisk contained in Symantec's Norton Utilities package. Another performance enhancing trick is to minimize the number of active system extensions. Usually, only the QuickTime extension is necessary. All other extensions take up valuable RAM and processor time without contributing to the output of the video. Deactivate all unnecessary extensions in the Extensions Manager and reboot the computer. Make sure there are no applications running other than the movie player application that is playing the movie. Turn off all networking. Give the movie player application as much RAM as possible. This allows the movie player to buffer movie data which results in smoother playback. Set the monitor's resolution to 640×480 if recording in NTSC and 512×768 if recording in PAL. If Apple's MoviePlayer is being used, choose "Present Movie" to play the movie. This option hides the Macintosh desktop and allows a delay to be set in the beginning and end of the movie. A shareware movie player called PetersPlayer also provides a presentation option and has slightly better memory buffering algorithms which allow it to play large movie files slightly smoother. The final step is to put a tape in the VCR, play the movie and press the record button.

6. There are two basic types of image formats: lossless and lossy. Lossless formats retain all image data, so they can be opened and resaved an infinite number of times without image degradation. Examples are TIFF, GIF (if it is an 8-bit image), uncompressed PICT, and Photoshop's proprietary format. The TIFF format provides a form of compression called LZW, but it is a lossless compression format. GIF is a lossless format that provides compression, but it is limited to 8-bit images. Twenty-four-bit color images usually should not be converted to GIF or a large amount of data will be lost. Lossy formats throw out image data in order to achieve smaller file sizes. Examples are JPEG and compressed PICT (which uses a form of JPEG compression). It is tempting to use these formats to save images because the files are much smaller, but each time the image is modified and resaved, more image data is lost. Eventually, there will be so many compression artifacts that the image will be unusable. It is a good idea to keep a master

copy of any images that will be used for multiple applications in a lossless format. The number of times an image is saved in a lossy format should be kept to a minimum, preferable one.

7. Presenting a movie directly from the computer is similar to recording to videotape. The difference is instead of connecting the computer to a VCR, the movie is simply played on the computer's monitor. A larger monitor, a TV or a video projector can be connected in addition to the standard monitor using the video out ports. The same issues regarding defragmentation, minimal extensions, and RAM buffering apply here.

8. Putting a digital movie file onto a CD-ROM is simply a matter of writing it to a CD-ROM using a special CD-ROM writing drive. Issues involving the playback of a movie from a CD-ROM are covered in the previous section on planning the morph project based on available hardware resources and end-product goals. The movie can be rendered in the morphing program at the desired resolution and frame rate, or it can be rendered at a high resolution and frame rate and then imported into a video editing program such as Premiere and recompiled at a lower resolution and frame rate. There are special CD-ROM authoring software packages designed to aid in creating fancy, interactive, multimedia CD-ROM's with pictures, sound, movies, and a snappy, custom user interface. Examples of CD-ROM authoring software packages are Macromedia's Director, and Terran Interactive's Media Cleaner Pro.

9. Access to a web server is necessary before any files can be put on the web. There are many ways to gain access to a web server, the details of which will not be discussed here. The most important issues involved in putting movies on the web are file size and file format. As mentioned previously, files should be as small as possible. The movie file can be rendered in the morphing program at a low resolution and frame rate, or it can be rendered at a high resolution and frame rate and then imported into a video editing program such as Premiere and recompiled at a lower resolution and frame rate. The issue of file formats is complicated. The web is platform-neutral. There are people browsing the web using every type of computer, every version of operating system and running every version of every software product available. It is essential to think in terms of compatibility and standards. If a movie is put onto the web in SGIs proprietary movie format, only people using SGIs will be able to view the movie. This may or may not be a problem. Usually it is. The most common movie formats found on the web are Apple's QuickTime format, MPEG, and Microsoft's AVI. Virtually anyone with an Apple computer, anyone with a recent version of Netscape Navigator or Microsoft Internet Explorer, most people running Microsoft Windows, and anyone with an SGI who isn't running ancient operating system software will be able to view QuickTime movie files. MPEG is a platform-neutral format. Anyone with an MPEG player application will be able to view an MPEG movie. The AVI format is especially popular with Windows users. If Windows users are the targeted audience, AVI may be the best choice. There are several applications designed to convert movie files from one format to another. The shareware util-

ity, Sparkle, is an MPEG player for the Macintosh and is also able to convert QuickTime movie files to MPEG. VfW Utilities are a set of utilities that allow QuickTime movies to be converted to AVI format and vice versa. For more information on creating animations for the web, refer to Lance Ladic's "Making Animations for the Web" at http://www.cs.ubc.ca/spider/ladic/animate.html.

References

1. Beier, T. and Neely, S. (1997) Feature-based image metamorphosis. URL: http://www.hammerhead.com/thad/morph.html
2. Palma, D. J. (1996) The Palma Morphing Paper URL: http://www.norcosoft.com/ncs/palma.html
3. Goh Han Tiong, A. (1997) Technology review paper–Morphing http://www.fit.qut.edu.au/Student/ITB235/papers/Moving/n2039389/morphing.html
4. Paddock, S. W., DeVries, P. J., Buth, E., and Carroll, S. B. (1994) Morphing: a new graphics tool for animating confocal images. *BioTechniques* **16,** 448–452.
5. Russ J. (1994) *Image Processing Handbook*, 2nd ed., CRC Press, Boca Raton, FL.
6. Paddock, S. W., Hazen, E. J., and DeVries, P. J. (1997) Methods and applications of three colour confocal imaging. *BioTechniques* **22,** 120–126.

Index